Bisher erschienen/Already published

Vol. 1: J. C. Erdmann. Wärmeleitung in Kristallen, theoretische Grundlagen und fortgeschrittene experimentelle Methoden. II, 283 Seiten. 1969.

Vol. 2: K. Hepp, Théorie de la renormalisation. III, 215 pages. 1969.

Vol. 3: A. Martin, Scattering Theory: Unitarity, Analyticity and Crossing. IV, 125 pages. 1969.

Vol. 4: G. Ludwig, Deutung des Begriffs „physikalische Theorie" und axiomatische Grundlegung der Hilbertraumstruktur der Quantenmechanik durch Hauptsätze des Messens. 1970. Vergriffen.

Vol. 5: Schaaf, The Reduction of the Product of Two Irreducible Unitary Representations of the Proper Orthochronous Quantummechanical Poincare Group. IV, 120 pages. 1970.

Vol. 6: Group Representations in Mathematics and Physics. Edited by V. Bargmann. V, 340 pages. 1970.

Vol. 7: R. Balescu, J. L. Lebowitz, I. Prigogine, P. Résibois, Z. W. Salsburg, Lectures in Statistical Physics. V, 181 pages. 1971.

Vol. 8: Proceedings of the Second International Conference on Numerical Methods in Fluid Dynamics. Edited by M. Holt. 1971. Out of print.

Vol. 9: D. W. Robinson, The Thermodynamic Pressure in Quantum Statistical Mechanics. V, 115 pages. 1971.

Vol. 10: J. M. Stewart, Non-Equilibrium-Relativistic Kinetic Theory. III, 113 pages. 1971.

Vol. 11: O. Steinmann, Pertubation Expansions in Axiomatic Field Theory. III, 126 pages. 1976.

Vol. 12: Statistical Models and Turbulence. Edited by C. Van Atta and M. Rosenblatt. Reprint of the First Edition. VIII, 492 pages. 1975.

Vol. 13: M. Ryan, Hamiltonian Cosmology. VII, 169 pages. 1972.

Vol. 14: Methods of Local and Global Differential Geometry in General Relativity. Edited by D. Farnsworth, J. Fink, J. Porter, and A. Thompson. V, 188 pages.

Vol. 15: M. Fierz, Vorlesungen zur Entwicklungsgeschichte der Mechanik. V, 97 Seiten. 1972.

Vol. 16: H.-O. Goergii, Phasenübergang 1. Art bei Gittergasmodellen. IX, 167 Seiten. 1972.

Vol. 17: Strong Interaction Physics. Edited by W. Rühl and A. Vancura. V, 405 pages. 1973.

Vol. 18: Proceedings of the Third International Conference on Numerical Methods in Fluid Mechanics, Vol. I. Edited by H. Cabannes and R. Temam. VII, 186 pages. 1973.

Vol. 19: Proceedings of the Third International Conference on Nemerical Methods in Fluid Mechanics, Vol. II. Edited by H. Cabannes and R. Temam. VII, 275 pages. 1973.

Vol. 20: Statistical Mechanics and Mathematical Problems. Edited by A. Lenard. VIII, 247 pages. 1973.

Vol. 21: Optimization and Stability Problems in Continuum Mechanics. Edited by P. K. C. Wang. V, 94 pages. 1973.

Vol. 22: Proceedings of the Europhysics Study Conference on Intermediate Processes in Nuclear Reactions. Edited by N. Cindro, P. Kulišic and Th. Mayer-Kuckuk. XIV, 329 pages. 1973.

Vol. 23: Nuclear Structure Physics. Proceedings 1973. Edited by U. Smilansky, I. Talmi, and H. A. Weidenmüller. XII, 296 pages. 1973.

Vol. 24: R. F. Snipes, Statistical Mechanical Theory of the Electrolytic Transport of Non-electrolytes. V, 210 pages. 1973.

Vol. 25: Constructive Quantum Field Theory. The 1973 "Ettore Majorana" International School of Mathematical Physics. Edited by G. Velo and A. Wightman. III, 331 pages. 1973.

Vol. 26: A. Hubert, Theorie der Domänenwände in geordneten Medien. XII, 377 Seiten. 1974.

Vol. 27: R. K. Zeytounian, Notes sur les Ecoulements Rotationnels de Fluides Parfaits. XIII, 407 pages. 1974.

Vol. 28: Lectures in Statistical Physics. Edited by W. C. Schieve and J. S. Turner. V. 342 pages. 1974.

Vol. 29: Foundations of Quantum Mechanics and Ordered Linear Spaces. Advanced Study Institute, Marburg 1973. Edited by A. Hartkämper and H. Neumann. VI, 355 pages. 1974.

Vol. 30: Polarization Nuclear Physics. Proceedings 1973. Edited by D. Fick. IX, 292 pages. 1974.

Vol. 31: Transport Phenomena. Sitges International Schools of Statistical Mechanics, June 1974. Edited by G. Kirczenow and J. Marro. XIV, 517 pages. 1974.

Lecture Notes in Physics

Edited by J. Ehlers, München, K. Hepp, Zürich,
H. A. Weidenmüller, Heidelberg, and J. Zittartz, Köln
Managing Editor: W. Beiglböck, Heidelberg

60

C. Gruber
A. Hintermann
D. Merlini

Group Analysis of Classical Lattice Systems

Springer-Verlag
Berlin Heidelberg GmbH 1977

Authors
C. Gruber
Lab. Phys. Théorique
Ecole Polytechnique Fédérale
Case Postale 1024
Lausanne/Schweiz

A. Hintermann
S.I.N.
Villigen

D. Merlini
C.R.P.P.
Ecole Polytechnique Fédérale
Case Postale 1024
Lausanne/Schweiz

Library of Congress Cataloging in Publication Data

Gruber, Christian, 1939-
Group analysis of classical lattice systems.

(Lecture notes in physics ; 60)
1. Lattice theory. 2. Groups, Theory of.
I. Hintermann, A., 1944- joint author. II. Merlini,
D., 1942- joint author. III. Title. IV. Series.
QC174.85.L38G78 530.1'32 77-2821

ISBN 978-3-540-08137-1 ISBN 978-3-540-37407-7 (eBook)
DOI 10.1007/978-3-540-37407-7

Originally published by Springer-Verlag Berlin · Heidelberg · New York in 1977

2153/3140-543210

FOREWORD

In 1969 a theoretical Physics Group was created at the Physics Department of the Federal Institute of Technology of Lausanne. In those days it was decided to initiate research on selected topics of Statistical Physics, in particular on the statistical mechanics of Lattice Systems. C. Gruber, a Ph. D. from Princeton University, who had joined the Group that same year, conducted this part of the research programme. In 1971 two graduate students from E.T.H. Zurich, A. Hintermann and D. Merlini started to work on their Ph.D. thesis with C. Gruber as thesis advisor.

In 1975, G. Gallavotti and G. Jona-Lasinio, who had been invited to be members of the thesis committee, suggested to present in one monograph the work of Gruber, Hintermann and Merlini. The idea was accepted with enthusiasm and an optimistic planning was laid down. After a finite sequence of lower bound estimates to its duration, the project came ultimatly to completion.

I would like to take this opportunity to congratulate the authors who have proved to successfully overcome the difficulties of all kind inherent to such enterprise and I wish for them an encouraging acceptance of their monograph by the audience interested in the subject.

Ph. Choquard

Lausanne, December 20, 1976

TABLE OF CONTENTS: Pages

PART II

SPIN $\frac{1}{2}$ LATTICE SYSTEMS WITH CONSTRAINTS

P A R T III

ARBITRARY SPIN LATTICE SYSTEMS

INTRODUCTION

The development of Statistical Mechanics has shown that the detailed analysis of simple systems (models) - even if they are not too realistic - is important to understand the general properties, as well as to give a possible mechanism, of certain physical phenomena. In particular the intensive study of Lattice Models such as the Ising Model was fundamental in the history of Statistical Mechanics to show that the basic principles were sufficiently general to describe phenomena such as phase transition; moreover these investigations have led to a better understanding on the nature of critical phenomena, coexistence of phase and metastability. More recently, these same models have been the starting points of researches in other fields such as mathematics or biology.

The first and maybe one of the most important results of these investigations on lattice systems was the fact that very interesting properties, which should not be allowed within the original framework, could appear in the so called "Thermodynamic Limit", which is the limit obtained by letting the dimension of the system become infinite in a suitable manner. Physically, this limit corresponds to the experimental fact that the size of the system, to which standard thermodynamic applies, is very large compared to the range of the effective forces, with the consequence that the surface effects may be neglected and certain physical quantities appear to be extensive. Mathematically, this limit implies the possibility of singularities in the thermodynamic functions or in the correlation functions, singularities whose physical consequences are precisely the properties observed in nature.

The investigation on Lattice Systems have been undertaken either from an analytic point of view or by means of computer simulation. In the first approach, which is the only one we have considered in our work, one can further introduce a distinction between "Rigorous Results" (existence theorem, analyticity properties) and "Approximate or Numerical Results" usually obtained by means of series expansion giving concrete results which can then be compared with experiment.

At the present time the study of lattice systems is especially directed towards understanding a certain number of fundamental properties for which only partial results if any, have been obtained up to now. Those open problems center mainly around the following questions :

1. What are the characteristics of model Hamiltonians giving rise to a phase transition ? Why some many body interactions give rise to a phase transition and others will not ?

2. Why, as observed in nature, phase transition occurs only on isolated lines in the p - T plane ?

3. If a critical point exists, why is it that some many body interactions, even if short ranged, can give rise to different critical behaviours, while others will not ?

4. What is the influence of dimensionality and "symmetry" on the phase coexistence phenomena ?

In this monograph we present a systematic analysis of classical spin systems on a lattice. This investigation is based on a simple mathematical structure which appears in a natural way in any study of lattice systems; it consists essentially of two abelian groups associated respectively with the configuration space and the interactions. As we shall show these two groups play a "dual" role and are related by two homomorphisms which are the basic tools of our whole analysis. Physically it appears that this framework reflects certain symmetries of the systems - either euclidean or internal - and gives therefore the natural setting to study phenomena such as symmetry breakdown and phase transition; moreover besides its simplicity it is also well adapted to take into account the local structure of the lattice.

We shall begin our discussion with the simplest case, namely the case of spin $\frac{1}{2}$ without constraints (Part I); for these systems we give a description of the group structure and then proceed to study the local properties of the lattice in terms of the group properties. This method allows us to suggest models, which in some cases have received further attention and interest, giving rise to new "solvable" models of statistical mechanics (see "triangular model" in our analysis).

In part II we extend this framework to include the case of spin $\frac{1}{2}$ with constraints such as hard-core models, ferro-electric models, vertex-models, where the effect of the constraints is to reduce the phase space by allowing only certain configurations. Essentially the only new idea which has to be introduced is to enlarge the group associated with the interactions to take explicitly into account the constraints imposed on the system.

Finally in Part III we present a generalisation of the group methods to arbitrary spin systems; the objective of this last part is to gain as much informations as possible on abstract systems using standard results of harmonic analysis and then to translate these properties into physical properties of real systems. In particular this analysis brings into evidence new symmetries which had not appeared in the case of spin $\frac{1}{2}$ systems; these symmetrics and their physical consequences are investigated and applied to certain spin 1 models.

Our main concern in writing this monograph has been to illustrate our general theory with simple applications yielding non trivial results, together with their physical implications. Furthermore for the sake of clarity we give at the beginning of each chapter a brief introduction which should situate the problems which we investigate into the general context of some interesting open problems of statistical mechanics.

Acknowledgment

We are much indebted to R. Calinon, W. Greenberg, J.L.Lebowitz, A. Messager, S. Miracle-Sole, B. Payandeh and J.Slawny for their collaboration in the derivations of certain results mentionned in these notes.

The suggestion to publish a synthesis of our work was made to us by G. Gallavotti and G. Jona-Lasinio and we express our gratitude to both of them for their support in this project.

Finally we wish to thank Ph. Choquard for his constant encouragements without which this monograph would have never appeared and all members of his "Laboratoire de Physique Théorique, EPF-L" for many fruitful discussions.

One of us (D.M.) has also the pleasure to thank the "Fonds National Suisse de la Recherche Scientifique".

Lausanne, Dec. 17, 1976

PART I

SPIN ½ LATTICE SYSTEMS WITHOUT CONSTRAINTS

CHAPTER 1 - DEFINITIONS AND GROUP STRUCTURE

1.1. Definition of Spin Lattice Systems [1]

Spin lattice systems, or equivalently lattice gas systems, are obtained most naturally by idealization of crystals constituted of subsystems (molecules) which may exist in a finite number of configurations.

We assume that the subsystems are parametrized by the points of a countable set $\mathcal{L}$ and in most cases of interest $\mathcal{L}$ will be either $\mathbb{Z}^{\nu}$, the set of ν-tuples of integers, (one lattice site per unit cell) or a sum of $\mathbb{Z}^{\nu}$ (several lattice sites per unit cell). The points of $\mathcal{L}$, called lattice sites, are denoted by small letters, x, y, z, ... We restrict ourselves to classical systems and in the following we adopt the spin language so that the system can be seen as a model of magnetic spins. In this first part we shall consider the case of spin ½ systems; therefore with each lattice site x in $\mathcal{L}$ is associated a spin variable s_x which takes the values ± 1, and describes the configurations of the subsystem at the site x. (one may think of these two values as the possible z-component of the spin variable of a particle of spin $\hbar/2$ located at the site x).

A finite system is defined by a finite subset Λ of $\mathcal{L}$ and a configuration may then be specified by the set of sites $X \subset \Lambda$ for which $s_x = -1$, $x \in X$. Therefore the configuration space is $\mathcal{P}(\Lambda) = \{X ; X \subset \Lambda\}$, the set of subsets of Λ (*).

The Hamiltonian of the system is defined by a function on the configuration space and its general form is given by :

$$H = -\sum_{X \subset \Lambda} J(X)\, \sigma_X \qquad (1.1)$$

where $\sigma_X = \prod_{x \in X} \sigma_x$ is the function defined by :

---

* We shall denote by capital letters X,Y,S,T,.. subsets of $\mathcal{L}$.

$$\sigma_X(Y) = (-1)^{|X \cap Y|} \tag{1.2}$$

and where for any $Z \subset \Lambda$, $|Z|$ denotes the cardinality of Z ; J is a complex function on $\mathcal{P}(\mathcal{L})$ which describes the many-body or multiple-spin interaction. To simplify the notation, we introduce $K(X) = \beta J(X)$ where $\beta = (kT)^{-1}$ is the reciprocal temperature.

In conclusion a general spin ½ lattice system is then defined by $\{\mathcal{L}, K\}$ with $\mathcal{L}$ a countable set of points and K a complex function on $\mathcal{P}(\mathcal{L})$.

1.2. Thermodynamics and Correlation Functions

The thermodynamics of the system is obtained from the study of the properties of the partition function and free energy density; for a finite system $\{\Lambda, K\}$ with $|\Lambda|$ points,(*) these quantities are defined by :

$$\begin{aligned} Z(\Lambda,K) &= \sum_{X\subset\Lambda} e^{\sum_{Y\subset\Lambda} K(Y)\sigma_Y(X)} \\ -\beta f_{(\Lambda,K)} &= |\Lambda|^{-1}. \ln Z(\Lambda,K) = -F_{(\Lambda,K)} \end{aligned} \tag{1.3}$$

Moreover, we shall also be interested in the correlation functions, functions on $\mathcal{P}(\Lambda)$, defined by :

$$\langle\sigma_X\rangle_{(\Lambda,K)} = Z^{-1} \sum_{Z\subset\Lambda} e^{\sum_{Y\subset\Lambda} K(Y)\sigma_Y(Z)} \sigma_X(Z) \tag{1.4}$$

In order to obtain a general and concise formulation, as well as to investigate some properties of such systems, it is

(*) In the following we shall always denote by Λ a finite set of sites; on the other hand $\mathcal{L}$ can be either finite or infinite.

convenient and useful to associate a group structure on the lattice, which we shall now introduce.

1.3. Group Structure for Finite Lattice Systems $\{\Lambda, K\}$ [2]

1.3.1. Group of Configuration on $\mathcal{P}(\Lambda)$

$\mathcal{P}(\Lambda)$ is taken to be an abelian(commutative) group with the product defined by the symmetric difference [3] , i.e. for any X , Y in $\mathcal{P}(\Lambda)$,

$$X \cdot Y = X \cup Y - X \cap Y$$

This group is of order $2^{|\Lambda|}$.

With any $X \subset \mathcal{P}(\Lambda)$, we associate the function σ_X on $\mathcal{P}(\Lambda)$ defined by Eq. (1.2) and having the following properties.

Property 1 :

a) $$\sigma_X(Y) = \sigma_Y(X)$$

b) $$\sigma_X(Y) \cdot \sigma_X(Z) = \sigma_X(Y \cdot Z)$$

c) $$\sigma_X(Z) \cdot \sigma_Y(Z) = \sigma_{X \cdot Y}(Z)$$

therefore the functions σ_X are the characters of $\mathcal{P}(\Lambda)$ and the mapping $X \mapsto \sigma_X$ is the isomorphism between $\mathcal{P}(\Lambda)$ and its character group $\hat{\mathcal{P}}(\Lambda)$. In this way to each spin lattice system, we can associate the groups $\mathcal{P}(\Lambda)$ and $\hat{\mathcal{P}}(\Lambda)$.

The orthogonality and completude relations read [3] :

d) $$\sum_{Z \subset \Lambda} \sigma_X(Z) \sigma_Y(Z) = 2^{|\Lambda|} \delta_{X,Y}$$

e) $$\sum_{Z\subset\Lambda} \sigma_Z(X)\,\sigma_Z(Y) = 2^{|\Lambda|}\,\delta_{X,Y}$$

Eq. (1.1) is then the Fourier decomposition of H in the basis $\{\sigma_X\}$ of $\hat{\mathcal{P}}(\Lambda)$, where the Fourier transform of a function f defined on $\mathcal{P}(\Lambda)$ is given by [4] :

$$f = \sum_{X\in\mathcal{P}(\Lambda)} \hat{f}(X)\cdot\sigma_X$$

and satisfies :

$$f^{\wedge} = 2^{-|\Lambda|} \sum_{X\in\mathcal{P}(\Lambda)} f(X)\cdot\sigma_X \qquad (1.5)$$

1.3.2. Definition of Some Subgroups of $\mathcal{P}(\Lambda)$

Set $\mathcal{B}$ of Bonds : subset of $\mathcal{P}(\Lambda)$ defined as the support of the function K, i.e. $\mathcal{B} = \{B\subset\Lambda;\ K(B)\neq 0\}$.

Interaction Subgroup $\bar{\mathcal{B}}$: subgroup of $\mathcal{P}(\Lambda)$ generated by $\mathcal{B}$.

Internal Symmetry Group $\mathcal{S}$: subgroup of $\mathcal{P}(\Lambda)$ defined by :
$\mathcal{S} = \{S\subset\Lambda;\ \sigma_B(S) = +1\ \forall B\in\mathcal{B}\}$.

Property 2 :

a) $\bar{\mathcal{B}} = \{X;\ \sigma_X(S) = 1\ \forall S\in\mathcal{S}\} = \mathcal{S}^{\perp}$

b) $\mathcal{P}(\Lambda)/\mathcal{S} \cong \mathcal{S}^{\perp} = \bar{\mathcal{B}}$, $\quad\mathcal{P}(\Lambda)/\bar{\mathcal{B}} \cong \mathcal{B}^{\perp} = \mathcal{S}$

c) With $|\bar{\mathcal{B}}| = 2^{N_i}$, $|\mathcal{S}| = 2^{N_s}$ then $|\Lambda| - N_i = N_s$.

d) If $K(B) > 0\ \forall B\in\mathcal{B}$, i.e. for ferromagnetic systems, then each element $S\in\mathcal{S}$ defines a groundstate.

Groundstate :

Y is called a groundstate if $H(X) \geq H(Y)$ for all $X \in \mathcal{P}(\Lambda)$. Notice that for any S in $\mathcal{S}$, $H(S \cdot Y) = H(Y)$ and assuming $K(B) > 0$ we thus have d).

Properties a), b) and c) are consequences of the following lemma.

Lemma 1 [4]

Let $\mathcal{G}$ be any locally compact abelian group, H a subset of $\mathcal{G}$, and $\hat{\mathcal{G}}$ the dual group of $\mathcal{G}$; with $H^{\perp} = \{ \chi \in \hat{\mathcal{G}} ; \chi(h) = +1 \ \forall h \in H \}$, then $H^{\perp} \cong (\mathcal{G}/\bar{H})^{\wedge}$

$$\text{and} \quad \bar{H} \cong (\hat{\mathcal{G}}/H^{\perp})^{\wedge}$$

where $\bar{H}$ is the closed subgroup of $\mathcal{G}$ generated by H. Moreover $H^{\perp\perp} = \bar{H}$.

In the above discussion we had $\mathcal{G} = \mathcal{P}(\Lambda)$, $H = \mathcal{B}$, $\mathcal{B}^{\perp} = \mathcal{S}$, therefore, $\mathcal{P}(\Lambda)/\bar{\mathcal{B}} \cong \mathcal{B}^{\perp} = \mathcal{S}$ and $\mathcal{P}(\Lambda)/\mathcal{S} \cong \mathcal{S}^{\perp} = \bar{\mathcal{B}}$ (since $\mathcal{P}(\Lambda)$ is isomorphic to its dual group).

The interest of the groups $\mathcal{S}$ and $\bar{\mathcal{B}}$ are given by the properties which we now discuss.

Symmetry Group $\{\tau_S\}_{S \in \mathcal{S}}$

Let $\mathfrak{A}_{\Lambda} = \{ A \}$ be the algebra of observables* [1] ;

---

* By definition $A \in \mathfrak{A}_{\Lambda}$ is a bounded function $A : X \mapsto A(X)$ defined on $\mathcal{P}(\Lambda)$; in particular for finite systems it follows from Eq. (1.5) that each observable of finite system has a Fourier decomposition of the form $A = \sum_{Y \subset \Lambda} A^{\wedge}(Y) \sigma_Y$.

$\mathcal{S}$ acts as a group of automorphisms on $\mathfrak{A}_\Lambda$ in the following way : for all S in $\mathcal{S}$ we define the automorphism τ_S by :

$$\begin{array}{rcl} \tau_S : \mathfrak{A}_\Lambda & \longrightarrow & \mathfrak{A}_\Lambda \\ \quad\ \ \, \cup\!\!\!\!\!\! & & \cup\!\!\!\!\! \\ A & \longmapsto & (\tau_S A)(X) = A(S\cdot X) \end{array} \tag{1.6}$$

in particular $\tau_S \sigma_Y = \sigma_Y(S)\, \sigma_Y$

τ_S induces a transformation τ'_S of the state* ω as [1] :

$$(\tau'_S \omega)[A] = \omega[\tau_S A]$$

which yields in particular :

$$(\tau'_S \omega)[\sigma_X] = \omega[\tau_S \sigma_X] = \sigma_X(S)\, \omega[\sigma_X] \tag{1.7}$$

<u>Definition</u> :

A state ω is said <u>invariant under $\mathcal{S}$</u> or <u>symmetric</u> if $\tau'_S \omega = \omega$ for all S in $\mathcal{S}$.

<u>Property 3</u> :

A state ω is invariant under $\mathcal{S}$ if and only if $\omega[\sigma_X] = 0$ for $X \notin \mathcal{S}^\perp = \bar{\mathcal{B}}$.

Indeed : $\tau'_S \omega = \omega$ implies $(1 - \sigma_X(S))\, \omega[\sigma_X] = 0$ and therefore $\sigma_X(S) = +1$ for all X such that $\omega[\sigma_X] \neq 0$. It thus follows that ω is invariant under $\mathcal{S}$ if and only if $\omega[\sigma_X] = 0$ for all $X \notin \mathcal{S}^\perp = \bar{\mathcal{B}}$.

Let us remark that this property will still be valid for infinite systems.

---

* A state ω is a probability measure P on $\mathcal{P}(\Lambda)$ and for all A in $\mathfrak{A}_\Lambda$, $\omega[A] = \sum_{X \subset \Lambda} A(X)\, P(X)$. For a finite system $\{\Lambda, K\}$, the state defined by the correlation function Eq. (1.4) is called the <u>Gibbs state</u>, and is denoted by $\omega_{\beta,\Lambda}$.

Property 4 :

The Hamiltonian H is invariant under $\mathcal{S}$, i.e. $\tau_S H = H$ $\forall S \in \mathcal{S}$. The Gibbs State $\omega_{\beta,\Lambda}$ is invariant under $\mathcal{S}$, i.e. $\tau'_S \omega_{\beta,\Lambda} = \omega_{\beta,\Lambda}$ $\forall S \in \mathcal{S}$.

1.3.3. Group of Graphs $\mathcal{P}(\mathcal{B})$ and Definition of Some Subgroups of $\mathcal{P}(\mathcal{B})$ [2]

The group of graphs $\mathcal{P}(\mathcal{B})$ is the group defined by the subsets β of $\mathcal{B}$ with the product defined by the symmetric difference. (Group defined by analogy with $\mathcal{P}(\Lambda)$ where the set of sites has been replaced by the set of bonds) .

Subgroup $\mathcal{K}$: subgroup of $\mathcal{P}(\mathcal{B})$ defined as the kernel of the surjective homorphism π of $\mathcal{P}(\mathcal{B})$ onto $\overline{\mathcal{B}} \subset \mathcal{P}(\Lambda)$, homomorphism defined by :

$$\begin{array}{cccc} \pi : & \mathcal{P}(\mathcal{B}) & \longrightarrow & \overline{\mathcal{B}} \\ & \cup & & \cup \\ & \beta = (B_1, \cdots B_n) & \mapsto & \pi\beta = \prod_{i=1}^{n} B_i \end{array}$$

$$\mathcal{K} = \ker \pi = \{ \beta \in \mathcal{P}(\mathcal{B}) ; \pi\beta = \phi \} \cong \mathcal{P}(\mathcal{B})/\overline{\mathcal{B}} \qquad (1.8)$$

By definition π is surjective and is an homomorphism since $B^2 = \phi$ for all B in $\mathcal{B}$.

$\mathcal{K}$ is therefore precisely the group of elements β , such that for each lattice site x of Λ there is an even number of bonds B in β containing x ; we shall then call "closed graph" the elements of $\mathcal{K}$, and $\mathcal{K}$ "group of closed graph" or "High Temperature Group" (see Ch. 2) ; it is a group of order $|\mathcal{K}| = 2^{|\mathcal{B}| - N_i}$ and $|\mathcal{B}| - N_i$ is the minimal number of generators of $\mathcal{K}$.

Subgroup Γ : subgroup of $\mathcal{P}(\mathcal{B})$ defined as the image of the homomorphism γ of $\mathcal{P}(\Lambda)$ into $\mathcal{P}(\mathcal{B})$:

$$\begin{array}{cccc} \gamma : & \mathcal{P}(\Lambda) & \longrightarrow & \mathcal{P}(\mathcal{B}) \\ & \Psi & & \Psi \\ & X & \longmapsto & \gamma(X) = \{B \in \mathcal{B}; \sigma_X(B) = -1\} \end{array} \quad (1.9)$$

The connection between the mappings π and γ is given by the relation [5]

$$<\beta ; \gamma(X)>_{\mathcal{B}} = <\pi\beta ; X>_{\Lambda} \quad (1.10)$$

where $<Y ; X>_{\Lambda} = \sigma_Y(X)$ and $<\beta ; \beta'>_{\mathcal{B}} = \sigma_{\beta}(\beta')$ are the standard bicharacter notations [4] .

From the above relation, it follows immediately that γ is an homomorphism since for all $\beta \subset \mathcal{B}$, $X_1, X_2 \in \mathcal{P}(\Lambda)$, we have :

$$<\beta ; \gamma(X_1X_2)>_{\mathcal{B}} = <\pi\beta ; X_1 \cdot X_2>_{\Lambda} = <\pi\beta ; X_1>_{\Lambda} \cdot <\pi\beta ; X_2>_{\Lambda} =$$
$$= <\beta ; \gamma(X_1)>_{\mathcal{B}} <\beta ; \gamma(X_2)>_{\mathcal{B}} = <\beta ; \gamma(X_1) \cdot \gamma(X_2)>_{\mathcal{B}}$$

and therefore $\gamma(X_1X_2) = \gamma(X_1) \cdot \gamma(X_2)$. Moreover $\ker \gamma = \{ S ; \sigma_B(S) = +1 \ \ \forall B \in \mathcal{B}\} = \mathcal{S}$, yields $\Gamma = \gamma(\mathcal{P}(\Lambda)) \cong \mathcal{P}(\Lambda)/\mathcal{S}$ and γ defines an injective homomorphism of $\mathcal{P}(\Lambda)/\mathcal{S}$ into $\mathcal{P}(\mathcal{B})$; we then have : $|\mathcal{P}(\Lambda)/\mathcal{S}| = |\Gamma| = 2^{|\Lambda| - N_S} = 2^{N_i}$. Moreover, since the image of a generating set for $\mathcal{P}(\Lambda)$ yields a subset generating Γ , Γ can be generated by the elements $\gamma(x) = \{ B ; \sigma_B(x) = -1\} = \{B \ni x\}$ $x \in \Lambda$. Notice that the elements of Γ are those subsets of $\mathcal{B}$ having an odd number of sites in common with some given subsets of Λ .

Using Lemma 1 we obtain the connection between the groups $\mathcal{K}$ and Γ ; indeed

$$\begin{aligned} \Gamma &\cong \mathcal{P}(\Lambda)/\mathcal{J} \cong \bar{\mathcal{B}} \\ \bar{\mathcal{B}} &\cong \mathcal{P}(\mathcal{B})/\mathcal{K} \cong \mathcal{K}^{\perp} \end{aligned} \qquad (1.11)$$

implies $\Gamma \cong \mathcal{K}^{\perp}$ and $\Gamma^{\perp} \cong \mathcal{K}$.

It should be noticed that these relations can be strengthened using the equation (1.10) $<\beta ; \gamma(X)>_{\mathcal{B}} = <\pi\beta ; X>_{\Lambda}$; indeed $\beta \in \mathcal{K} \Leftrightarrow <\pi\beta ; X>_{\Lambda} = 1 \ \forall X$, i.e. $<\beta ; \gamma(X)>_{\mathcal{B}} = 1 \ \forall \gamma(X)$, therefore $\beta \in \mathcal{K} \Leftrightarrow \beta \in \Gamma^{\perp}$, which gives(*)

$$\Gamma = \mathcal{K}^{\perp} \quad \text{and} \quad \Gamma^{\perp} = \mathcal{K} \ . \qquad (1.12)$$

This structure represents the basic mathematical tool which we need; the interest of these abelian groups for lattice systems will be shown already in the next chapter, where we define the generalized duality transformation and investigate the thermodynamic equivalence between a large class of lattice systems.

(*) We identify the group $\mathcal{P}(\mathcal{B})$ with its character group $\mathcal{P}(\mathcal{B})^{\wedge}$ using the isomorphism $\beta \leftrightarrow \sigma_{\beta}$.

1.4. Group Structure for Infinite Systems

To conclude this section we recall that the modern approach to some problems in Statistical Mechanics consists in studying directly infinite systems rather than considering first a finite system and then take the thermodynamic limit [see also Ch. 4, 6-9]. We shall now show that the group structure which we have introduced so far for finite systems $\{\Lambda, K\}$, can also be introduced for infinite systems with only minor modifications; moreover we shall introduce this group structure in a way which makes straightforward the generalization to arbitrary spin (Part III).

A finite or infinite spin $\frac{1}{2}$ Lattice System Σ is abstractly defined by $\Sigma = \{\mathcal{L}, \mathcal{B}, K, \pi\}$, where :

$\mathcal{L}$ is a countable set of points x, called the "lattice sites"

$\mathcal{B}$ is a countable set of indices b, called the "bonds"

K is a real or complex function on $\mathcal{B}$

π is a mapping $b \mapsto \sigma_b$ from $\mathcal{B}$ into the Dual Group $\hat{\mathcal{P}}(\mathcal{L})$ where

$$\hat{\mathcal{P}}(\mathcal{L}) \cong \mathcal{P}_f(\mathcal{L}) = \{X \subset \mathcal{L};\ |X| < \infty\} \tag{1.13}$$

For any $\Lambda \in \mathcal{P}_f(\mathcal{L})$ and $Y \subset \mathcal{L}/\Lambda$, the energy of the configuration $X \subset \Lambda$ given the configuration Y outside of Λ, $H_{\Lambda,Y}(X)$ is defined by :

$$H_{\Lambda,Y}(X) = -kT \sum_{b \in \mathcal{B}_\Lambda} K(b)\,\sigma_b(Y)\,\sigma_b(X) \tag{1.14}$$

$$\mathcal{B}_\Lambda = \{ b \in \mathcal{B};\ \sigma_b(x) = -1 \quad \text{for some } x \text{ in } \Lambda \}$$

Although in the previous discussion, we had considered the set $\mathcal{B}$ to be defined by the support of K and therefore $b \mapsto \sigma_b$ was an injection, it will not always be the case;

in particular to define LT-HT duality (Ch. 2 example 4 and Ch. 3) and to discuss higher spin systems (Part III) it will be necessary to consider cases where $b \mapsto \sigma_b$ is not an injection.

With the system $\Sigma = \{\mathcal{L}, \mathcal{B}, K, \Pi\}$ is associated the group structure defined by the groups $\mathcal{P}(\mathcal{L})$, $\hat{\mathcal{P}}(\mathcal{L}) \cong \mathcal{P}_f(\mathcal{L})$, $\mathcal{P}(\mathcal{B})$, $\hat{\mathcal{P}}(\mathcal{B}) \cong \mathcal{P}_f(\mathcal{B})$, together with the homomorphism π and γ defined by :

$$\begin{array}{cccc} \pi : & \mathcal{P}(\mathcal{B}) & \longrightarrow & \mathcal{P}(\mathcal{L}) \\ & \psi & & \psi \\ & \beta & \longmapsto & \pi\beta \\ & & & \sigma_{\pi\beta} = \prod_{b\in\beta} \sigma_b \end{array} \qquad \begin{array}{cccc} \gamma : & \mathcal{P}(\mathcal{L}) & \longrightarrow & \mathcal{P}(\mathcal{B}) \\ & \psi & & \psi \\ & X & \longmapsto & \gamma(X) = \{b ; \sigma_b(X) = -1\} \end{array}$$

and related by(*) : $\langle \pi\beta ; X \rangle_{\mathcal{L}} = \langle \beta ; \gamma(X) \rangle_{\mathcal{B}}$ (1.15)

for all $\{\beta \in \mathcal{P}_f(\mathcal{B}), X \in \mathcal{P}(\mathcal{L})\}$, and for all $\{\beta \in \mathcal{P}(\mathcal{B}), X \in \mathcal{P}_f(\mathcal{L})\}$

The <u>Kernel $\mathcal{S}$ of γ</u> and the <u>image $\bar{\mathcal{B}}$ of π restricted to $\mathcal{P}_f(\mathcal{B})$</u> are subgroups of $\mathcal{P}(\mathcal{L})$ and $\mathcal{P}_f(\mathcal{L})$ given by :

$$\mathcal{S} = \{ S \in \mathcal{P}(\mathcal{L}) ; \sigma_b(S) = +1 \quad \forall b \in \mathcal{B} \}$$

$$\bar{\mathcal{B}} = \{ \bar{B} \in \mathcal{P}_f(\mathcal{L}) ; \bar{B} = \pi\beta \quad \beta \in \mathcal{P}_f(\mathcal{B}) \}$$

and related by : $\mathcal{S} = \bar{\mathcal{B}}^{\perp}$ $\quad \bar{\mathcal{B}} = \mathcal{S}^{\perp}$

The <u>kernel $\mathcal{K}$ of π</u> and the <u>image Γ of γ</u> are subgroups of $\mathcal{P}(\mathcal{B})$ given by :

$$\mathcal{K} = \{ \beta \in \mathcal{P}(\mathcal{B}) ; \prod_{b\in\beta} \sigma_b(X) = +1 \quad \text{for all } X \in \mathcal{P}_f(\mathcal{L}) \}$$

$$\Gamma = \{ \gamma(X) ; \quad X \in \mathcal{P}(\mathcal{L}) \}$$

(*) Eq. (1.15) could also be considered as the definition of γ since $\langle \{b\} ; \gamma(X) \rangle_{\mathcal{B}} = \langle \pi\{b\} ; X \rangle_{\mathcal{L}} = \sigma_b(X)$ implies that $b \in \gamma(X) \Leftrightarrow \sigma_b(X) = -1$ [5].

and are related by [6]

$$\mathcal{K} = \Gamma^{f\perp} \qquad \text{and} \qquad \mathcal{K}^{\perp} = \Gamma^{f}$$

$$\mathcal{K}_f = \Gamma^{\perp} \qquad \text{and} \qquad \mathcal{K}_f^{\perp} = \Gamma$$

where $\mathcal{K}_f = \mathcal{K} \cap \mathcal{P}_f(\mathcal{B})$ and $\Gamma^f = \{\gamma(X)\ ;\ X \in \mathcal{P}_f(\mathcal{L})\}$

relation which are obtained directly from Eq. (1.15).

Moreover :

$$(\mathcal{P}(\mathcal{B})/\mathcal{K})^{\wedge} \cong \Gamma^{f} \cong \mathcal{I}m\,\pi^{\wedge}\ ;\ (\mathcal{P}_f(\mathcal{B})/\mathcal{K}_f)^{\wedge} \cong \Gamma \cong \bar{\mathcal{B}}^{\wedge}$$

$$(\mathcal{P}(\mathcal{L})/\mathcal{S})^{\wedge} \cong \bar{\mathcal{B}} \cong \Gamma^{\wedge}\ ;\ (\mathcal{P}_f(\mathcal{L})/\mathcal{S}_f)^{\wedge} \cong (\Gamma^{f})^{\wedge} \cong \mathcal{S}_f^{\perp} = \mathcal{I}m\,\pi$$

where $\mathcal{S}_f = \mathcal{S} \cap \mathcal{P}_f(\mathcal{L})$

Property 5 :

$\sigma_{b_1} = \sigma_{b_2} \iff (b_1, b_2) \in \mathcal{K} \iff$ for any $\beta \in \Gamma$ $\ b_1 \in \beta$ implies $b_2 \in \beta$

Summary of the Group Structure

In conclusion, for any general lattice system $\Sigma = \{\mathcal{L}, \mathcal{B}, K, \pi\}$ we have the following Group Structure :

$$\mathcal{P}(\mathcal{L}) \underset{\pi}{\overset{\gamma}{\rightleftarrows}} \mathcal{P}(\mathcal{B})$$

$\mathcal{S} = \mathcal{B}^{\perp}$	$\Gamma = \mathcal{K}_f^{\perp}$
$\bar{\mathcal{B}} = \mathcal{S}^{\perp}$	$\mathcal{K} = \Gamma^{f\perp}$

$$\bar{\mathcal{B}} \cong (\mathcal{P}(\mathcal{L})/\mathcal{S})^{\wedge} \cong \Gamma^{\wedge}$$

$$\Gamma \cong (\mathcal{P}_f(\mathcal{B})/\mathcal{K}_f)^{\wedge} \cong \bar{\mathcal{B}}^{\wedge}$$

CHAPTER 2 - THE DUALITY TRANSFORMATION

2.1. The Duality Relation

In the history of the two-dimensional Ising Model without external field, the duality relation was introduced by Kramers and Wannier [7] . This duality relation appears as a relation between the high and the low temperature expansion of the partition function of the model [7,8,9] . In fact Kramers and Wannier started from the expansion of $Z(\Lambda, K)$ where each non-vanishing term in the expansion can be seen as a topological object, a closed graph, i.e. a set of multipolygons on the square lattice Λ . On the other hand the low temperature expansion was obtained by counting all the contributions from the perturbed ordered state at $T=0$, each contribution corresponding to the energy associated with the presence of droplets of overturned spin in Λ , i.e. a configuration in Λ . To specify this low temperature expansion, Kramers and Wannier associated with each configuration a set of multipolygons on the "dual lattice", lattice obtained by drawing a unit segment perpendicular to the center of each bond having opposite spins at its extreme. In this way, they were able to relate the value of the partition function at temperature T to value of the partition function at a temperature $T^* = T^*(T)$. Assuming the existence of a unique critical point, this duality relation was sufficient to obtain the critical temperature for this model (Wannier argument) before the solution was published by Onsager [10] . Later, it was proved that both expansions are essentially related by a Poisson Formula for some abelian groups associated with the lattice [11, 5].

This duality relation reflects some symmetry of the model and one can expect that this is not a particularity of the Ising Model; in fact the duality concept can be extended to arbitrary systems and we shall introduce the general concept of duality transformation from one model to another model, called the dual model. For such a generalization the idea is to use the group structure considered in Chapter 1, from which the duality concept emerges naturally.[2]

2.2. The High and Low Temperature Expansion

High Temperature Expansion : with the definition of $Z(\Lambda,\mathcal{K})$ and of the group $\mathcal{K}$, we have immediately :

$$Z(\Lambda,K) = 2^{|\Lambda|} \prod_{B\in\mathcal{B}} \cosh K(B) \sum_{\mathcal{T}\beta=\phi} \prod_{B\in\beta} \tanh K(B) =$$

$$= 2^{|\Lambda|} \prod_{B\in\mathcal{B}} \cosh K(B) \sum_{\beta\in\mathcal{K}} \prod_{B\in\beta} \tanh K(B) \tag{2.1}$$

where we have used the orthogonality of the characters (Property 1, Ch. 1). The expansions parameters are given by $\{\tanh K(B)\}$ which are precisely small at high temperature (small K); therefore with each closed graph β in $\mathcal{K}$ is associated the weight $w(\beta) = \prod_{B\in\beta} \tanh K(B)$ which is smaller than 1 for real interaction $K(B)$. The group $\mathcal{K}$ of closed graphs is thus the group associated with the HT expansion, which justifies the definition "High Temperature Group" introduced in Chapter 1.

Low Temperature Expansion : in the same way, using the definition of Γ and $\mathcal{J}$ we have the following expansion :

$$Z(\Lambda,K) = \prod_{B\in\mathcal{B}} e^{K(B)} \sum_{X\subset\Lambda} \prod_{B\in\mathcal{B}} e^{-K(B)[1-\sigma_B(X)]} =$$

$$= \prod_{B\in\mathcal{B}} e^{K(B)} \cdot |\mathcal{J}| \cdot \sum_{\beta\in\Gamma} \prod_{B\in\beta} e^{-2K(B)} \tag{2.2}$$

The expansion parameters are given by $\{e^{-2K(B)}\}$ and are small at low temperature (large K) when $K(B)>0$ $\forall B \in \mathcal{B}$.

We shall then call "Low Temperature Graph", the elements β of Γ and "Low Temperature Group", the group Γ ; with each low temperature graph β is associated a weight $\omega(\beta)=\prod_{B\in\beta} e^{-2K(B)}$. which is smaller than 1 for purely ferromagnetic systems (i.e. $K(B) > 0$ for all B in $\mathcal{B}$).

Let us remark that for systems which are not purely ferromagnetic, we can also write the Low Temperature expansion as a perturbation expansion with respect to the groundstate Y (as defined on p.5 and which corresponds to an ordered state at $T=0$), in the following manner :

$$Z(\Lambda,K) = 2^{N_S} \prod_{B\in\mathcal{B}} e^{K(B)\sigma_B(Y)} \sum_{\beta\in\Gamma} \prod_{B\in\beta} e^{-2K(B)\sigma_B(Y)}$$

and in this case, the expansion parameters are $\{e^{-2K(B)\sigma_B(Y)}\}$.

2.3. Duality Transformation for Finite Systems and Duality Relation for the Partition Function

We now discuss further applications of the group structure for general lattice systems; notice that beside the Kramers-Wannier duality transformation, several examples exist; F. Wegner has given a first generalization; explicit examples are given in [12], but the role of the boundary conditions is not investigated. On the other hand, it must be noted that the usual definition of duality assumes the existence of a bijection between $\mathcal{B}$ and $\mathcal{B}^*$ where $\mathcal{B}$ and $\mathcal{B}^*$ are respectively the set of bonds for the lattice system $\{\Lambda,K\}$ and its dual $\{\Lambda^*,K^*\}$. However, as we shall see later, this definition is not sufficiently general since it is possible to find dual transformations such that $\mathcal{B} \longrightarrow \mathcal{B}^*$ is not necessarily a bijection [2].

Definition :

Let $\{\Lambda,K\}$ be a finite lattice system; the lattice system $\{\Lambda^*,K^*\}$ is called a High Temperature - Low Temperature (HT-LT) dual for $\{\Lambda,K\}$ if there exists a surjective mapping d of $\mathcal{B}$ onto $\mathcal{B}^*$ such that $\mathcal{D}: \mathcal{P}(\mathcal{B}) \rightarrow \mathcal{P}(\mathcal{B}^*)$ defined as $\mathcal{D}\beta = \bigcup_{B\in\beta} (dB)$ induces a surjection of $\mathcal{K}$ onto Γ^* satisfying the condition $\beta = \bigcup_{B\in\beta} d^{-1}(dB)$ and if the interaction satisfies

$$e^{-2K^*(B^*)} = \prod_{B\in d^{-1}B^*} \tanh K(B). \tag{2.3}$$

Remark :

As we shall see below, from the above condition and the fact that $\mathcal{D}$ is a surjection, it follows that the mapping $\mathcal{K} \longrightarrow \Gamma^*$ is an isomorphism. Therefore, we could also define the HT-LT duality transformation as a surjection $d: \mathcal{B} \mapsto \mathcal{B}^*$ which induces a bijection of $\mathcal{K}$ onto Γ^* satisfying Eq. (2.3).

Similarly (LT-HT), (HT-HT) and (LT-LT) duals are defined by replacing $(\mathcal{K},\Gamma^*)$ by $(\Gamma,\mathcal{K}^*)$, $(\mathcal{K},\mathcal{K}^*)$ and (Γ,Γ^*) respectively with corresponding change in the role of the exponential and tanh functions in Eq. (2.3). For example $\{\Lambda^*,K^*\}$ is a (LT-HT) dual for $\{\Lambda,K\}$ if there exists a surjection $d: B\mapsto B^*$ of $\mathcal{B}$ onto $\mathcal{B}^*$ such that $\Gamma\to\mathcal{K}^*$ is a surjection,

$$\tanh K^*(B^*) = \prod_{B\in d^{-1}B^*} e^{-2K(B)} \quad \text{and such that} \quad \forall\beta\in\Gamma \quad ,$$

$\beta = \bigcup_{B\in\beta} d^{-1}(dB)$. In all cases we call the mapping $\{\Lambda,K\}\to$ $\to\{\Lambda^*,K^*\}$ a <u>generalized duality transformation</u>. From this definition we obtain immediately the following relations :

<u>Proposition 1</u> :

If $\{\Lambda^*,K^*\}$ is a HT-LT, LT-HT, HT-HT, LT-LT dual lattice system for $\{\Lambda,K\}$ then :

i)

$$|\mathcal{B}| - N_i = |\Lambda^*| - N_s^* = N_i^* \qquad \text{HT-LT}$$

$$|\Lambda| - N_s = N_i = |\mathcal{B}^*| - N_i^* \qquad \text{LT-HT}$$

$$|\mathcal{B}| - N_i = |\mathcal{B}^*| - N_i^* \qquad \text{HT-HT}$$

$$|\Lambda| - N_s = |\Lambda^*| - N_s^* \qquad \text{LT-LT}$$

ii) The partition functions of dual systems are related by the following <u>Duality Relations</u> :

$$Z(\Lambda,K) = 2^{|\Lambda|}|\mathcal{S}^*|^{-1}\prod_{B\in\mathcal{B}}\cosh K(B)\prod_{B^*\in\mathcal{B}^*} e^{-K^*(B^*)} Z(\Lambda^*,K^*) \qquad \text{HT-LT}$$

$$Z(\Lambda,K) = |\mathcal{S}|\,2^{-|\Lambda^*|}\prod_{B\in\mathcal{B}} e^{K(B)}\prod_{B^*\in\mathcal{B}^*}\cosh K^*(B^*)^{-1} Z(\Lambda^*,K^*) \qquad \text{LT-HT} \tag{2.4}$$

$$Z(\Lambda,K) = 2^{|\Lambda|-|\Lambda^*|}\prod_{B\in\mathcal{B}}\cosh K(B)\prod_{B^*\in\mathcal{B}^*}\cosh K^*(B^*)^{-1} Z(\Lambda^*,K^*) \qquad \text{HT-HT}$$

$$Z(\Lambda,K) = |\mathcal{S}|\cdot|\mathcal{S}^*|^{-1}\prod_{B\in\mathcal{B}} e^{K(B)}\prod_{B^*\in\mathcal{B}^*} e^{-K^*(B^*)} Z(\Lambda^*,K^*) \qquad \text{LT-LT}$$

The proof of these relations is immediate; for example in the HT-HT case, we have :

$$
\begin{aligned}
Z(\Lambda,K) &= 2^{|\Lambda|}\prod_{B\in\mathcal{B}}\cosh K(B)\sum_{\beta\in\mathcal{K}} W(\beta) = \\
&= 2^{|\Lambda|}\prod_{B\in\mathcal{B}}\cosh K(B)\sum_{\beta^*\in\mathcal{K}^*}\prod_{B^*\in\beta^*}\prod_{B\in d^{-1}B^*}\tanh K(B) = \\
&= 2^{|\Lambda|-|\Lambda^*|}\prod_{B\in\mathcal{B}}\cosh K(B)\prod_{B^*\in\mathcal{B}^*}\cosh K^*(B^*)^{-1} Z(\Lambda^*,K^*)
\end{aligned}
$$

These relations can be written in a more concise form; for example HT-LT duality has the form :

$$
Z(\Lambda,K) = \frac{2^{(|\Lambda|+N_S)\frac{1}{2}}}{2^{(|\Lambda^*|+N_S^*)\frac{1}{2}}}\cdot\prod_{B\in\mathcal{B}}\left(\sinh 2K(B)\right)^{1/2} Z(\Lambda^*,K^*) \qquad (2.5)
$$

To obtain a general method to construct a dual model, we need the following lemma :

Lemma : [2]

Let $\{\Lambda,K\}$ and $\{\Lambda^*,K^*\}$ be two spin lattice systems and G be a subgroup of $\mathcal{P}(\mathcal{B})$. Given a surjective mapping $d:\ \mathcal{B}\rightarrow\mathcal{B}^*$ such that the induced mapping :

$$
\begin{aligned}
D:\quad &\mathcal{P}(\mathcal{B}) \longrightarrow \mathcal{P}(\mathcal{B}^*) \\
&\beta=\{B\} \mapsto \beta^*=\{B^*\}
\end{aligned}
$$

satisfies the condition below, then $D|_G$ is an injective group homomorphism.

Condition :

With $(\beta_1,\cdots\beta_n)$ a subset generating G then $\beta_i = D^{-1}(D\beta_i)$ where $D^{-1}\beta^* = \{B\,;\,dB\in\beta^*\}$, i.e. $B\in\beta_i$ implies that β_i contains the whole fiber over B .

Proof :

Let $\beta_i^* = D\beta_i = \{B_{ij}^*\}_{j=1,\dots n_i}$; for all β in G

we have the decomposition $\beta = \prod_{i\in I} \beta_i$ with $I \subset \{1,2,\cdots n\}$.

With the above condition, we have $\beta = \prod_{i\in I} \{d^{-1} B_{ij}^*\}_{j=1,2,\dots n_i} = \{d^{-1}B_\alpha^*\}$ where $\{B_\alpha^*\}$ are those B_{ij}^* which appear an odd number of times in the set $\{B_{ij}\}_{\substack{i\in I\\ j=1,2,\cdots n_i}}$; moreover $\beta^* = D\beta = \{B_\alpha^*\} = \prod_{i\in I}\{B_{ij}\}_{j=1,2,\dots n_i}$ therefore :

$$D(\prod \beta_i) = \prod_{i\in I}(D\beta_i)$$

thus $D|_G$ is a group homomorphism; it is injective, since from the condition, it follows that $\beta_i \neq \beta_j \Rightarrow \beta_i^* \neq \beta_j^*$.
It must be remarked that the above condition, i.e. $\beta = \bigcup_{B\in\beta} d^{-1}(dB)$ is trivially satisfied if d is a bijection; this assumption, which is usually tacitely assumed in the literature is not always satisfied; we note that $D: \mathcal{P}(\mathcal{B}) \rightarrow \mathcal{P}(\mathcal{B}^*)$ is a group isomorphism if and only if $d: \mathcal{B} \rightarrow \mathcal{B}^*$ is a bijection.

2.4. General Method to Construct a Dual Lattice [2]

In this section, we give a general method to construct HT-LT as well as LT-LT duals $\{\Lambda^*, K^*\}$ for any finite lattice system $\{\Lambda, K\}$. To do this, we only need to remember that Γ is generated by $\gamma(x) = \{B \,;\, B \ni x\}, x \in \Lambda$ (see sec. 1.3.3.).

Construction of Dual Lattices

With $\{\Lambda, K\}$ a lattice system and G any subgroup of $\mathcal{P}(\mathcal{B})$ (we shall be interested in $G = \mathcal{K}$ or $G = \Gamma$) we define a new lattice system $\{\Lambda^*, K^*\}$ in the following manner.

Let $\beta_j = (B_{j1}, \cdots B_{jn_j})$ $j = 1, \cdots n$ be any subset of G generating G . With each β_j we associate a point $r^*_{\beta_j} \in \Lambda^*$ and define the mapping :

$$\begin{array}{cccc} d & : & \mathcal{B} & \longrightarrow \mathcal{B}^* \\ & & \cup\!\!\!\!\! \shortmid & \cup\!\!\!\!\! \shortmid \\ & & B & \longmapsto B^* = \{ r^*_{\beta_j} \,;\, \beta_j \ni B \} \end{array}$$

$$\begin{array}{cccc} D & : & \mathcal{P}(\mathcal{B}) & \longrightarrow \mathcal{P}(\mathcal{B}^*) \\ & & \cup\!\!\!\!\! \shortmid & \cup\!\!\!\!\! \shortmid \\ & & \beta_j = (B_{j1} \cdots B_{jn_j}) & \longmapsto \beta_j^* = \bigcup_\alpha B^*_{j\alpha} \end{array}$$

By construction, we have :

$$B^* \ni r^*_{\beta_j} \iff B \in \beta_j$$

which yields $\beta_j^* = D\beta_j = \{ B^* \,;\, B^* \ni r^*_{\beta_j} \} = \gamma^*(r^*_{\beta_j})$

and therefore the mapping d induces a surjection of G into Γ^* (see Ch. 1).

Finally $B^* \in \beta_j^* \Rightarrow B^* \ni r^*_{\beta_j} \Rightarrow B \in \beta_j$ gives $\beta_j = D^{-1}\beta_j^*$.

In conclusion, the lattice system $\{\Lambda^*, K^*\}$ obtained with $G = \mathcal{K}$ (resp. $G = \Gamma$) yields a HT-LT (resp. LT-LT) dual lattice for $\{\Lambda, K\}$.

It must be noticed that :

1) Choosing different subsets of $\mathcal{K}$ (resp. Γ) generating $\mathcal{K}$ (resp. Γ) one can obtain, with the above method different dual lattices.

2) Since Λ is finite, $N_s^* = n - (|\mathcal{B}| - N_i)$; in particular, choosing a minimal set of generators for $\mathcal{K}$, (resp. Γ), the dual lattice will have $\mathcal{S}^* = \{\phi\}$.

3) This method can be applied to find HT-HT, LT-HT; in fact if $\{\Lambda, K\}$ is such that $\mathcal{K}$ separates $\mathcal{B}$ and if $\{\Lambda^*, K^*\}$ is a HT-LT dual for $\{\Lambda, K\}$ then $\{\Lambda^*, K^*\}$ is also LT-HT dual for $\{\Lambda, K\}$ (See Sec. 2.5). Combining a HT-LT and a LT-HT duality transformation, we obtain the construction of a HT-HT dual lattice.

2.5. General Properties of the Mapping d

In this Section we want to show that it follows from our definitions of LT-HT duality and LT-LT duality that the map d of $\mathcal{B} \to \mathcal{B}^*$ is always a bijection. Moreover, it will always be a bijection for HT-LT and HT-HT duals of a certain class of systems. This is expressed by the following properties [14] :

Property 1 :

If $\{\Lambda^*, K^*\}$ is a LT-HT or LT-LT dual for $\{\Lambda, K\}$ then the map d of $\mathcal{B}$ onto $\mathcal{B}^*$ is a bijection.

Indeed, from the condition (page 18), it follows that $dB_1 = dB_2$ implies that for any $\beta \in \Gamma$, $B_2 \in \beta$ if and only if $B_1 \in \beta$. Therefore $dB_1 = dB_2$ implies $\sigma_X(B_1) = \sigma_X(B_2)$, $\forall X \subset \Lambda$ and thus $B_1 = B_2$.

To study the map d associated with HT-HT and HT-LT duality, we introduce, for all B in $\mathcal{B}$ the set $\delta B = \bigcap_{\beta \in \mathcal{K}, \beta \ni B} \beta$. If $\{\Lambda^*, K^*\}$ is a HT-HT or a HT-LT dual for $\{\Lambda, K\}$ defined by $d : \mathcal{B} \to \mathcal{B}^*$ then from condition on p. 18, two elements B_1 and B_2 in $\mathcal{B}$ may have the same image under d, only if $B_2 \in \delta B_1$. In particular if $\mathcal{K}$ separates $\mathcal{B}$, that is, if for any $B_1, B_2 \in \mathcal{B}$ there exists $\beta \in \mathcal{K}$ such that $B_1 \in \beta$, $B_2 \notin \beta$, then for any $B \in \beta$ we have $\delta B = B$. We thus have the following properties.

Property 2 :

Let $\{\Lambda^*, K^*\}$ be a HT-LT dual for $\{\Lambda, K\}$, then $d : B \mapsto B^*$ is a bijection if and only if $\mathcal{K}$ separates $\mathcal{B}$.

Property 3 :

If $\{\Lambda, K\}$ is such that $\mathcal{K}$ separates $\mathcal{B}$, then any HT-LT dual $\{\Lambda^*, K^*\}$ constructed by the method of Section 2.4 is also a LT-HT dual for $\{\Lambda, K\}$.

2.6. Duality Transformation for General Systems

Generalized HT-LT Duality Transformation (resp. LT-HT, HT-HT, LT-LT) for abstract systems as described in Sec. 1.4 can be simply defined by means of a surjection $b \mapsto b^*$ of $\mathcal{B}$ onto $\mathcal{B}^*$ which induces a bijection between the group $\mathcal{K}_f$ and Γ^{*f} (resp. $\Gamma^{*f} \leftrightarrow \mathcal{K}_f^*$, $\mathcal{K}_f \leftrightarrow \mathcal{K}_f^*$, $\Gamma^f \leftrightarrow \Gamma^{*f}$) and such that

$$e^{-2K^*(b^*)} = \prod_{b \in d^{-1} b^*} \tanh K(b)$$

(with the corresponding changes of exp. and tanh functions)

Property 1 :

If Σ^* is any Dual System for Σ such that the subgroup G of $\mathcal{P}(\mathcal{B})$ ($G = \mathcal{K}_f$ or Γ^f) is mapped onto the subgroup H^* of $\mathcal{P}(\mathcal{B}^*)$ [$H^* = \mathcal{K}_f^*$ or Γ^{*f}] then for any g in G, the relation

$b \in g$ implies $d^{-1}b^* \subset g$; moreover for finite systems the duality relations Eq. (2.4) remain valid.

In other words the condition page 18 is entirely equivalent to the condition that the surjection of G onto H^* is a bijection. Moreover the generalization of properties 1-2, Sec. 2.5 are given by the following results :

Property 2 :

i) If Σ^* is a LT-HT dual for Σ then $db_1 = db_2$ implies $\sigma_{b_1} = \sigma_{b_2}$.

ii) If Σ^* is a LT-LT dual for Σ then $\sigma_{db_1} = \sigma_{db_2}$ implies $\sigma_{b_1} = \sigma_{b_2}$.

Proof :

i) From Property 1 follows that $db_1 = db_2$ implies that for any $\beta \in \Gamma^f$ $b_1 \in \beta \Leftrightarrow b_2 \in \beta$ and therefore $\sigma_{b_1} = \sigma_{b_2}$. (See Sec. 1.3.4.)

ii) $\sigma_{b_1^*} = \sigma_{b_2^*}$ implies that for any $\beta^* \in \Gamma^{*f}$, $b_1^* \in \beta^* \Leftrightarrow b_2 \in \beta^*$ therefore for any $\beta \in \Gamma^f$, $b_1 \in \beta \Leftrightarrow b_2 \in \beta$, which yields $\sigma_{b_1} = \sigma_{b_2}$.

Property 3 :

If Σ^* is a HT-LT dual for Σ then $\sigma_{dB_1} = \sigma_{dB_2}$ implies that for any $\beta \in \mathcal{K}_f$, $b_1 \in \beta \Leftrightarrow b_2 \in \beta$.

Proof :

$\sigma_{db_1} = \sigma_{db_2}$ implies that for any $\beta^* \in \Gamma^{*f}$, $db_1 \in \beta^* \Leftrightarrow db_2 \in \beta^*$ which concludes the proof.

In conclusion, we have generalized the notion of duality transformation in such a way to recover the results of Sec. 2.3 and 2.5 and which will be applicable to higher spins (Part III). Moreover, the general construction (Sec. 2.4) can still be used to construct dual systems.

Let us remark that in all cases of interest, we will have an action of $\mathbb{Z}^\nu$ on $\mathcal{L}$, i.e. for all $a \in \mathbb{Z}^\nu$ and $x \in \mathcal{L}$, $T_a x \in \mathcal{L}$ and an action of a subgroup $\mathcal{T}$ of $\mathbb{Z}^\nu$ with $|\mathbb{Z}^\nu/\mathcal{T}| < \infty$ on $\mathcal{B}$, i.e. for all $a \in \mathcal{T}$ and $b \in \mathcal{B}$, $T_a b \in \mathcal{B}$ and $\pi(T_a b) = T_a(\pi b)$. In those cases we shall define Duality Transformations in such a way that it commutes with $\mathcal{T}$, i.e. $d[T_a b] = T_a^* [db]$. Finally we introduce fundamental set of bonds $\mathcal{B}_0$ as minimal subsets of $\mathcal{B}$ such that for all b in $\mathcal{B}$ there exists b_0 in $\mathcal{B}_0$ and a in $\mathbb{Z}^\nu$ such that $b = T_a b_0$ [15].

Remarks :

1. If $\{\mathcal{L}, K\}$ and $\{\mathcal{L}, K'\}$ are two lattice systems whose interactions have the same support $\mathcal{B} = \mathcal{B}'$ then any mapping d of $\mathcal{B}$ onto $\mathcal{B}^*$ defining a dual $\{\mathcal{L}^*, K^*\}$ for $\{\mathcal{L}, K\}$ will automatically define a dual $\{\mathcal{L}^*, K'^*\}$ for $\{\mathcal{L}, K'\}$; in fact it is the support of the interactions, rather than the exact form of the Hamiltonian, which plays the fundamental role in duality. This is the reason we shall also denote concrete systems by $\{\mathcal{L}, \mathcal{B}, K\}$ although $\mathcal{B}$ is defined by the support of K. This remark will be useful in the next chapter.

2. For self-dual models, i.e. models which are mapped onto themselves by a HT-LT or LT-HT duality transformation, or for weakly self-dual models which are mapped onto themselves up to boundary terms, Eq. (2.5) yields the following Duality Relation for the free energy density of the translationally invariant infinite system :

$$-F[S] = -F[S'] + \frac{\alpha}{2} \ln \prod_{B \in \mathcal{B}_0} S(B) \tag{2.6}$$

where α is a geometrical constant.

$$S(B) = \sinh 2K(B) \qquad S' = \frac{1}{S(d^{-1}B)}$$

and $-F[S] = \lim_{\Lambda \to \mathcal{L}} \frac{1}{|\Lambda|} \ln Z(\Lambda, K)$

As was shown recently by J. Slawny [16] the class of (weakly) self-dual models is a large class since it contains in particular all translationally invariant systems whose fundamental family of bonds $\mathcal{B}_0$ contains exactly two elements; we should remark that all examples found prior to this general result falls into this class and it might be that this is a general characterization of all (weakly) self-dual models. Indeed for (weakly) self-dual models we must have $|\mathcal{B}| - N_i = |\Lambda| + O(|\partial\Lambda|)$ (= minimal number of generators for $\mathcal{K}$) from which follows that $\mathcal{K}$ should be generated by the translates of only one element β of $\mathcal{K}$; on the other hand, if the fundamental family of bonds contains more than two elements, we will need more than one element β of $\mathcal{K}$ to obtain the generators of $\mathcal{K}$ by translations.

3. For models which are mapped onto themselves by a LT-LT or HT-HT duality transformation, the duality relation yields a Symmetry Relation of the free energy density.

4. The interest of LT-HT duality transformation will be illustrated in the next chapter (see also example 4 of the next section) where we consider the duality transformation for the correlation function and where it is essential to fix the boundary conditions in order to define the equilibrium state of the infinite systems in the Low Temperation region.

5. For general systems property 3 of Sec. 2.5 does not always hold, i.e. if d is a bijection such that $\mathcal{K}_f$ is mapped onto

Γ^{*f} it does not always imply that Γ^f is mapped onto $\mathcal{K}_f^*$. In fact Γ^f is mapped onto $\mathcal{K}_f^*$ if and only if $\Gamma^f = \Gamma \cap \mathcal{P}_f(\mathcal{B})$ [6].

2.7. Some Examples

In this Section, we illustrate the method discussed and construct a dual lattice for some lattice systems without and with magnetic field. To obtain a simple dual lattice, we choose a simple generating set for the group $\mathcal{K}$ (or Γ), i.e. we choose a set $\{\beta_1, \cdots \beta_n\} \subset \mathcal{K}$ such that $|\beta_i|$ is minimal and the β_i are obtained by the translations of a given β_0 . From the above result it is clear that the dual lattice Λ^* can be easily constructed considering the "barycenter" of the generators; the mapping $d : \mathcal{B} \rightarrow \mathcal{B}^*$ defines the dual interaction as the set of barycenters associated with the generators β_i containing B . We restrict ourselves here to give some simple examples of interest which illustrate some general features concerning the thermodynamic equivalence between lattice systems related by a duality transformation. The method allows us to find new systems having the interesting property to be self-dual; in this case, the internal symmetry such as self-duality, allows us to locate the critical point exactly assuming the existence and their unicity (Wannier-Kramers argument); the same symmetry gives us some properties about the locus of zeros of the partition function (see also Ch. 8). Notice that a very interesting implication of self-duality in the two-dimensional Ising Model has been given recently by G. Benettin, G. Jona Lasinio and A. Stella [17] .

Example 1 : HT-LT Duality Transformation for Finite Systems

We consider a NxN square lattice Λ such that $|\Lambda| = N^2$, with pure 4 body ferromagnetic interaction $\beta J = K$ as represented in Fig. 1. The generators of $\mathcal{K}$ can be choosen as the translates of $\beta_0 = (B_1, B_2, B_3, B_4)$. (Fig. 1). We assume open boundary conditions such that the spins of the boundary do not interact with the outside $\mathbb{Z}^2/\Lambda$.

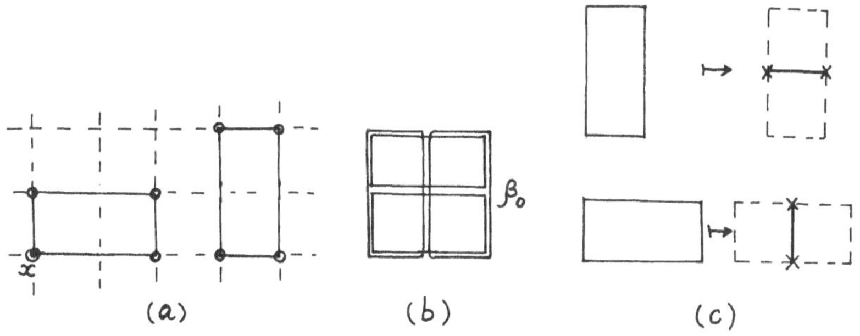

Fig. 2.1 : (a) Fundamental set of bonds
(b) Generator β_0 for $\mathcal{K}$
(c) Duality mapping

With this choice of $\{\beta_i\} \subset \mathcal{K}$ one obtains a dual lattice $\{\Lambda^*, K\}$ with $\Lambda^* \subset \Lambda$ and K^* a two-body interaction between nearest neighbours of strength $K^* = \beta^* J^* = -\frac{1}{2} \ln \tanh K$. Additional magnetic fields appear on the boundary of Λ^* which is the same as to fix the spins outside Λ^* to be + 1 (positive boundary conditions). This is the usual Ising Model with + boundary conditions.

We thus have an example of a model with pure 4-body forces which gives rise to a phase transition. Some of the properties known for the Ising Model can be directly translated onto the model. In particular the critical point (singularity of the free energy in K) is unique and is given by $\sinh 2K = 1$, and the specific heat diverges at T_c as $\ln(T - T_c)$. Notice that if instead of the above interaction one chooses a 4-body force between the 4 spins in each unit square, then the thermodynamics in zero field is trivial [18] ; in nonzero field the model is selfdual [13] .

It should be noticed that for infinite system $\Gamma^f \neq \Gamma \cap \mathcal{P}_f(\mathcal{B})$ and the Ising model is not a LT-HT dual for this model (see remark 5, sec. 2.6.).

Example 2 : LT-HT Duality Transformation for Infinite Systems.

As it has been suggested [19] interesting properties are expected in the study of models containing three-body interactions and whose Hamiltonian does not possess the up-down spin reversal symmetry. A simple two-dimensional model having this property is that defined by an infinite triangular lattice with translationally invariant 3-body forces $K = \beta J$ between nearest neighbours. This model has four groundstates $(\phi, \{S_i = \mathbb{E}_i\}_{i=1,2,3})$ where $\mathbb{E}_i$ are the three hexagonal sub-lattices of the triangular lattice $\mathbb{Z}^2$ and are obtained one from each others by some translation. Let d map a 3-body interaction K onto itself, then $\forall x \in \mathbb{Z}^2$, $\gamma(x) \in \Gamma$ is mapped onto $\gamma^*(x) \in \mathcal{K}$ the elementary hexagon with center $x \in \Lambda$ (Fig. 2.2). We thus obtain the same model with $\mathbb{Z}^2 = \mathbb{Z}^{2*}$, $K^* = -\frac{1}{2} \ln \tanh K$ and the model is self-dual $(\mathcal{K}_f = \Gamma^f)$ as the Ising model. Assuming the existence of one critical point only, it must be located at $S = \sinh 2K = 1$. From Eq. (2.6), Section 2.6, the free energy density satisfies the duality relation :

$$- F(S) = \ln S - F(S^{-1})$$

Other results such as the proof of the existence of a phase transition and symmetry breaking of the state will be obtained in the sequel (Ch. 3-9).

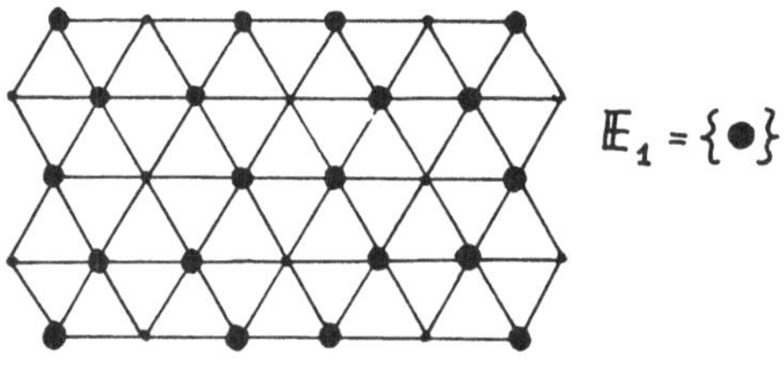

Fig. 2.2 : the Triangular Model

Example 3 : HT-LT Duality Transformation

Let us consider a triangular lattice $\mathbb{Z}^2$ with diluted anisotropic 3-body forces (K_1, K_2, K_3) and homogeneous magnetic field h (fig.2.3). From Section 2.4 we conclude easily that this model is self-dual (see Fig. 2.3); from Eq. (2.6), Section 2.6, we obtain a relation for the free energy, which reads, in the isotropic case $K_i = K$:

$$-F(S_K, S_h) = \frac{1}{2} \ln S_K \cdot S_h - F(S_h^{-1}, S_K^{-1})$$

where $S_K = \sinh 2K$ and $S_h = \sinh 2h$. The symmetry line is given by $S_K \cdot S_h = 1$. Since both, the low and high temperatures expansions are up to some unimportant factors polynominals in e^{-2K}, e^{-2h} , resp. $\tanh K, \tanh h$, the locus of zeros in the complex plane $z = e^{-2h}$ for a given K coincides with the locus of zeros in the plane $\tilde{z} = \tanh K$ for $h = -\frac{1}{2} \ln \tanh K$. The model is strongly singular at $h = 0$, for this value of h the thermodynamics is trivial ($\mathcal{K} = \{\phi\}$ for $h = 0$).

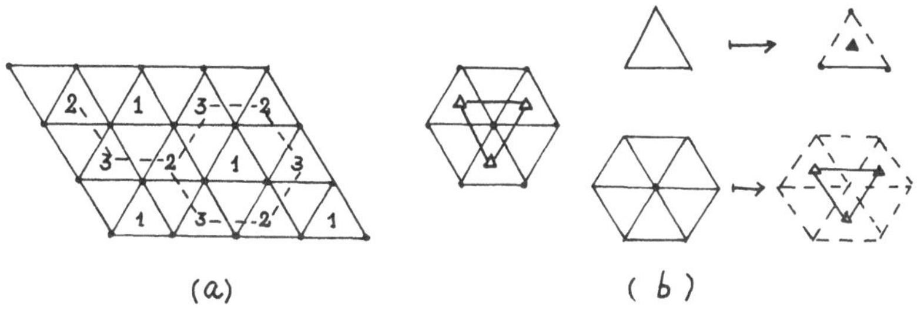

Fig. 2.3 : (a) The diluted anisotropic model and its dual for $K_1 = 0$
(b) Dual system and duality mapping for $K_1 \neq 0$.

Moreover, it is interesting to consider the anisotropic case $K_1 = 0$ and $K_2 = K_3 = K$. Let $\Lambda \subset \mathbb{Z}^2$ be finite. Since we shall be interested in the limit $K \to \infty$, it is necessary to translate the zero point energy to coincide with the ground-state, by subtraction of the corresponding energy $\sum_{B \in \mathcal{B}} K(B)$ (assume $K(B) > 0 \ \forall B \in \mathcal{B}$). In such cases the partition function is replaced by the reduced partition function defined as :

$$Z^{red}(\Lambda, K) = Z(\Lambda, K) \prod_{B \in \mathcal{B}} e^{-K(B)}$$

and the reduced free energy is defined similarly as :

$$-F^{red}(\Lambda, K) = \frac{1}{|\Lambda|} \ln Z^{red}(\Lambda, K)$$

(see Part II). Then a HT-LT dual is given by the hexagonal Ising Model with field $h^* = -\frac{1}{2} \ln \tanh K$ and 2-body forces between nearest neighbours $K^* = -\frac{1}{2} \ln \tanh h$ (Fig. 2.3). We then have the following results :

1. For h real, the reduced partition function $Z^{red}(\Lambda, K, h)$ has zeros whose locus in the $\tilde{z} = \tanh K$ plane coincides with the unit circle (this follows from the Lee-Yang unit circle theorem [20] ; by assumption $K^* > 0$). Thus a physical singularity could occur only in the infinite interaction limit $K \to \infty$ $(h^* = 0)$.

2. For fixed real h , the free energy of $\{\Lambda, K, h\}$ is analytic in $\tilde{z} = \tanh K$ inside the unit circle. From Eq. (2.5), Section 2.3, we have :

$$\frac{1}{|\Lambda|} \cdot \ln Z^{red}(\Lambda, K, h) = \frac{1}{6} \ln 2 + \frac{1}{3} \ln S_K e^{-2K} + \frac{1}{2} \ln S_h e^{-2h} + \\ + \frac{1}{|\Lambda|} \cdot \ln Z^*(K^*, h^*) + O(|\Lambda|^{-1/2})$$

3. In the limit $K \to \infty$ we obtain :

$$-F^{red} = \frac{1}{6}\ln 2 + \frac{1}{3}\ln\frac{1}{2} + \frac{1}{2}\ln Sh.e^{-2h} - \frac{2}{3}F^{*}(K^{*},0)$$

and thus a phase transition occurs (in the sense of a singularity of F^{red}). The critical field is given by $e^{-2h} = \frac{\sqrt{3}}{3}$ since the critical point of the hexagonal lattice is given by $\tanh K = \frac{\sqrt{3}}{3}$ [21] .

4. The magnetization $m_0(h)$ is continuous; defining

$$m_0(h) = \lim_{\Lambda \to \mathbb{Z}^2} \left(\frac{1}{|\Lambda|}\cdot\frac{\partial}{\partial h}\ln Z^{red}(\Lambda, K=\infty, h) + 1\right)$$ and

$$<\sigma_{12}>_0(K^*) = \lim_{\Lambda \to \mathbb{Z}^2} \frac{1}{|\Lambda|}\ln Z^{*}(K^{*},0)$$ one has the relation :

$$m_0(h) = \frac{\cosh 2h - <\sigma_{12}>(K^{*},0)}{\sinh 2h}$$

5. The susceptibility χ and specific heat C diverge both with indices (α,α') of the nearest neighbours hexagonal Ising Model; defining $\chi = \left(\frac{\partial m_0}{\partial h}\right)_T$ and $C = \left(\frac{\partial U}{\partial T}\right)(J=\infty, T)$ then, near T_c they behave as :

$$\chi(T \sim T_c) \cong a\frac{\partial <\sigma_{12}>}{\partial h^*} + b \sim a \ln\left(\frac{T-T_c}{T_c}\right) + b$$

$$C(T \sim T_c) \cong a'\frac{\partial <\sigma_{12}>}{\partial h^*} + b' \sim a' \ln\left(\frac{T-T_c}{T_c}\right) + b'$$

We thus have an example of phase transition (in the sense of a singularity of some thermodynamic variables) without symmetry breaking of the state.

Moreover, we remark that for $K = \infty$ a HT-HT dual for the model is given by the triangular lattice with 2-body nearest neighbours in zero field. The mapping d is represented in Fig. 2.4, where the lattice Λ^* is given by the center of the K_1 bonds. From the same argument as in Part II we conclude that this model has a unique equilibrium state for all temperature although it has a phase transition at $T \neq 0$. (Part II, Sec. 4.2).

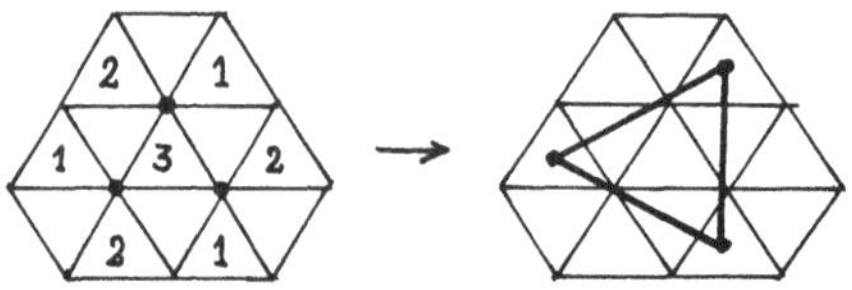

Fig. 2.4 : The HT-HT mapping d

The general discussion of systems with infinite interactions will be given in Part II. To conclude this example we remark that analogous properties were obtained by F.Y. Wu [22] for a model on a square lattice with 4-body forces and external field, which is the dual system for the usual two-dimensional Ising Model with field [2] .

Example 4 : LT-HT Duality Transformation for Finite Systems with + Boundary Conditions.

With this example, we want to illustrate the role played by the Boundary Conditions (B.C.) for LT-HT duality transformations; on the other hand, it will also show the interest of introducing the abstract set of indices $\mathcal{B}$ rather than the support of the interactions.

We consider the finite system defined by a finite portion Λ of the infinite model discussed in example 2; this portion can be in any configuration while the outside is fixed with all spins having the value + 1.

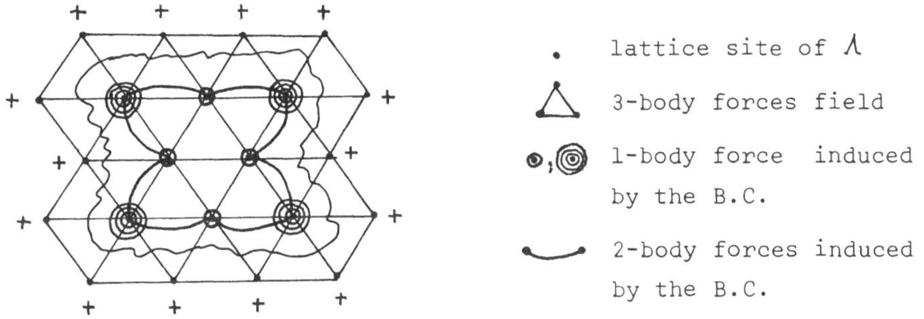

Fig. 2.4 : Triangular Model with 3-Body Forces and Positive Boundary Conditions.

With $\mathcal{B}$ the family of bonds of the infinite systems, the Hamiltonian of the finite system with positive boundary conditions is given by :

$$H_\Lambda = - \sum_{\substack{B \in \mathcal{B} \\ B \cap \Lambda \neq \phi}} J \sigma_B \qquad (\text{ considered as a function on } \mathcal{P}(\Lambda) \subset \mathcal{P}(\mathbb{Z}^2))$$

In that case, the set of indices $\mathcal{B}_\Lambda$ is defined by those B in $\mathcal{B}$ which intersect Λ and the mapping $\pi : b \mapsto \sigma_b$ of $\mathcal{B}_\Lambda$ onto $\widehat{\mathcal{P}(\Lambda)}$ is simply given by $B \mapsto \sigma_B$; it is immediately clear that because of the boundary effect $\mathcal{B}_\Lambda$ is not the support of the interactions and the map $B \mapsto \sigma_B$ is not injective since $\sigma_B = \sigma_{B\cap\Lambda}$ on $\mathcal{P}(\Lambda)$. (Recall that by definition the support of J is a family of subsets of Λ).

Let us then consider the mapping $d : \mathcal{B}_\Lambda \to \mathcal{B}^*_{\Lambda^*}$ defined by $B^* = B$, with $\Lambda^* = \bigcup_{B \in \mathcal{B}_\Lambda} B$. The system $\{\Lambda^*, \mathcal{B}^*_{\Lambda^*}\}$ is therefore the triangular system with open or free boundary conditions; moreover by definition the mapping d induces a bijection between the groups Γ and $\mathcal{K}^*$. In conclusion the system with free B.C. is a LT-HT dual for the system with positive B.C.

Let us remark that this situation is rather general; i.e. the LT-HT dual for a system with positive B.C. is a system with open B.C. In fact we shall show in Ch. 7 that for a large class of ferromagnetic systems, the LT-HT duals of a system with "+" (resp. "open") boundary condition is a system with "open" (resp. "+") boundary condition.

2.8. Concluding Remarks and Summary of Results

The idea to employ the simple properties of the groups which can be associated with any general lattice systems has been useful already to give a simple and concise general way to construct and relate dual models. This transformation has been given for finite as well as for infinite systems. It must be noticed that the necessity of considering these relationships for the exact duals, rather than "weak duals", i.e. duals in the thermodynamic limit, or for duals with prudently chosen boundary conditions only, is of central importance in studying the correlation functions. Indeed, although the free energy may not in general be affected in the thermodynamic limit by boundary terms in the Hamiltonian, the same is not true for the correlation functions; in fact, below criticality one expects the state of the system to be determined precisely by these terms (Ch. 3,4).

Another fundamental problem is the following : what are the characteristics of Model Hamiltonians, which give rise to such transition ? Why some interactions give rise to a phase transition and others will not ? From the analysis of above examples, one can remark the following : Systems such that the group $\mathcal{H}$ of closed graphs is of small order or empty does not give rise to any phase transition (for $h=0$), since the corresponding free energy density is analytic in β ; but such systems (see example 3) may be strongly singular in some limit of interactions parameter and if $h\neq 0$, it is possible to have a phase transition without symmetry breaking of the state ($\mathcal{S}=\{\phi\}$) . Moreover, if $|\mathcal{S}|>1$ and if the group $\mathcal{H}$ is of the order $2^{\alpha|\Lambda|}$ with α a positive number, we expect and shall prove (Sec. 4.3) for a large class of interesting systems that there exists a symmetry breakdown of the state at low temperature, which is precisely one of the definitions of a phase transition.

As we know from Baxter's model [23] addition of a 4 body force to 2 body interactions yields a critical behaviour depending on the values of the 4 body interactions. Moreover, example 1 shows that this situation is not typical of systems having 4-body forces. On the other hand, such a change of critical behaviour will also appear in the example 2 with zero field and anisotropic 3-body forces K , as will be discussed in Chapters 5 and 6 ; another model with 3-body forces which exhibits similar critical behaviour has been discussed in [24] .

From the relation between the free energy densities of dual systems and the facts that the dual parameters K^* are, except for some points, analytic functions of the interaction K , it follows that the critical exponent α of the specific heat of the model around T_c coincides with the exponent α^* of any dual systems in the corresponding region around T_c^*.
For HT-LT or LT-HT dual systems, an asymmetry in the specific heat for $\{\Lambda,K\}$ around T_c is also apparent in the specific heat of $\{\Lambda^*,K^*\}$, but the low and high temperature regions are interchanged.

In particular those self-dual models which have a unique critical point exhibit a symmetric singularity of the specific heat around T_c . For these models, this is an additional information to the scaling laws.

Another interesting property, which is evident from the above analysis, is the relationship between the locus of the zeros of the partition functions of two dual models [13.22]. This additional property for self-dual models can be employed to relate property of the state above and below critical regions (Ch. 8).

Finally, it must be noticed that the abstract structure discussed in Sections 1.3.4 and 2.6 can be naturally generalized to investigate properties of higher spin models; our generalization will be presented in Part III.

We now summarize the results of this Chapter :

Result 1 :

For any lattice system $\{\mathcal{L}, K\}$ there exists many dual systems which can be constructed by the method of section 2.3 for finite systems, the influence of boundary conditions can be given explicitly.

Result 2 :

For a class of systems, the critical indices α and α^* of the system and its dual are equal. In particular for self-dual lattices, $\alpha = \alpha'$. (Symmetric behaviour around T).

Result 3 :

The locus of the zeros of the partition function (up to some factors) in the corresponding parameters, is the same, for two models related by a duality transformation.

Result 4 :

There exists models having a phase transition without symmetry breakdown of the state and without local order parameter.

2.9. Summary of Duality Transformation

A general spin $\frac{1}{2}$ lattice system was defined by $\Sigma = (\mathcal{L}, \mathcal{B}, K, \Pi)$ with K a real or complex function on $\mathcal{B}$ and Π a mapping from $\mathcal{B}$ into $\widehat{\mathcal{P}(\mathcal{L})}$. With $\mathcal{S}$, $\mathcal{K}$, Γ defined in Chapter 1, we have obtained a concrete realisation of the following scheme :

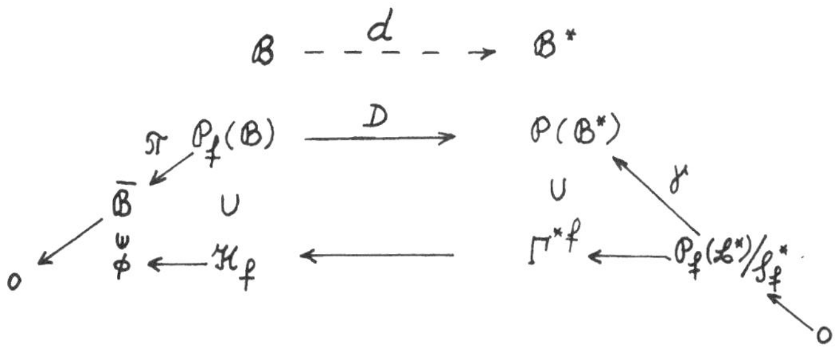

Analogue scheme can be given with the role of $(\mathcal{K}, \Gamma^*)$ replaced by $(\Gamma, \mathcal{K}^*)$, $(\mathcal{K}, \mathcal{K}^*)$ and (Γ, Γ^*) .

We have already remarked at the end of Section 2.6, that the equilibrium states at low temperature depend essentially on the boundary conditions; it is therefore important to investigate the effect of duality transformation on quantities which depend on the boundary conditions. This is in particular the case for certain correlation functions and it is the purpose of next Chapter to study the effect of duality transformation for the correlation functions. (See also Ch. 7)

CHAPTER 3 - DUALITY RELATION FOR THE CORRELATION FUNCTIONS

3.1. Equilibrium State and Boundary Conditions

As we have seen in Chapter 1, a complete statistical description of the properties of a lattice system is provided by the correlation functions $\{<\sigma_X>\}$; for finite system $\{\Lambda, K\}$ they were defined by Eq. (1.4). If H_Λ is real, one just refers to these correlation functions as the equilibrium state or Gibbs state of the system in the box Λ for the interaction K.

As we have also seen in Chapter 1 (Property 4), the Gibbs states of finite systems are always invariant under the internal symmetry group $\mathcal{S}$; therefore to investigate the possible breakdown of this internal symmetry (Ch. 4) it will be necessary to start in such a way that the symmetry is already broken at finite volume and then investigate the thermodynamic limit. To do so we shall adopt the approach introduced by Dobrushin in the study of Gibbs Random Field and introduce the notion of "Gibbs State for finite system Λ with boundary condition Y".

Let $\{\mathcal{L}, \mathcal{B}, K\}$ be an infinite system and Y be any subset of $\mathcal{L}$. A finite system Λ with boundary condition Y is defined as a finite subset Λ of $\mathcal{L}$ which can be in any configurations while the outside $\Lambda^c = \mathcal{L}/\Lambda$ is in the given configuration $Y^c = Y \cap \Lambda^c$; introducing as before the set $\mathcal{B}_\Lambda$ of bonds which intersect Λ , $\mathcal{B}_\Lambda = \{B \in \mathcal{B} \ ; \ B \cap \Lambda \neq \phi\}$, the Hamiltonian of the system Λ with boundary condition Y is then defined by :

$$H_{\Lambda,Y} = -\sum_{B \in \mathcal{B}_\Lambda} J(B)\sigma_B(Y^c)\sigma_B = -\sum_{\substack{B \in \mathcal{B} \\ B \subset \Lambda}} J(B)\sigma_B - \sum_{\substack{B \in \mathcal{B}_\Lambda \\ B \cap \Lambda^c \neq \phi}} J(B)\sigma_B(Y^c)\sigma_B =$$

$$= H_\Lambda + \delta H_{\Lambda,Y} \qquad (3.1)$$

where H_Λ describes the energy of the portion (corresponding to "open" or "free" boundary conditions); and $\delta H_{\Lambda,Y}$ describes the interaction energy between Λ and the outside Λ^c [19,25].

The Gibbs state defined by $H_{\Lambda,Y}$ will be denoted by $\omega^Y_{\beta,\Lambda}$ or $\langle\sigma_X\rangle^Y_{\beta,\Lambda}$ while the Gibbs state defined by H_Λ will be denoted by $\omega_{\beta,\Lambda}$ or $\langle\sigma_X\rangle_{\beta,\Lambda}$ *. An <u>equilibrium state</u> of the infinite system is then defined [19] as a convex combination of limiting sequence of finite volume Gibbs states with suitable boundary conditions, i.e.

$$\omega^P_\beta = \sum_Y P_Y \lim_{\Lambda\to\infty} \omega^Y_{\beta,\Lambda} \qquad \langle\sigma_X\rangle^P_\beta = \sum_Y P_Y \lim_{\Lambda\to\mathcal{L}} \langle\sigma_X\rangle^Y_{\beta,\Lambda} \tag{3.2}$$

with $\sum_Y P_Y = 1 \quad , \quad 0 \le P_Y \le 1 \quad \forall Y$

where P is a probability measure on the configuration space. Moreover, it is assumed that a sequence of boundary conditions satisfies the requirement, (constraint of physical nature) that the energy of the boundary terms, should not be of the same order as the volume energy of the box Λ, i.e. it is a surface term, and

$$\lim_{\Lambda\to\mathcal{L}} \max_X \frac{\delta H_{\Lambda,Y}(X)}{|\Lambda|} = 0$$

The situation is not any more as simple as for quantities, such as the energy density, which are not affected in the thermodynamic limit by these boundary terms in the Hamiltonian; as we know, this is not always true for the correlation functions and in each case, one must investigate what happens in the thermodynamic limit. From this point of view, we adopt the following definition:

* This state corresponds to open boundary condition and will also be denoted by $\omega^{op}_{\beta\Lambda}$.

Definition [19]

A phase transition is said to take place at the value K_c of the interaction K, if the system is unstable with respect to boundary perturbations, i.e. if there are at least two boundary conditions Y_1 and Y_2 such that for some $X \in \mathcal{P}_f(\mathcal{L})$, we have :

$$\lim_{\Lambda \to \mathcal{L}} \langle \sigma_X \rangle_{\beta,\Lambda}^{Y_1} \neq \lim_{\Lambda \to \mathcal{L}} \langle \sigma_X \rangle_{\beta,\Lambda}^{Y_2} \tag{3.3}$$

Remark :

If Eq. (3.3) holds, we have a macroscopic instability; some quantities are sensitive to the boundary conditions; for instance, if for x in $\mathcal{L}$

$$\lim_{\Lambda \to \infty} \langle \sigma_x \rangle_{\beta,\Lambda}^{Y_1} \neq \lim_{\Lambda \to \infty} \langle \sigma_x \rangle_{\beta,\Lambda}^{Y_2}$$

then the local magnetization depends on the boundary condition even if this boundary is very far away. On the other hand, some quantities, such as the free energy, do not depend on the boundary condition, since we have :

$$|\beta f_{Y_2} - \beta f_{Y_1}| < \lim_{\Lambda \to \infty} \max_{X \subset \Lambda} \frac{1}{|\Lambda|} (H_{\Lambda,Y_2}(X) - H_{\Lambda,Y_1}(X)) = 0$$

In this regard, it is then important to know how the boundary conditions change by duality transformation, if one is interested in the relationship between the properties of the states of dual models; the relationship must be considered first for a finite volume Λ, i.e. for every sequence of box Λ_i, such that $\lim \Lambda_i \to \mathcal{L}$ in the manner defined above.

We should also remark at this point that the above definition of phase transition is not the only possible definition; by analogy with thermodynamics, we could also define a phase transition by means of the singularities of the free energy. Although in most cases these definitions are equivalent, there exists models which are stable with respect to boundary perturbations but which have a phase transition associated with a singularity of the free energy ; the study of such phase transitions can also be undertaken at least in some cases within the framework of this Chapter [26] (Example 2, Sec. 2.7 and Part II Sec. 4.2).

Boundary conditions of special interest in the study of lattice systems are [19] :

1. Open Boundary Conditions (perfect walls);
2. Boundary Conditions defined by Y in $\mathcal{S}$, in particular for ferromagnetic systems $Y = \phi \in \mathcal{S}$ and is called "Positive Boundary Condition";
3. Periodic Boundary Conditions ;
4. Boundary Conditions defined by $Y = (S_1 \cap \mathcal{L}_1)\cdot(S_2 \cap \mathcal{L}_2)$ where $(\mathcal{L}_1, \mathcal{L}_2)$ is a partition of $\mathcal{L}$ and $S_1, S_2 \in \mathcal{S}$, also called "Mixed Boundary Conditions".

The physical meaning of these different boundary conditions and their interest in the study of phase coexistence and surface tension in spin lattice systems is given in [27,28,29] .

It is clear that in order to prove the existence of a phase transition, it will be sufficient to find any two boundary conditions such that Eq. (3.3) is satisfied. However, in general it will be possible to choose some special boundary conditions which yields the <u>Extremal Equillbrium States</u>, where an equilibrium state ω_P is extremal if for any $\alpha \in [0,1]$, the decomposition

$$< \sigma_X >_P = \alpha < \sigma_X >_{P_1} + (1-\alpha) < \sigma_X >_{P_2} \quad \forall X \in \mathcal{P}_f(\mathbb{Z}^\nu)$$

implies $\alpha = 0$ or 1. A knowledge of all the extremal equilibrium states will then define all equilibrium states by convex combination.

As a first step in the study of equilibrium states of general lattice systems and in order to investigate their behaviour under duality transformation, we shall now discuss the duality relation for the correlation functions. The simple relationships which we shall derive are direct consequences of remark 1, Section 2.6 and is an application of the general scheme given in Section 2.9.

3.2. Duality Relations for Correlation Functions of Finite Systems

The aim of this Section is to relate the correlation functions of the finite system $\{\Lambda, K\}$ to those of the dual system $\{\Lambda^*, K^*\}$. These relations are easily obtained if we make use of a simple identity of exponential function, which allows to write any correlation function as a ratio between two partition functions; using the group structure, we generalize a method originally given by H. Ceva and L. Kadanoff [30] which yields a simple relation between the functions $\{\langle \sigma_X \rangle_{(\Lambda,K)}\}$ and $\{\langle \sigma_{X^*} \rangle_{(\Lambda^*,K^*)}\}$ with $X \in \overline{\mathcal{B}}$ and $X^* \in \overline{\mathcal{B}}^*$ [14].

Since for any Gibbs state of $\{\Lambda, K\}$, $\langle \sigma_X \rangle_{(\Lambda,K)} = 0$ if $X \notin \overline{\mathcal{B}}$ (Sec. 1.3, Properties 3-4), it is sufficient to consider $\langle \sigma_X \rangle_{(\Lambda,K)}$ with $X \in \overline{\mathcal{B}}$, i.e. $\langle \prod_{B \in \beta} \sigma_B \rangle_{(\Lambda,K)} = \langle \sigma_{\pi\beta} \rangle_{(\Lambda,K)}$ with β in $\mathcal{P}(\mathcal{B})$. It then follows from the general discussion of Chapter 1 that the function on $\mathcal{P}(\mathcal{B})$ given by $\sigma(\beta) = \langle \prod_{B \in \beta} \sigma_B \rangle_{(\Lambda,K)}$ is in fact a function on the factor group $\mathcal{P}(\mathcal{B})/\mathcal{K} \cong \overline{\mathcal{B}}$, i.e. $\sigma(\beta_1) = \sigma(\beta_2)$ if $\pi\beta_1 = \pi\beta_2$.

Definition :

With $\bar{B}$ in $\bar{\mathcal{B}}$, any subset β of $\mathcal{B}$ such that $\pi\beta = \bar{B}$ will be called a path associated with $\bar{B}$. We notice that with any path β associated with $\bar{B}$ and any $\varkappa \in \mathcal{K}$, $\beta \cdot \varkappa$ is also a path associated with $\bar{B}$.

Using the identity $\sigma_X = (-i)\, e^{i\frac{\pi}{2}\sigma_X}$ for all $X \subset \Lambda$, we obtain for any β in $\mathcal{P}(\mathcal{B})$:

$$< \prod_{B\in\beta} \sigma_B >_{(\Lambda,K)} = (-i)^{|\beta|} Z^{-1} \sum_{Y\subset\Lambda} e^{\sum_{B\in\mathcal{B}} K(B)\sigma_B(Y) + i\frac{\pi}{2}\sum_{B\in\beta}\sigma_B(Y)}$$

$$\text{i.e. } < \prod_{B\in\beta} \sigma_B >_{(\Lambda,K)} = (-i)^{|\beta|} \frac{Z(\Lambda,K')}{Z(\Lambda,K)} = \frac{\sum_{\gamma\in\Gamma} \sigma_\beta(\gamma) \prod_{B\in\gamma} e^{-2K(B)}}{\sum_{\gamma\in\Gamma} \prod_{B\in\gamma} e^{-2K(B)}} \tag{3.4}$$

$$\begin{aligned} \text{where } \quad K'(B) &= K(B) && \text{if } B \notin \beta \\ K'(B) &= K(B) + \frac{i\pi}{2} && \text{if } B \in \beta \end{aligned} \tag{3.5}$$

and $\sigma_\beta(\varkappa) = (-1)^{|\beta\cap\varkappa|}$ (Ch. 1).

The support $\mathcal{B}'$ of K' being identical with the support $\mathcal{B}$ of K, any dual $\{\Lambda^*, K^*\}$ for $\{\Lambda, K\}$ will define a corresponding dual for $\{\Lambda^*, K'^*\}$, (see remark 1, Sec. 2.6). We now consider the four types of duality.

3.2.1. HT-LT Duality

Let $\{\Lambda^*, K^*\}$ be a HT-LT dual for $\{\Lambda, K\}$ and $\{\Lambda^*, K'^*\}$ the corresponding dual for $\{\Lambda, K'\}$. From Eq. (2.3) follows that :

$$K'^*(B^*) = K^*(B^*) + \varphi(B^*) \tag{3.6}$$

where

$$\varphi(B^*) = \sum_{B\in d^{-1}B^*\cap\beta} \ln \tanh K(B) \tag{3.7}$$

From Eq. (3.4-3.6) and the proposition of Section 2.3, we obtain :

$$< \sigma_{\pi\beta} >_{(\Lambda,K)} = Z(\Lambda^*, K'^*)/Z(\Lambda^*,K^*) = < \prod_{B^* \in \beta^*} e^{\varphi(B^*)\sigma_{B^*}} >_{(\Lambda^*,K^*)}$$

and the following holds :

Proposition 1 [14]

If $\{\Lambda^*,K^*\}$ is a HT-LT dual for $\{\Lambda,K\}$ then for any $\beta \subset \mathcal{B}$,

$$< \sigma_{\pi\beta} >_{(\Lambda,K)} = \prod_{B^* \in \beta^*} \cosh \varphi(B^*) \sum_{\bar{\beta}^* \subset \beta^*} \prod_{B^* \in \bar{\beta}^*} \tanh \varphi(B^*) < \sigma_{\pi\bar{\beta}^*} >_{(\Lambda^*,K^*)}$$

Corollary

If $\{\Lambda^*,K^*\}$ is a HT-LT dual for $\{\Lambda,K\}$, then for any β such that d restricted to β is one-one,

$$< \prod_{B \in \beta} \sigma_B >_{(\Lambda,K)} = < \prod_{B^* \in \beta^*} e^{-2K^*(B^*)\sigma_{B^*}} > = < \prod_{B^* \in \beta^*} \mu_{B^*} >_{(\Lambda^*,K^*)}$$

with $\mu_{B^*} = e^{-2K^*(B^*)\sigma_{B^*}}$ (3.8)

3.2.2. LT-HT Duality

Let $\{\Lambda^*,K^*\}$ be a LT-HT dual for $\{\Lambda,K\}$ and $\{\Lambda^*,K'^*\}$ be the corresponding dual for $\{\Lambda,K'\}$. Then from property 1 of Section 2.5 the mapping $d: B \mapsto B^*$ is a bijection , $\tanh K'^*(B^*) = e^{-2K'(B)} =$ $= e^{-2K(B)}(-1)^{|d^{-1}B^* \cap \beta|}$, and $K'^*(B^*) = K^*(B^*)\,\sigma_{d^{-1}B^*}(\beta)$.

We thus obtain :

Proposition 2

If $\{\Lambda^*, K^*\}$ is a LT-HT dual for $\{\Lambda, K\}$ then for any $\beta \subset \mathcal{B}$

$$< \prod_{B \in \beta} \sigma_B >_{(\Lambda, K)} = < \prod_{B^* \in \beta^*} e^{-2K^*(B^*)\sigma_{B^*}} >_{(\Lambda^*, K^*)} = < \prod_{B^* \in \beta^*} \mu_{B^*} >_{(\Lambda^*, K^*)}$$

3.2.3. LT-LT Duality

If $\{\Lambda^*, K^*\}$ is a LT-LT dual for $\{\Lambda, K\}$ then $K'^*(B^*) = K'(B)$ and $< \sigma_{\pi\beta} >_{(\Lambda, K)} = Z(\Lambda^*, K'^*) / Z(\Lambda^*, K^*)$; we thus have :

Proposition 3

If $\{\Lambda^*, K^*\}$ is a LT-LT dual for $\{\Lambda, K\}$ then for any $\beta \subset \mathcal{B}$

$$< \prod_{B \in \beta} \sigma_B >_{(\Lambda, K)} = < \prod_{B^* \in \beta^*} \sigma_{B^*} >_{(\Lambda^*, K^*)}$$

3.2.4. HT-HT Duality

Let $\{\Lambda^*, K^*\}$ be a HT-HT dual for $\{\Lambda, K\}$ and $\{\Lambda, K'^*\}$ the corresponding dual for $\{\Lambda, K'\}$; suppose that the path β is such that $d^{-1}(dB) \subset \beta$ for all B in β ; from definition (Sec. 2.3) we have :

$$\tanh K'^*(B) = \prod_{B \in d^{-1}B^*} \tanh K'(B) = \prod_{B \in d^{-1}B^*} \tanh K(B) . \prod_{B \in d^{-1}B^* \cap \beta} [\tanh K'^*(B)]^{-2}$$

From the condition imposed on β we have :

$$\forall B^* \in \beta^* : \tanh K'^*(B^*) = \prod_{B \in d^{-1}B^*} (\tanh K(B))^{-1} = \tanh K^*(B^*)^{-1}$$

thus

$$<\sigma_X>_{(\Lambda, K)} = \prod_{B^* \in \beta^*} \tanh K^*(B^*) . \frac{\cosh K^*(B^*)}{\cosh K'^*(B^*)} . \frac{Z(\Lambda^*, K'^*)}{Z(\Lambda^*, K^*)} = (-i)^{|\beta^*|} . \frac{Z(\Lambda^*, K'^*)}{Z(\Lambda^*, K^*)}$$

With $K'^{*}(B^{*}) = K^{*}(B^{*}) + \frac{i\pi}{2}$ we obtain :

Proposition 4

If $\{\Lambda^{*}, K^{*}\}$ is a HT-HT dual for $\{\Lambda, K\}$ then for any path $\beta \subset \mathcal{B}$ such that $d^{-1}(dB) \subset \beta$ for all B in β we have :

$$< \sigma_{\pi\beta} >_{\{\Lambda, K\}} = < \prod_{B\in\beta} \sigma_{dB} >_{\{\Lambda^{*}, K^{*}\}}$$

Remark :

The above propositions will remain true for any finite abstract lattice system $\{\Lambda, \mathcal{B}, K, \pi\}$ as introduced in Section 1.4. However, it is already clear that in general for infinite systems for which there may exist equilibrium states which are not invariant under $\mathcal{S}$, the above transformation of the correlation functions will not yield the values of $< \sigma_{X^{*}} >_{(\mathcal{L}^{*}, K^{*})}$ from the values of $\{ < \sigma_{X} >_{(\mathcal{L}, K)} \}$ if $X^{*} \notin \overline{\mathcal{B}^{*}}$, (except for LT-LT duality transformations [31]). Moreover, it may also happen that there exists no states of the dual systems such that $< \sigma_{X^{*}} >_{(\mathcal{L}^{*}, K^{*})}$, as defined by the duality transformations, represents a correlation function of the dual lattice (see Ch. 6). However, as in Section 2.8, for most ferromagnetic systems, the open boundary conditions on the original model correspond positive boundary conditions on the HT-LT dual and vice-versa; in this case $< \prod_{B\in\beta} \sigma_{B} >_{op} = < \prod_{B^{*}\in\beta^{*}} \mu_{B^{*}} >_{+}$ and $< \prod_{B\in\beta} \mu_{B} >_{op} = < \prod_{B^{*}\in\beta^{*}} \sigma_{B^{*}} >_{+}$. (see Chapter 7).

3.3. Symmetric States and Symmetric Algebra

Propositions 1 and 2 of Section 3.2 have shown that for LT-HT and HT-LT duality transformations, where the range of temperature is interchanged, the role played by the family of functions $\{\prod_{B\in\beta} \sigma_B\}_{\beta\subset\mathcal{B}}$ and $\{\prod_{B\in\beta} \mu_B\}_{\beta\subset\mathcal{B}}$ is also interchanged; on the other hand, for LT-LT and HT-HT transformations, where the range of temperature is not interchanged, the role of these families of functions is also not interchanged. In this section we want to discuss further the properties of these families of functions, their connection and their interest.

The family of functions $\{\prod_{B\in\beta} \mu_B\}$ were introduced in [30] for the 2-dimensional Ising Model and within the context of duality, they appeared naturally in propositions 1 and 2; however these functions may be considered independently of any duality transformations [32] . Moreover, in the study of Equilibrium States defined by linear equations (Ch. 6), the family of functions $\{<\prod_{B\in\beta} \sigma_B>\}$ appears well adapted to the high temperature region, while the family of functions $\{<\prod_{B\in\beta} \mu_B>\}$ appears more adapted to low temperature region; we remark already at this point that these expectation values are expected to be small in the corresponding temperature region.

For any B in β we have defined (sec. 3.2) the function μ_B on $\mathcal{P}(\Lambda)$ by :

$$\mu_B = e^{-2K(B)\sigma_B} = \cosh 2K(B) - \sigma_B \sinh 2K(B) \tag{3.9}$$

which yields :

$$\sigma_B = \cosh 2K_*(B) - \mu_B \sinh 2K_* \tag{3.10}$$

where $e^{-2K_*(B)} = \tanh K(B)$.

For any $\beta \in \mathcal{P}_f(\mathcal{B})$ we shall furthermore define the functions $\sigma_\beta = \sigma_\beta(X)$ and $\mu_\beta = \mu_\beta(X)$ on $\mathcal{P}(\Lambda)$ by :

$$\sigma_\beta = \prod_{B\in\beta} \sigma_B \tag{3.11}$$

$$\mu_\beta = \prod_{B\in\beta} \mu_B \tag{3.12}$$

and we have immediately the following result :

Proposition 1 :

For any Gibbs states of $\{\Lambda, K\}$, and any $\beta \in \mathcal{P}(\mathcal{B})$:

$$\mu(\beta) = \langle \mu_\beta \rangle_{(\Lambda,K)} = \frac{\sum_{\bar\beta\in\mathcal{K}} \sigma_\beta(\bar\beta) \prod_{B\in\bar\beta} \tanh K(B)}{\sum_{\bar\beta\in\mathcal{K}} \prod_{B\in\bar\beta} \tanh K(B)}$$

$$\sigma(\beta) = \langle \sigma_\beta \rangle_{(\Lambda,K)} = \frac{\sum_{\bar\beta\in\Gamma} \sigma_\beta(\bar\beta) \prod_{B\in\bar\beta} e^{-2K(B)}}{\sum_{\bar\beta\in\Gamma} \prod_{B\in\bar\beta} e^{-2K(B)}}$$

(We insist on the fact that we have two distinct functions denoted by σ_β , one on $\mathcal{P}(\Lambda)$, $\sigma_\beta(X) = (-1)^{|\pi\beta\cap X|}$, the other on $\mathcal{P}(\mathcal{B})$ $\sigma_\beta(\bar\beta) = (-1)^{|\beta\cap\bar\beta|}$).

Using proposition 1, the result of Section 3.2 appears now as a direct consequence of the definition of duality transformation (indeed $\sigma_B(\beta) = \sigma_{B^*}(\beta^*)$ for any duality transformation).

Proposition 2 [32] :

With any state ω of $\{\Lambda, K\}$, the function on $\mathcal{P}(\mathcal{B})$, $\sigma(\beta) = \omega[\sigma_\beta]$ and $\mu(\beta) = \omega[\mu_\beta]$ are related by the transformation :

$$\sigma = D_{K_*}\mu \qquad \text{and} \qquad \mu = D_K\,\sigma$$

where for any (real or complex) function K on $\mathcal{B}$, D_K is the linear operator on the space of bounded functions $\varphi = \varphi(\beta)$ defined by :

$$(D_K\,\varphi)(\beta) = \prod_{B\in\beta} \cosh 2K(B) \sum_{\bar\beta\subset\beta} \prod_{B\in\bar\beta} \tanh[-2K(B)]\cdot\varphi(\bar\beta) \tag{3.13}$$

which has a unique inverse, if for all B in $\mathcal{B}$, $K(B) \neq 0$, $\pm\frac{i\pi}{2}$, given by :

$$D_K^{-1} = D_{K_*} \tag{3.14}$$

with $K_* = K_*(B)$ defined by $e^{-2K_*(B)} = \tanh K(B)$. Indeed this proposition follows immediately from Eqs. (3.9) and (3.10); on the other hand the equation $D_K^{-1} = D_{K_*}$ is obtained directly by computation, i.e. for all φ, $D_K D_{K_*}\varphi = D_{K_*} D_K\varphi = \varphi$.

We remark that the relation given by proposition 2, being independent of the volume Λ will remain valid for infinite systems.

As was recalled at the beginning of Section 3.2, the Gibbs states are invariant under $\mathcal{S}$ and the introduction of the family of observables $\{\sigma_\beta\}$ and $\{\mu_\beta\}$ appears naturally in the study of any state invariant under $\mathcal{S}$. From the definition of Chapter 1, a state ω is invariant under $\mathcal{S}$, or symmetric, if $\tau_S'\omega = \omega$ for all S in $\mathcal{S}$, from which followed that ω is invariant under $\mathcal{S}$ if and only if $\omega[\sigma_X] = 0$ if $X \notin \mathcal{S}^\perp = \bar{\mathcal{B}}$. We thus have :

Proposition 3 :

For any finite or infinite system $\{\mathcal{L}, K\}$, then :

1) A state ω is invariant under any arbitrary group $\mathcal{S}$ if and only if

$$\omega[A] = 0 \qquad \text{for} \quad A \in \mathfrak{A}_{\mathcal{S}}^{\perp}$$

where $\mathfrak{A}_{\mathcal{S}} = \{ A \in \mathfrak{A} \; ; \; \tau_s A = A \quad \forall \; s \in \mathcal{S} \}$

$$\mathfrak{A}_{\mathcal{S}}^{\perp} = \{ A \in \mathfrak{A} \; ; \; |\mathcal{S}|^{-1} \sum_{s \in \mathcal{S}} \tau_s A = 0 \}$$

2) Any symmetric state ω is uniquely specified either by

$$\{ \sigma(\beta) = \omega[\sigma_\beta] \}_{\beta \subset \mathcal{B}} \quad \text{or by} \quad \{ \mu(\beta) = \omega[\mu_\beta] \}_{\beta \subset \mathcal{B}}$$

Definition :

With $\mathcal{S}$ the internal symmetry group of the system the subalgebra $\mathfrak{A}_{\mathcal{S}}$ is called the "Symmetric Algebra".

The symmetric algebra $\mathfrak{A}_{\mathcal{S}}$ can be defined as the (closure of) linear span of the family of observables $\{ \sigma_\beta \}_{\beta \subset \mathcal{B}}$ or $\{ \mu_\beta \}_{\beta \subset \mathcal{B}}$, i.e. for any A in $\mathfrak{A}_{\mathcal{S}}$

$$A = \sum_{\beta \subset \mathcal{B}} \tilde{A}(\beta)\, \sigma_\beta = \sum_{\beta \subset \mathcal{B}} \tilde{\tilde{A}}(\beta)\, \mu_\beta \tag{3.15}$$

In conclusion any symmetric state is a state defined on the symmetric algebra $\mathfrak{A}_{\mathcal{S}}$. We shall analyse further the properties of symmetric equilibrium states in Ch. 7 ; in the next section we shall now discuss some simple applications of the previous methods and results.

3.4. Applications

Example 1 : Solution of the Two-Dimensional Ising Model with External Field $h = \frac{i\pi}{2}$.

In this example, we apply the above method to find the free energy and magnetization of the two-dimensional Ising Model $\{\Lambda, K, h\}$ at $e^{-2h} = z = -1$, which corresponds to a pure imaginary magnetic external field $h = \frac{i\pi}{2}$. As was pointed out by Lee and Yang, the case of non-zero external magnetic field is related to the complete solution of the lattice gas model outside the transition region, but the explicit solution for $h \neq 0$ is not known. However, a result has been obtained for the special value $h = \frac{i\pi}{2}$. Notice that the solution of this problem was announced by Lee and Yang in one of their two fundamental papers on the circle theorem on zeros and the spontaneous magnetization of the Ising Model [20]. Nevertheless, the proof was not published. Later a quasi-proof was first given by G.Baxter [33]; the solution was then also found by B. MacCoy and T.T. Wu by a different approach using the transfer matrix method [34] . We shall now give another very simple proof of this result based on the properties of duality transformations [35] .

Let us first consider the infinite system $\{\mathbb{Z}^2, \mathcal{B}\}$ defined by a square lattice $\mathbb{Z}^2$ and the family $\mathcal{B}$ of nearest neighbour pair of points (i.e. the usual 2-dimensional Ising model without external field); the finite system we want to study is defined by $\Lambda \subset \mathbb{Z}^2$, a square with N^2 sites, and the Hamiltonian :

$$\beta H_\Lambda = - K \sum_{B \in \mathcal{B}_\Lambda} \sigma_B - h \sum_{x \in \Lambda} \sigma_x$$

where $\mathcal{B}_\Lambda = \{B, B \cap \Lambda \neq \phi\}$ and we consider positive boundary conditions ($s_x = +1$ for all $x \in \mathbb{Z}^2/_\Lambda$).

For the special value $h = \frac{i\pi}{2}$ of the external field, it is possible to eliminate this external field in the interior of Λ ; indeed there exists $\beta \subset \mathcal{B}_\Lambda$ such that $\pi\beta \cap \Lambda = \Lambda$ (cf. Fig. 3.1) and therefore

$$\sigma_\Lambda = \prod_{B\in\beta} \sigma_B = \prod_{B\in\beta}(-i\, e^{i\frac{\pi}{2}\sigma_B}) = (-i)^{|\Lambda|}\, e^{i\frac{\pi}{2}\sum_{x\in\Lambda}\sigma_x} \text{ on } \mathcal{P}(\Lambda) .$$

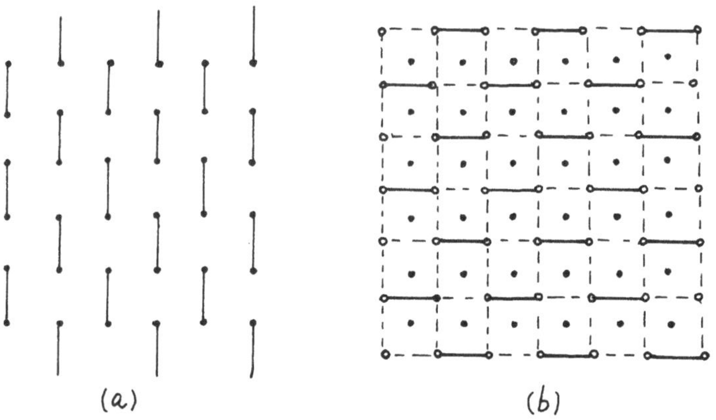

Fig. 3.1 : (a) Lattice site $\Lambda = \{\cdot\}$ and path $\beta \subset \mathcal{B}_\Lambda$
(b) LT-HT Dual systems $\{\Lambda^*, \tilde{K}^*\}$ and dual path β^* .

It then follows that the partition function of the model with field $h = \frac{i\pi}{2}$ is up to a numerical factor, equal to the partition function of the model without external field except on the boundary, defined by the Hamiltonian :

$$\beta \tilde{H}_\Lambda = -\sum_{B\in\mathcal{B}_\Lambda} \tilde{K}_B \quad \sigma_B$$

where $\tilde{K}_B = K + i\frac{\pi}{2}$ if $B \in \beta$

$\tilde{K}_B = K$ if $B \notin \beta$

In fact, we have :

$$Z^+_\Lambda(K, h = \tfrac{i\pi}{2}) = (-i)^{|\beta| - |\Lambda|} Z_\Lambda(\tilde{K})$$

Using then a LT-HT duality transformation such as given in example 4 (Ch. 2), we obtain :

$$Z_\Lambda(K, h=\tfrac{i\pi}{2})_+ = (i)^{|\Lambda|}(\sinh 2K)^{|\Lambda|+\sqrt{|\Lambda|}} \cdot 2^{-(\sqrt{|\Lambda|}+1)} Z_{\Lambda^*}(\tilde{K}^*)_{op}$$

where $Z_{\Lambda^*}(K^*)_{op}$ is the partition function of the dual lattice Λ^* with open boundary conditions and coupling $\tilde{K}^*_{B^*} = K^*$ except on the dual path β^* where $\tilde{K}^*_{B^*} = -K^*$ ($K^* = -\frac{1}{2}\ln\tanh K$) (see Fig. 3.1 (b)). In the thermodynamic limit :

$$-F\,(K, i\tfrac{\pi}{2}) = \frac{i\pi}{4} - F\,(\tilde{K}) = \frac{i\pi}{2} + \ln\sinh 2K - F\,(\tilde{K}^*)$$

We now remark that $-F(\tilde{K})$ resp. $-F(\tilde{K}^*)$ are the free energy density of the generalized anisotropic square lattice of Utiyama for the cases $\tilde{K}_1 = K + \frac{i\pi}{2}$, $\tilde{K}_i = K$, i = 2,3,4 resp. $\tilde{K}_2^* = -K^*$, $\tilde{K}_i^* = K^*$, i = 1,3,4. The solution of this model is known [21] and we obtain :

$$\text{(I)} \quad -\beta f\,(K, i\tfrac{\pi}{2}) = \ln 2 + \frac{i\pi}{2} + \frac{1}{4(2\pi)^2} \cdot \iint_0^{2\pi} dk_1\,dk_2 \ln S^2\left(1 + S^2 + \frac{\cos(k_1+k_2) - \cos(k_1-k_2)}{2}\right)$$

which is the first Lee-Yang formula in the variables $S = \frac{1-x^2}{2x} = \sinh 2K$, $x = e^{-2K}$. In the same way the "magnetization" at $z = -1$ can be related to that of the anisotropic lattice of Utiyama for which an expression has been conjectured by Syozi and Naya [21] ; we then obtain for the "magnetization" at $h = \frac{i\pi}{2}$:

$$\text{(II)} \qquad I\,(z=-1) = \left[\frac{(1+x^2)^2}{1-x^2} \cdot \left(1 + 6x^2 + x^4\right)^{-1/2}\right]^{1/4}$$

which is the second Lee-Yang formula [20] . Notice that $I\,(z=-1)$ is always greater than 1.

Moreover, in the case $h=0$ the zeros of Z in the variable $S = \sinh 2K$ are known to lie asymptotically on the unit circle $|S| = |\sinh 2K| = 1$ [36] . In the case $h = \frac{i\pi}{2}$, and asymptotically, the zeros in the same variable S are on the imaginary axis $|S| < \sqrt{2}$ (see Eq. (I)). This illustrates the changing of the locus of the zeros in the S -plane depending on the values of the magnetic field. On the other hand, Eq. (II) has a direct bearing on the distribution function $g(\theta)$ [20] , of the zeros of the partition function on the unit circle in the $z = e^{-2h}$ plane. Notice that $(2\pi)^{-1} g(\theta = 0)$ (spontaneous magnetization) is less than unity, while $(2\pi)^{-1} g(\theta = \pi)$ is always greater than unity (unphysical domain). $g(0)$ increases with decreasing temperature, while $g(\pi)$ decreases with decreasing temperature. This shows the motion of the roots towards the right along the unit circle as the temperature decreases.

To conclude the discussion of this application, we give two explicit equivalences of the model at $z = -1$ with other problems :

1. The Lee-Yang Formula Eq. (I) is, up to the factor $i^{|\Lambda|} (\cosh K)^{|2\Lambda|}$ the generating function of all graphs with odd valency (1 or 3) covering Λ [35]. (With weight $\tanh K$ for each line). This follows directly from the structure of the high temperature expansion of Z (Sec. 2.2). In Figure 3.2

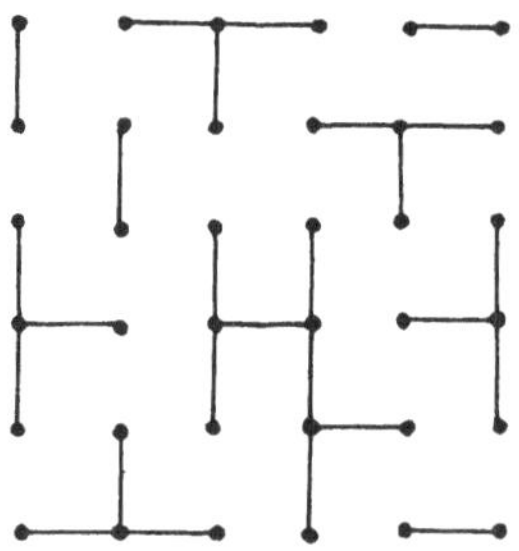

Fig. 3.2

we represent a typical graph entering into the expansion and consisting of 2,4 and 8 point subsets in the case of open boundary conditions for $|\Lambda|$ even.

2. Another interesting equivalence is given by the relation of the above problem with the correlation function of the whole lattice in zero field. For open boundary conditions and if $|\Lambda|$ is even, we have, in fact, from definition 3.2 :

$$< \sigma_\Lambda >_{K,0} = i^{-|\Lambda|} Z(\tilde{K},0) / Z(K,0)$$

As expected $\lim_{|\Lambda| \to \infty} < \sigma_\Lambda >_{op} (K,0) = 0$ exponentially with $|\Lambda|$.

The same method can also be considered for the hexagonal lattice and by duality for the triangular lattice yielding analogous results. The expression for **I(z=1)** will be also useful in the discussion of the example of Sec.4.2 Part II.

Example 2 : Phase Transition for a Ferromagnetic Model with 3-Body Interactions. An Application of HT-LT Duality and GKS Inequality.

In this example, we consider a simple application of HT-LT duality transformation to prove the existence of a phase transition for a ferromagnetic system. The idea, which is quite general, is to combine HT-LT duality transformation together with GKS inequalities [37] . As example, we choose the triangular lattice with 3-body forces which we have already defined in Chapter 2 (example 2).

As we have discussed at the beginning of this chapter, to prove the existence of a phase transition, it is sufficient to prove the existence of a symmetry breaking of the state as expressed by Eq. (3.4); in the following, we shall consider the thermodynamic limit respectively for open and positive boundary conditions and show that the limiting states are not identical.

Consider first the case of open boundary conditions. In this case, the symmetry group $\mathcal{S}$ is of order 4; since $x \notin \bar{\mathcal{B}}$, Property 1 of Chapter 2 yields for all finite $\Lambda \subset \mathbb{Z}^2$, $\langle \sigma_x \rangle_{(\Lambda,K)_{op}} = 0$ for all x in Λ and therefore :

$$\langle \sigma_x \rangle_{op} = \lim_{\Lambda \to \mathbb{Z}^2} \langle \sigma_x \rangle_{(\Lambda,K)_{op}} = 0 \quad .$$

Let us then consider the same model in the presence of an external field $h_x = h$ for all x in Λ ; to analyze this model in field, we consider also the "diluted" model $(\Lambda, \mathcal{B}')$ defined as above except that some of the 3-body forces $K(B)$ and some of the external fields h_x are taken to be zero (Fig. 3.3); Λ^* will denote the set of x where $h_x \neq 0$, it is an hexagonal sublattice of Λ .

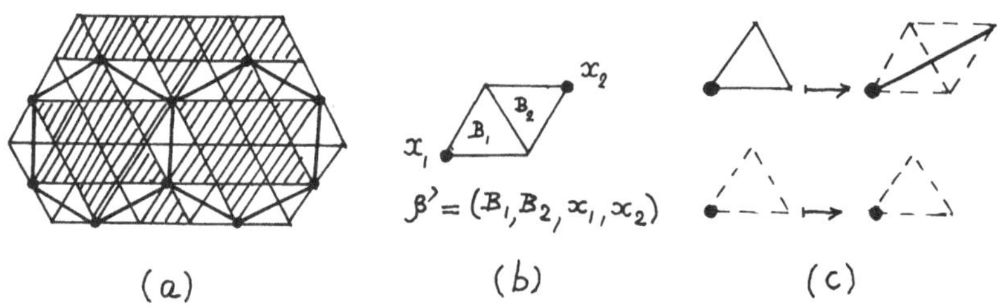

Fig. 3.3 : (a) Diluted Model $\{\Lambda, \mathcal{B}'\}$: $\Lambda = \{\cdot\}$, $\mathcal{B}' = \{\bullet, \blacktriangle\}$ and its HT-HT dual $\{\Lambda^*, \mathcal{B}'^*\}$: $\Lambda^* = \{\bullet\}$, $\mathcal{B}'^* = \{\cdot, \text{—}\}$

(b) Generator of the group $\mathcal{K}'$ for the diluted model:

(c) Duality map d

We shall now show that at sufficiently low temperature :

$$\langle \sigma_x \rangle_+ = \lim_{h \to 0} \lim_{|\Lambda| \to \infty} \langle \sigma_x \rangle_{(\Lambda, K, h)} \neq 0 \quad .$$

Using GKS inequalities, we have :

$$\langle \sigma_x \rangle_{(\Lambda, K, h)} \geq \langle \sigma_x \rangle^{diluted}_{(\Lambda, K, h)} \qquad \forall x \in \Lambda$$

and for any x in Λ^* , the right hand side of this inequality can easily be obtained by means of a HT-HT duality transformation, applied to the model $(\Lambda, \mathcal{B}')$. A generating set for the group $\mathcal{K}'$ is given by the elements $\beta' = (B_1, B_2, x_1, x_2)$ (Fig. 3.3. (b)), i.e. a pair of 3-body forces and fields. Let (Λ^*, K^*) be the isotropic hexagonal Ising model with lattice sites Λ^* and let $d : \mathcal{B}' \mapsto \mathcal{B}'^*$ map the fields onto themselves and 3-body forces onto 2-body forces between nearest neighbours on Λ^* (Fig. 3.3. (c)). The mapping d (which is not a bijection) maps the generators of $\mathcal{K}'$ onto the generators of $\mathcal{K}'^*$ and satisfies the condition of Proposition 1 (Ch. 3). Therefore, the hexagonal Ising model with external magnetic field $h^* = h$ and 2-body isotropic interactions K^* defined by $\tanh K^* = (\tanh K)^2$ is a HT-HT dual for the diluted model $(\Lambda, \mathcal{B}')$. Application of Proposition 4 (Sec. 3.2.4) gives :

$$\langle \sigma_x \rangle^{diluted}_{(\Lambda, K, h)} = \langle \sigma_x \rangle^{Ising\ hex.}_{(\Lambda^*, K^*, h)}$$

and therefore, for any $x \in \Lambda^*$:

$$\langle \sigma_x \rangle_{(\Lambda, K, h)} \geq \langle \sigma_x \rangle^{Ising\ hex.}_{(\Lambda^*, K^*, h)}$$

In particular, in the limit $|\Lambda| \to \infty$, followed by $h \to 0_+$ we have :

$$\langle \sigma_x \rangle (K, 0) > m_0^{hex.}(K^*)$$

where $m_0^{hex}(K^*)$ is the spontaneous magnetization of the isotropic hexagonal Ising model with 2-body interactions K^* i.e. $m_0^{hex}(K^*) = [C^3(C^2+1)(C-2)(C^2-1)^{-3}]^{-1}$, $C = \cosh 2K^*$ The same argument used in [38] yiedls :

$$\lim_{h\to 0}\lim_{|\Lambda|\to\infty} <\sigma_x>_{(\Lambda,K,h)} = \lim_{|\Lambda|\to\infty}\lim_{h\to 0} <\sigma_x>_{(\Lambda,K,h)_+}$$

where $<\cdots>$ denotes the state defined by means of positive boudary conditions. We finally obtain :

$$<\sigma_x>_+(K) \geqslant m_0^{hex}(K^*) \quad \forall x \in \mathbb{Z}^2$$

which proves the instability of the local mangetization $<\sigma_x>$ with respect to boundary perturbations and the existence of a phase transition. A lower bound for the critical temperature is obtanied by solving the equation $C = 2$ which yields

$e^{-2K_0} = e^{-2J/kT_0} = \sqrt{3}+2 - \sqrt{3\cdot 2+4\sqrt{3}} \cong 3{,}73 - 3{,}6 = 0{,}13$ which is not too far from the critical point (Ch. 2), $e^{-2J/kT_c} = 0{,}41$.

Moreover, from Griffith's inequality , $<\sigma_x>_{(\Lambda,K,h)}$ is an increasing function of the field; consider the limit

$h(y) \longrightarrow \infty \quad \forall y \in \mathbb{T}$ where $\mathbb{T}$ is a triangular sublattice which does not contain the point x . An upper bound for $<\sigma_x>$ can then be found [39] in terms of the magnetization $m_0(2K)$ of the hexagonal Ising model, yielding an upper bound $X = e^{-2J/kT_c} < 0{,}5176$ for the critical temperature T_c . Further results concerning the minimal number of extremal equilibrium states and thermodynamic phases will be obtained in the Chapters 4 and 6.

To conclude this exemple, we remark that better bounds for this model can be obtained by diletting fewer interactions from the model in fields [40] .

CHAPTER 4 - PHASE TRANSITIONS WITH SPONTANEOUS SYMMETRY BREAKDOWN. ERGODIC DECOMPOSITION

4.1. Introductory Remarks

In this chapter we want to investigate the problem of the "existence of a phase transition associated with a spontaneous symmetry breakdown", i.e. the existence of equilibrium states for the infinite systems which are not invariant under the full symmetry group $\mathcal{G}$ of the system. This same problem can also be stated as the "non-uniqueness of the Gibbs Random Field associated with the interaction K". Following the discussion of Sec.3.1 we have to study in more details the influence of the boundary conditions on the definition of the state and we shall show that, at low temperature, different boundary conditions may define different states in the thermodynamic limit.

To attack this problem we shall use the group structure and the LT-HT duality transformation. It allows us to transform and generalise graphical arguments first given by Peierls[41] into group properties. This generalized Peierls arguments, or method of contours, yield then lower bounds for the critical temperature of ferromagnetic systems (sec. 4.3). Below this critical temperature there will exist several equilibrium states ω_β whose symmetry group is a subgroup of the full group $\mathcal{G}$.

The idea of "contours" introduced by Peierls for the Ising model with two-body forces was made rigourous by Griffiths [37,42] and Dobrushin [43], and even today it is almost the only general technique that one has to study phase transitions. This same method was then applied to several other models either discrete [44] or continuous [45] to obtain bounds on the critical temperature. In the case of lattice system it was shown by C.Gruber that Peierls argument can be generalised by means of the group structure to a large class of lattice systems [25]. Along the same line, B. Payandeh and D.Merlini proved the existence of a phase transition in the triangular model[46,39]. This work was recently extended by W.Holsztyński and J. Slawny [6] who gave, for $\mathbb{Z}^{\nu}$ invariant systems, explicit conditions on the groups, as well as on the hamiltonian, for the existence of a phase transition ; furthermore, they also derived explicit bounds on the critical temperature for any ferromagnetic systems satisfying these conditions.

The basic idea of the Peierls argument consists in showing that there exists some boundary condition Y and some finite $X \notin \bar{\mathcal{B}}$, for example $x \in \mathcal{L}$, such that $|\omega^{Y}_{(\beta,\Lambda)}[\sigma_X]| \geq \eta > 0$ with η independent of Λ. It follows that what appears in Peierls argument is not a discontinuity of the thermodynamic functions, but rather the fact that the properties at a given point inside the system can be strongly influenced by what happens at the boundary even if this boundary is infinitely remote.

4.2. Internal and Euclidean Symmetry Group for Infinite Systems. Ergodic Decomposition

Besides the internal symmetry group $\mathcal{S}$ which may be broken, there exist other symmetries which could as well be broken in the thermodynamic limit. In particular, the euclidean symmetry $\mathcal{E}$ which we shall now introduce is one of the most important group one has to consider together with $\mathcal{S}$.

With $\{\mathcal{L}, \mathcal{B}, K\}$ an infinite system defined by lattice sites x in $\mathbb{R}^\nu$, $\mathcal{B} \subset \mathcal{P}_f(\mathcal{L})$ and $\mathcal{E}_\mathcal{L}$ the subgroup of the euclidean group in $\mathbb{R}^\nu$ which leaves $\mathcal{L}$ invariant, i.e. $\{ T_e x \}_{x\in\mathcal{L}} \equiv \mathcal{L}$ for all $e \in \mathcal{E}_\mathcal{L}$, we define :

$$\mathcal{S} = \{ S \subset \mathcal{L} \; ; \; \sigma_B(S) = +1 \quad \forall B \in \mathcal{B} \} = \mathcal{B}^\perp = \text{"\underline{Internal symmetry group}"}$$

$$\mathcal{E} = \{ e \in \mathcal{E}_\mathcal{L} \; ; \; T_e B \in \mathcal{B},\; K(T_e B) = K(B) \;\; \forall B \in \mathcal{B} \} = \text{"\underline{Euclidean symmetry group}"}$$

Property 1.

$\mathcal{S}$ is invariant under $\mathcal{E}$, i.e. $T_e S \in \mathcal{S}$ for all $S \in \mathcal{S}$, and $e \in \mathcal{E}$.

The full symmetry group $\mathcal{G} = \mathcal{S} \circledS \mathcal{E}$ of the system is then defined as the semi-direct product of $\mathcal{S}$ and $\mathcal{E}$:

$$\mathcal{G} = \{ g = (S, e) \; ; \; S \in \mathcal{S}, \; e \in \mathcal{E} \}$$

$$g_1 \cdot g_2 = (S_1, e_1)\cdot(S_2, e_2) = (S_1 \cdot T_{e_1} S_2 \, , \; e_1 e_2)$$

This group $\mathcal{G} = \mathcal{S} \circledS \mathcal{E}$ acts as group of automorphism on the observables $A \in \mathfrak{A}$:

$$(\tau_g A)(X) = A(T_e^{-1}(SX)) \qquad \forall g = (S,e) \in \mathcal{G}$$

which yields
$$(\tau_{g^{-1}} A)(X) = A(S \cdot T_e X) \quad , \quad \tau_{g^{-1}} \sigma_Y = \sigma_S(Y) \cdot \sigma_{T_e^{-1} Y}$$

Moreover $\mathcal{G}$ acts as a group of linear transformations on the states by :

$$(\tau'_g \omega)[A] = \omega[\tau_{g^{-1}} A]$$

which implies
$$(\tau'_g \omega)[\sigma_X] = \sigma_S(X)\, \omega[\sigma_{T_e^{-1} X}] \qquad (4.1.)$$

and the state ω <u>is said "invariant" if $\tau'_g \omega = \omega$ for all $g \in \mathcal{G}$</u>

We recall that from Property 3 of sec. 1.3.2 the internal symmetry group $\mathcal{S}_\omega$ of any state ω is given by :

$$\mathcal{S}_\omega = \{ X \, ; \, \omega[\sigma_X] \neq 0 \}^\perp \qquad (4.2)$$

Property 2

If ω is invariant under $\mathcal{E}_\omega \subset \mathcal{E}$, then $\mathcal{S}_\omega$ is also invariant under $\mathcal{E}_\omega$, i.e. $T_e S \in \mathcal{S}_\omega$ for all $S \in \mathcal{S}_\omega$ and $e \in \mathcal{E}_\omega$

Indeed : $\tau'_e \tau'_S \omega = \omega = \tau'_{T_e S} \tau'_e \omega = \tau'_{T_e S} \omega$

The interest of the property lies in the fact that for $\mathbb{Z}^\nu$-invariant ferromagnetic systems, it is possible to show that the state ω^+ is $\mathbb{Z}^\nu$- invariant [47] ; it then follows that $\mathcal{S}_{\omega^+}$ is $\mathbb{Z}^\nu$- invariant. Therefore to find all possible internal symmetry groups of ω_β^+ it is sufficient to consider all the subgroups of $\mathcal{S}$ which are left invariant by action of $\mathbb{Z}^\nu$ (See application).

Theorem 1

1) $\tau'_{S_1} \omega^{S_2}_{(\beta,\Lambda)} = \omega^{S_1 S_2}_{(\beta,\Lambda)}$ for all S_1, S_2 in $\mathcal{S}$

2) For any [extremal] equilibrium state ω and any $g \in \mathcal{G}$ then $\tau'_g \omega$ is also an [extremal] equilibrium state.

3) For any [extremal] equilibrium state ω of $\{\mathcal{L}, \mathcal{B}, K\}$ and any $Y \subset \mathcal{L}$ then $\tau'_Y \omega$ is an [extremal] equilibrium state of $\{\mathcal{L}, \mathcal{B}, K'\}$ where $K'(B) = \sigma_B(Y) K(B)$.

The proof of (1) follows from the definition of $\omega^Y_{(\beta,\Lambda)}$ (sec.3.1); on the other hand the proof of (2) and (3) follows from the fact that if ω is a solution of eq.(6.10) (Ch.6) satisfying the positivity condition, then $\tau'_g \omega$ is also a solution of the same equation satisfying also the positively condition, while $\tau'_Y \omega$ is a solution of the equation with K replaced by K' ; the results mentioned in Ch. 6 on the equivalence of the definitions of equilibrium states conclude the proof.

It thus follows from the theorem that to investigate the Gibbs states of $\{\mathcal{L},\mathcal{B},K\}$ with boundary condition S in $\mathcal{S}$, it is in fact sufficient to study the "+" boundary condition (i.e. $S=\phi$) and for all S in $\mathcal{S}$ we have $\omega_\beta^S = \tau'_S \omega_\beta^+$. Moreover, the results which we shall obtain for ferromagnetic systems can be extended at once to the "essentially ferromagnetic systems", systems defined by the condition that the ground state (sec. 1.3.2) Y satisfy the condition

$$\sigma_B(Y) = \text{sign } J(B) \quad \forall B \in \mathcal{B}$$

and we have for these systems

$$\omega_K^Y[\sigma_X] = \sigma_Y(X) \cdot \omega_{|K|}^+[\sigma_X] \tag{4.3}$$

for any ground state Y(*).

Moreover, the following results is a direct consequence of the definitions.

Property 3

i) $\mathcal{S}_{\tau'_S \omega} \equiv \mathcal{S}_\omega$ for all S in $\mathcal{S}$

ii) $\mathcal{S}_{\tau'_e \omega} = \{T_e S ; S \in \mathcal{S}_\omega\}$ for all e in $\mathcal{E}$

$\mathcal{E}_{\tau'_e \omega} = \mathcal{E}_\omega$

(*) For essentially ferromagnetic systems, the ground states are given by $\{Y \cdot S\}_{S \in \mathcal{S}}$ where Y is any ground state.

Theorem 2

1) If ω is an extremal equilibrium state invariant under $\mathcal{S}_\omega \subset \mathcal{S}$, then there exists at least $|\mathcal{S}/\mathcal{S}_\omega|$ distinct extremal equilibrium states.
2) If ω_1 and ω_2 are two extremal equilibrium states which coincide on the symmetrie algebra $\mathcal{O}\!\mathcal{L}_{\mathcal{S}}$, i.e $\omega_1[\sigma_X] = \omega_2[\sigma_X]$ for all $X \in \mathcal{S}^\perp = \bar{\mathcal{B}}$, then there exists some S' in $\mathcal{S}$ such that $\omega_2 = \tau'_S\,\omega_1$.

Proof :
The proof of (1) follows from the definitions; to establish (2) we note that $\omega_1[\sigma_X] = \omega_2[\sigma_X]$ for all $X \in \mathcal{S}^\perp$ implies

$$|\mathcal{S}|^{-1} \sum_{S \in \mathcal{S}} \tau'_S\,\omega_1 = |\mathcal{S}|^{-1} \sum_{S \in \mathcal{S}} \tau'_S\,\omega_2$$

Hence, since the equilibrium states form a Choquet simplex [48] the decomposition of any states into extremal states is unique and theorem 1 concludes thus the proof.

Corollary
If $\omega_\beta^+[\sigma_X] \neq 0$ for some $X \notin \bar{\mathcal{B}}$ there exists at least two equilibrium states and the systems has a phase transition, moreover if $\omega_\beta^+[\sigma_x] \neq 0$ for all x in $\mathcal{L}$ there exists at least $|\mathcal{S}|$ distinct extremal equilibrium states.

This corollary follows from theorem 2 and the fact that $\omega_\beta^+[\sigma_X] \neq 0$ for $X \notin \bar{\mathcal{B}}$ implies that ω_β^+ is not invariant under $\mathcal{S}$ i.e $\tau'_S\,\omega_\beta^+ \neq \omega_\beta^+$ for some S in $\mathcal{S}$; therefore this state is unstable with respect to boundary perturbation, which is one of the definition of phase transition ; finally $\omega_\beta^+[\sigma_x] \neq 0, \forall x \in \mathcal{L}$ implies $\mathcal{S}_\omega = \{\phi\}$ which conclude the proof.

In the following chapters we shall show that there exists domains of temperature where there exists a unique equilibrium states invariant under the full symmetry group $\mathcal{G}$; in such cases we can then apply the following result :

Theorem 3

If there exists a unique equilibrium state invariant under $\mathcal{G} = \mathcal{S} \circledS \mathcal{E}$, then

a) all extremal equilibrium states ω_i invariant under some subgroup $\mathcal{G}_i$ of $\mathcal{G}$ with $\mathcal{G}/\mathcal{G}_i$ finite are of the form

$$\omega_i = \tau'_g \omega_0 \quad \text{with } g \in \mathcal{G}$$

b) all equilibrium states ω invariant under some $\mathcal{G}_\omega \subset \mathcal{G}$ with $\mathcal{G}/\mathcal{G}_\omega$ finite are of the form

$$\omega = \sum_{g \in \mathcal{G}/\mathcal{G}_\omega} \mu(g) \tau'_g \omega_0$$

where $\mu(g)$ is a probability measure on $\mathcal{G}/\mathcal{G}_\omega$

c) if ω_0 is invariant under $\mathcal{E}$, then all extremal states ω_i with $\mathcal{G}/\mathcal{G}_i$ finite are of the form

$$\omega_i = \tau'_s \omega_0 \quad \text{with } s \in \mathcal{S}$$

and any equilibrium state ω is given by

$$\omega = \int_{\mathcal{S}} (\tau'_s \omega_0) \mu[ds]$$

where μ is a probability measure on $\mathcal{S}$.

In such a case all extremal equilibrium state ω_i have the same internal symmetry group $\mathcal{S}_{\omega_i}$ which is a subgroup of $\mathcal{S}$ invariant under $\mathcal{E}$.

The proof of this theorem relies on the same arguments as the proof of theorem 2.

We shall conclude this general discussion with the following result which is a direct consequence of G.K.S inequalities:

Theorem 4

For ferromagnetic systems $T_1 < T_2$ implies $\mathcal{S}_{\omega^+_{T_1}} \subseteq \mathcal{S}_{\omega^+_{T_2}}$

Applications

Let us consider $\mathbb{Z}^\nu$- invariant ferromagnetic systems for temperature β such that there exists a unique equilibrium state invariant under $\mathcal{S} \circledS \mathbb{Z}^\nu$, for such systems ω^+_β is extremal $\mathbb{Z}^\nu$ - invariant and it follows from theorem 3 that all extremal equilibrium states ω_i with $\mathcal{G}/\mathcal{G}_i$ finite are given by $\omega_i = \tau'_S \omega^+$ with $S \in \mathcal{S}/\mathcal{S}_{\omega^+}$; moreover all these ω_i will be invariant under $\mathcal{S}_{\omega^+}$ which is a subgroup of $\mathcal{S}$ invariant under $\mathbb{Z}^\nu$, while the ω_i's will not necessarily be invariant under $\mathbb{Z}^\nu$.

1) Systems on $\mathbb{Z}^\nu$ with two-body forces only.
In this case $\mathcal{S} = \{\phi, \mathbb{Z}^\nu\} \supset \mathcal{S}_0 = \{\phi\}$
implies that there exists :
either one pure phase (*) invariant under $\mathcal{S}$, namely ω^+,
or two pure phases invariant under $\mathcal{S}_0$, namely ω^+ and $\omega^- = \tau'_{\mathbb{Z}^\nu}\omega^+$.

2) Triangular model with 3-body forces only (Example 2, sec. 2.7). In this case

$$\mathcal{S} = \{\phi, S_1, S_2, S_3\} \supset \begin{bmatrix} \mathcal{S}_1 = \{\phi, S_1\} \\ \mathcal{S}_1' = \{\phi, S_2\} \\ \mathcal{S}_1'' = \{\phi, S_3\} \end{bmatrix} \supset \mathcal{S}_0 = \{\phi\}$$

Since $\mathcal{S}_1, \mathcal{S}_1', \mathcal{S}_1''$, are not invariant under $\mathbb{Z}^\nu$, it follows that there exists either one pure phase invariant under $\mathcal{S}$, namely ω^+ , or two pure phases invariant under $\mathcal{S}_0$, ω^+ and $\frac{1}{3}(\tau'_{S_1} + \tau'_{S_2} + \tau'_{S_3})\omega^+$, which decompose further into 4 extremal states.

(*) By definition, ω is a "pure phase" if it is an extremal invariant equilibrium state [1] , i.e. if it cannot be decomposed into two $\mathbb{Z}^\nu$- invariant equilibrium states.

3) The Baxter model, defined by a square lattice $\mathbb{Z}^2$ with 4 body forces on a unit square and two-body forces between next nearest neighbour. (Example a, sec. 6.4).

In this case $\mathcal{S} = \{\phi, \mathbb{Z}^\nu, S_1, S_2\} \supset \begin{bmatrix} \mathcal{S}_1 = \{\phi, \mathbb{Z}^\nu\} \\ \mathcal{S}_1' = \{\phi, S_1\} \\ \mathcal{S}_1'' = \{\phi, S_2\} \end{bmatrix} \supset \mathcal{S}_0 = \{\phi\}$.

Since S_1 is invariant under $\mathbb{Z}^2$ there exists either one pure phase invariant under $\mathcal{S}$, ω^+
or two pure phases invariant under $\mathcal{S}_1$, ω^+ and $\omega^1 = \tau'_{S_1}\omega^+$
or three pure phases invariant under $\mathcal{S}_0$, ω^+, $\omega^- = \tau'_{\mathbb{Z}^\nu}\omega^+$, $\omega' = \frac{1}{2}(\tau'_{S_1} + \tau'_{S_2})\omega^+$
which decompose further into 4 extremal states.

4) The Ashkin-Teller model defined by $\mathcal{L} = \mathbb{Z}_A^2 \times \mathbb{Z}_B^2$ with 2 and 4 body forces (Example b, sec.6.4).
In this case
$$\mathcal{S} = \{\phi, \mathbb{Z}_A^2, \mathbb{Z}_B^2, \mathcal{L}\} \supset \begin{bmatrix} \mathcal{S}_1 = \{\phi, \mathcal{L}\} \\ \mathcal{S}_1' = \{\phi, \mathbb{Z}_A^2\} \\ \mathcal{S}_2' = \{\phi, \mathbb{Z}_B^2\} \end{bmatrix} \supset \mathcal{S}_0 = \{\phi\}$$
and thus all subgroups of $\mathcal{S}$ are invariant under $\mathbb{Z}^2$; therefore there exists either one pure phase invariant under $\mathcal{S}$, ω^+
or two pure phases invariant under either $\mathcal{S}_1$, $\mathcal{S}_1'$, $\mathcal{S}_1''$, given by ω^+ and either $\tau'_{\mathbb{Z}_A^2}\omega^+$, $\tau'_{\mathbb{Z}_B^2}\omega^+$ or $\tau'_{\mathcal{L}}\omega^+$
or three pure phases invariant under $\mathcal{S}_0$, ω^+, $\omega^- = \tau'_{\mathcal{L}}\omega^+$, $\omega' = \frac{1}{2}(\tau'_{\mathbb{Z}_A^2} + \tau'_{\mathbb{Z}_B^2})\omega^+$
which decompose further into 4 extremal states.

Let us note that if there exists only two pure phases, then only one of the symmetry groups $\mathcal{S}_1, \mathcal{S}_1', \mathcal{S}_1''$ will appear because of theorem 4. The symmetry group which does appear depends upon the values of the coupling constant $K(B)$.

In these last two examples, we could thus have in principle two phases transitions associated with a spontaneous symmetry breakdown, as shown in fig. 1 and 2.

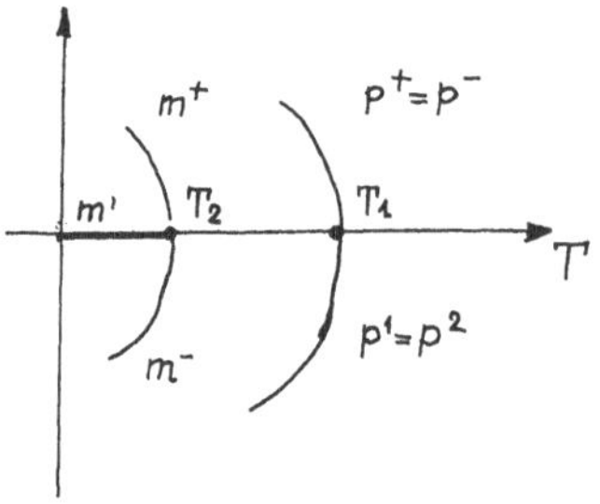

$$|\mathcal{S}| = 4 \quad m^+ = \omega^+[\sigma_x] = 0 \,;\; p^+ = \omega^+[\sigma_{<xy>}] = 0$$

$$|\mathcal{S}| = 2 \quad m^+ = 0 \qquad p^+ > 0$$

$$|\mathcal{S}| = 1 \quad m^+ > 0 \qquad p^+ > 0$$

Fig. 1 Possible order parameters for example 3.

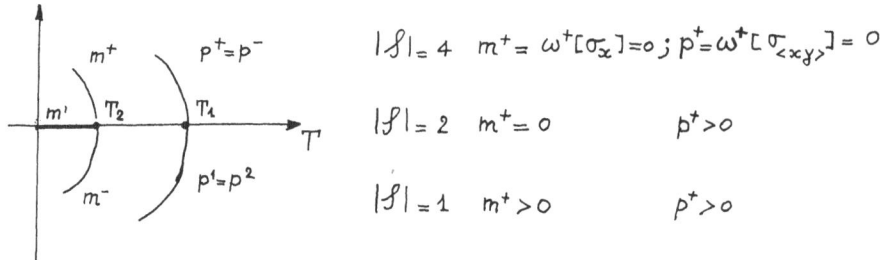

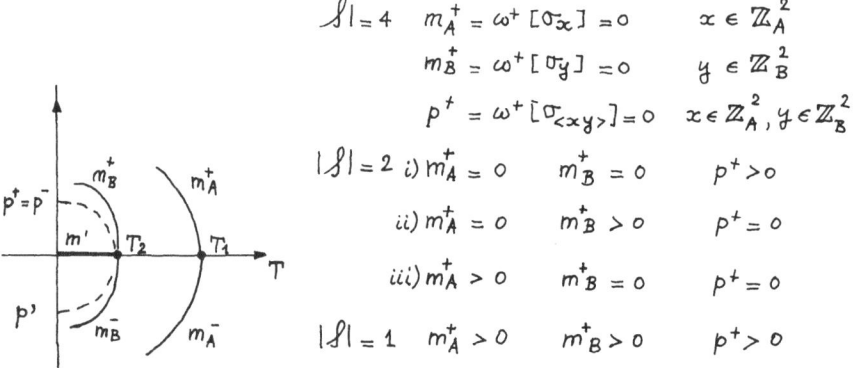

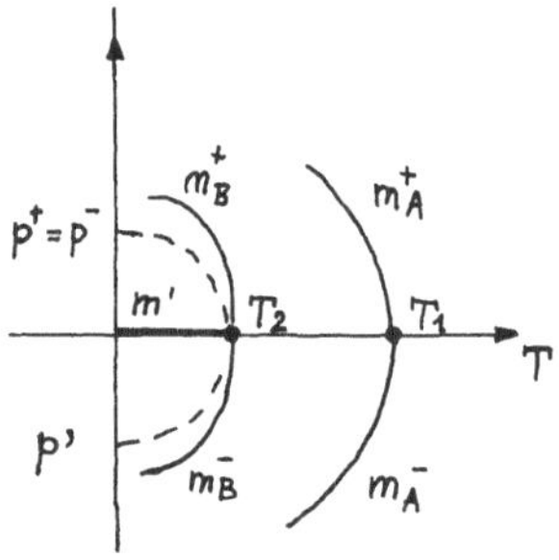

$$|\mathcal{S}| = 4 \quad m_A^+ = \omega^+[\sigma_x] = 0 \qquad x \in \mathbb{Z}_A^2$$

$$m_B^+ = \omega^+[\sigma_y] = 0 \qquad y \in \mathbb{Z}_B^2$$

$$p^+ = \omega^+[\sigma_{<xy>}] = 0 \qquad x \in \mathbb{Z}_A^2,\ y \in \mathbb{Z}_B^2$$

$$|\mathcal{S}| = 2 \quad i)\ m_A^+ = 0 \qquad m_B^+ = 0 \qquad p^+ > 0$$

$$ii)\ m_A^+ = 0 \qquad m_B^+ > 0 \qquad p^+ = 0$$

$$iii)\ m_A^+ > 0 \qquad m_B^+ = 0 \qquad p^+ = 0$$

$$|\mathcal{S}| = 1 \quad m_A^+ > 0 \qquad m_B^+ > 0 \qquad p^+ > 0$$

Fig. 2 Possible order parameters for the case $\mathcal{S} \supset \mathcal{S}_1'' \supset \mathcal{S}$

To conclude the discussion of these two models, we recall that the exact solution of Baxter [23] indicates that $T_1 = T_2$ for example 3 ; on the other hand the arguments of Wu and Lin[49] as well as the renormalization group techniques of Knops[50] indicates that $T_1 \neq T_2$ for example 4, and for certain values of the coupling constants (see sec. 7.4 and Part III sec. 4.4).

4.3 Generalized Peierls Argument

In this section, we generalise Peierls argument to arbitrary systems and give general conditions in terms of the group structure for the existence of a phase transition associated with a spontaneous breakdown of $\mathcal{G}$. For any finite Λ and x in Λ

$$\omega_{\Lambda}^{Y}[\sigma_x] = \sigma_x(Y)\{1 - 2\,\mathrm{Prob}_{\Lambda,Y}(s_x = -\sigma_x(Y))\} \qquad (4.4)$$

where $\mathrm{Prob}_{\Lambda,Y}(s_x = -\sigma_x(Y))$ denotes the probability that the spin at site x has the value $s_x = -\sigma_x(Y)$ opposite to that of the boundary condition Y.

It is thus sufficient to show that there exists some boundary condition Y and $\eta < \frac{1}{2}$ independent of Λ such that

$$\mathrm{Prob}_{\Lambda,Y}(s_x = -\sigma_x(Y)) \leq \eta < 1/2 \qquad (4.5)$$

If such a bound can be found then the influence of the boundary condition on the site x will remain even in the limit of infinite volume.

In the following we shall use the LT-HT duality transformation to obtain a generalization of Peierls argument in terms of closed contours on the dual lattice ; from theorem 2 of Sec.4.3 it appears at once that the same analysis could be carried out directly on the original lattice in terms of the "low temperature graphs". The advantage to go over the dual lattice is that in general it gives an intuitive picture of the decomposition into phases (Sec. Fig.4).

4.3.1 Graphical Structure

Given a lattice $\mathcal{L}$ and $\mathcal{B}_0$ <u>any family</u> of finite subset of $\mathcal{L}$

then if the group $\mathcal{K}_o$ associated with $\{\mathcal{L}, \mathcal{B}_o\}$ separate $\mathcal{B}_o$ it is always possible to construct a HT-LT dual $\{\mathcal{L}^*, \mathcal{B}_o^*\}$ by means of a bijection d between $\mathcal{B}_o$ and $\mathcal{B}_o^*$. (Sec. 2.4, 2.5). However, property 3 of sec. 2.5 does not always hold for infinite $\mathcal{L}$ and we can only conclude that d induces a bijection of Γ_o onto $\mathcal{K}_o^*$.

Property 1

The HT-LT dual $\{\mathcal{L}^*, \mathcal{B}_o^*\}$ defined by a bijection of $\mathcal{B}_o$ onto $\mathcal{B}_o^*$ is also a LT-HT dual for $\{\mathcal{L}, \mathcal{B}_o\}$ if and only if $\Gamma_o^f \equiv \Gamma_o \cap \mathcal{P}_f(\mathcal{L})$.

In conclusion for any $\{\mathcal{L}, \mathcal{B}_o\}$ satisfying the following conditions

$$\text{c.1} \quad \mathcal{K}_o \text{ separates } \mathcal{B}_o \tag{4.5}$$

$$\text{c.2} \quad \Gamma_o^f = \Gamma_o \cap \mathcal{P}_f(\mathcal{B}_o) \tag{4.6}$$

It is possible to construct a LT-HT dual $\{\mathcal{L}^*, \mathcal{B}_o^*\}$ by the method of sec. 2.4.

As we have already discussed the elements β^* of $\mathcal{K}_o^*$ are "closed graphs" and we shall call "Length of β^*" the cardinality $|\beta^*|$ of $\beta^* \subset \mathcal{B}_o^*$. Moreover any closed graph β^* can be decomposed into "non intersecting connected closed graphs", i.e.

$$\beta^* = \prod_j \beta_j^* ; \quad \beta_i^* \cap \beta_j^* = \phi , \quad \beta_i^* \text{ connected}^{(+)} \tag{4.7}$$

Theorem 1

Let $\{\mathcal{L}^*, \mathcal{B}_o^*\}$ be a HT-LT dual for $\{\mathcal{L}, \mathcal{B}_o\}$ where $\mathcal{B}_o \subset \mathcal{P}_f(\mathcal{L})$ satisfies the condition

$$\text{c.3}' \quad \mathcal{B}_o^\perp \cap \mathcal{P}_f(\mathcal{L}) = \{\phi\} \text{ i.e. } \mathcal{S}_{o,f} = \{\phi\}$$

(+) A graph $\beta^* \in \mathcal{P}(\mathcal{B}^*)$ is said "connected" if the graph defined by the vertices B_i^* in β^* and the lines l_{ij} between those vertices B_i^*, B_j^* such that $B_i^* \cap B_j^* \neq \phi$ in $\mathcal{P}(\Lambda^*)$ is connected in the usual sense of graph theory[51].

then for any finite $\Lambda \subset \mathcal{L}$ and any fixed $Y \subset \mathcal{L}$, the mapping $X \mapsto \beta_X^* = D[\gamma_0(X\, Y_\Lambda)]$, $Y_\Lambda = Y \cap \Lambda$,

is an injective group homomorphism from $\mathcal{P}(\Lambda)$ into $\mathcal{K}^*_{0,\Lambda^*} = \mathcal{K}^*_0 \cap \mathcal{P}(\mathcal{B}^*_{0,\Lambda^*})$,

where $\mathcal{B}_{0,\Lambda} = \{ B \in \mathcal{B}_0 \,;\, B \cap \Lambda \neq \phi \}$.

Furthermore, if $\mathcal{B}_0$ satisfies the condition c.3 below, the mapping $X \mapsto \beta_X^*$ is an isomorphism.

c.3 For any finite $\Lambda \subset \mathcal{L}$ and $X \in \mathcal{P}_f(\mathcal{L})$ the condition $\sigma_B(X) = +1$ for all $B \notin \mathcal{B}_{0,\Lambda}$ implies $X \subset \Lambda$ (+) (4.8)

i.e. if the conditions c.1 - c.3 are satisfied, we have the following sequence of isomorphisms.

$$\begin{array}{ccccccc} \mathcal{P}(\Lambda) & \overset{Y}{\longleftrightarrow} & \mathcal{P}(\Lambda) & \overset{\gamma_0}{\longleftrightarrow} & \Gamma_0 & \overset{D}{\longleftrightarrow} & \mathcal{K}^*_{0,\Lambda^*} \\ \Psi & & \Psi & & \Psi & & \Psi \\ X & \longleftrightarrow & X \cdot Y_\Lambda & \longleftrightarrow & \gamma_0(X \cdot Y_\Lambda) & \longleftrightarrow & \gamma_0(X Y_\Lambda)^* \end{array}$$

<u>Proof</u>

c.3 implies that γ_0 restricted to $\mathcal{P}(\Lambda)$ is injectif since for all $X, X' \subset \Lambda$ $\gamma_0(X) = \gamma_0(X')$ implies $X \cdot X' \in \mathcal{B}_0^\perp \cap \mathcal{P}_f(\mathcal{L})$ i.e. $X = X'$. The definition of LT-HT duality implies that for any $\beta^* \in \mathcal{K}^*_{0,\Lambda^*}$ there exists $X \in \mathcal{P}_f(\mathcal{L})$ such that $\gamma_0(X)^* = \beta^*$. Furthermore the condition c.3 implies that $X \subset \Lambda$ since $\sigma_B(X) = +1$ for all $B \notin \mathcal{B}_{0,\Lambda}$.

In conclusion for any $\{\mathcal{L}, \mathcal{B}_0\}$ where $\mathcal{B}_0$ satisfies c.1-3 we can always construct a LT-HT dual $\{\mathcal{L}^*, \mathcal{B}_0^*\}$ such that for any finite $\Lambda \subset \mathcal{L}$ and any $Y \subset \mathcal{L}$ there exists an isomorphism $X \mapsto \beta_X^*$ of the configuration space $\mathcal{P}(\Lambda)$ onto closed graphs $\mathcal{K}^*_{0,\Lambda^*}$ of the dual; we are thus led to introduce the following :

(+) Taking $\Lambda = \phi$ c.3 $\mapsto$ c.3'

Definition :

With any closed graph $\beta^* \in \mathcal{K}^*_{o,f}$, the "Interior of β^* " is the finite subset of $\mathcal{L}$ defined by :

$$\mathrm{Int}\,\beta^* = \gamma_o^{-1} D^{-1} \beta^*$$

Property 2

1. $\mathrm{Int}\,\beta^* = \mathrm{Int} \prod_i \beta_i^* = \prod_i \mathrm{Int}\,\beta_i^*$

i.e. $x \in \mathrm{Int}\,\beta^*$ where $\beta^* = \prod_i \beta_i^*$ if and only if it is in the interior of an odd number of closed graphs β_i^*.

2. $\mathrm{Int}\,\beta_X^* = X \cdot Y_\Lambda$

i.e $\mathrm{Int}\,\beta_X^*$ is precisley the set of sites z in Λ where $\sigma_z(X) = -\sigma_z(Y)$.

3. $\beta_X^* = \prod_{z \in \mathrm{Int}\,\beta_X^*} \gamma(z)^*$ is the decomposition of the closed graph β_X^* into generators of $\mathcal{K}_o^*$.

Remarks

1. The above definitions are the immediate generalisations of the graphical definition introduced by R. Minlos and Y. Sinaï [52] for the Ising model ; indeed for the Ising model $\gamma(x)^*$ defines a "small square" covering the site x which will then be in the interior of the contour defined by the "small square" which is precisely $\beta^* = \gamma(x)^*$.

2. We have introduced Y for the essentially ferromagnetic systems where the ground states differs from $\mathcal{S}$; in this case the contours are defined by means of generators $\gamma(x)^*$ placed at these sites x where $s_x = -\sigma_x(Y)$.

3. We have introduced $\mathcal{B}_o$ rather then the set of bonds $\mathcal{B}$, since

in the study of phase transition without breakdown of $\mathcal{S}$ we shall consider LT-HT duality transformation with respect to a subfamily $\mathcal{B}_0$ of $\mathcal{B}$ (see sec. 4.6.)

Theorem 2

For any lattice system $\{\mathcal{L},\mathcal{B},K\}$, $\mathcal{B}_0$ a subset of $\mathcal{P}_f(\mathcal{L})$ such that $\mathcal{B}_0^{\perp} \cap \mathcal{P}_f(\mathcal{L}) = \phi$ and $\{\mathcal{L}^*,\mathcal{B}_0^*\}$ a LT-HT dual for $\{\mathcal{L},\mathcal{B}_0\}$ then

$$H_{\Lambda,Y}(X) = H_{\Lambda,Y}[\beta_X^*]$$

where $H_{\Lambda,Y}[\cdot]$ is the function defined on $\mathcal{P}_f(\mathcal{B}_0^*)$ by

$$H_{\Lambda,Y}[\beta^*] = H_{\Lambda,Y}[\phi^*] + 2\sum_{B_0 \in D^{-1}\beta^*} J(B_0)\sigma_{B_0}(Y) + 2\sum_{\substack{B \notin \mathcal{B}_0 \\ \sigma_B(Int\,\beta^*) = -1}} J(B)\sigma_B(Y) \quad (4.9)$$

Proof :

$$H_{\Lambda,Y}(X) = -\sum_{B\in\mathcal{B}_\Lambda} J(B)\sigma_B(Y^c)\sigma_B(X) = -\sum_{B\in\mathcal{B}_\Lambda} J(B)\sigma_B(Y) + 2\sum_{\substack{B\in\mathcal{B}_\Lambda \\ \sigma_B(X\cdot Y_\Lambda) = -1}} J(B)\sigma_B(Y)$$

$$H_{\Lambda,Y}(X) = -\sum_{B\in\mathcal{B}_\Lambda} J(B)\sigma_B(Y) + 2\sum_{B\in\gamma_0(X\cdot Y_\Lambda)} J(B)\sigma_B(Y) + 2\sum_{\substack{B\notin\mathcal{B}_0 \\ \sigma_B(X\cdot Y_\Lambda) = -1}} J(B)\sigma_B(Y)$$

4.3.2. Phase Transition with Spontaneous Breakdowns of the Internal Symmetry Group

In this section, we generalize Peierls argument to arbitrary essentially ferromagnetic systems and give general conditions for the existence of a phase transition associated with a breakdown of $\mathcal{S}$. As it was shown in sec. 4.1, we can restrict ourselves to the case of ferromagnetic systems with + boundary conditions, i.e. $S = \phi$; furthermore, from the discussions at the beginning of this section 4.3, it will be sufficient to show that there exists η independent of Λ such that

$$\mathrm{Prob}_{\Lambda,+}(s_x = -1) \leq \eta < 1/2 \quad (4.10)$$

In the following we shall thus apply the general graphical structure of sec. 4.3.1 with $\mathcal{B}_o \equiv \mathcal{B}$ and $\mathcal{Y} = \Phi$.

Lemma

Let $\{\mathcal{L}, \mathcal{B}, K\}$ be any ferromagnetic system such that

c.1 $\mathcal{K}$ separates $\mathcal{B}$.

c.2 $\Gamma^f \equiv \Gamma \cap \mathcal{P}_f(\mathcal{B})$

c.3 for any $\Lambda, X \in \mathcal{P}_f(\mathcal{L})$, $\sigma_B(X) = +1$, $\forall B \notin \mathcal{B}_\Lambda$ implies $X \subset \Lambda$.

then :

$$Prob_{\Lambda,+}(s_x = -1) \leq \sum_{\substack{\text{connected graph } \beta^* \in \mathcal{K}^* \\ \text{with } Int\,\beta^* \ni x}} e^{[-2\beta \sum_{B \in D^{-1}\beta^*} J(B)]}$$

Proof :

From theorems 1,2 and property 2 of sec 4.3.1, it follows that

$$Prob_{\Lambda,+}(s_x = -1) = \frac{\sum_{\beta^* \in \mathcal{K}^*,\, Int\,\beta^* \ni x} e^{[-2 \sum_{B \in D^{-1}\beta^*} K(B)]}}{\sum_{\beta^* \in \mathcal{K}^*} e^{[-2 \sum_{B \in D^{-1}\beta^*} K(B)]}} = \sum_{\substack{\beta^* \in \mathcal{K}^*_\Lambda \\ x \in Int\,\beta^*}} Prob_{\Lambda,+}[\beta^*]$$

Moreover since for any $\beta^* \in \mathcal{K}^*_\Lambda$ such that $Int\,\beta^* \ni x$ there exists at least one $\tilde{\beta}_i^*$ (Eq. 4.7) such that $Int\,\tilde{\beta}_i^* \ni x$, then

$$Prob_{\Lambda,+}(s_x = -1) \leq \sum_{\substack{\text{connected} \\ \tilde{\beta}^* \in \mathcal{K}^*_{\Lambda^*} \\ Int\,\tilde{\beta}^* \ni x}} P[\tilde{\beta}^*]$$

where

$$P[\tilde{\beta}^*] = \sum_{\substack{\beta^* \in \mathcal{H}^*_{\Lambda^*} \\ \beta^* \supset \tilde{\beta}^* \\ \beta^* \cdot \tilde{\beta}^* \cap \beta^* = \phi}} \frac{\prod_{B \in D^{-1}\beta^*} e^{-2K(B)}}{\sum_{\beta^* \in \mathcal{H}^*_{\Lambda^*}} \prod_{B \in D^{-1}\tilde{\beta}^*} e^{-2K(B)}} \tag{4.11}$$

But to each term β^* of the numerator, we can associate the term $\tilde{\beta}^* \cdot \beta^*$ in the denominator and $D^{-1}(\beta^*) = D^{-1}(\tilde{\beta}^* \beta^*) \cup D^{-1}(\tilde{\beta}^*)$ yields

$$P[\tilde{\beta}^*] \leq \prod_{B \in D^{-1}\tilde{\beta}^*} e^{-2K(B)}$$

Combining the condition Eq. (4.10) with the above lemma we obtain the following conclusion [25]:

Main theorem

Let $\{\mathcal{L}, \mathcal{B}, K\}$ be any essentially ferromagnetic system with an action of $\mathbb{Z}^\nu$ on $\mathcal{L}$ and bonds $\mathcal{B}$ satisfying the conditions c.1 - c.3 of the lemma, and let $\{\mathcal{L}^*, \mathcal{B}^*\}$ be a LT-HT dual for $\{\mathcal{L}, \mathcal{B}\}$

If $\sum_{n=0}^{\infty} e^{-2\beta |J| \cdot n} N_n s_n \leq \eta < 1/2$

where $|J| = \min_{B \in \mathcal{B}} |J(B)|$

N_n = number of direct connected graphs $\tilde{\beta}^*$ of length n which are not congruent

s_n = maximum number of sites in the interior of any closed connected graph of length n,

then there exists a phase transition associated with a spontaneous breakdown of $\mathcal{S}$. Moreover there exists at least $|\mathcal{S}|$

distinct extremal equilibrium states although not necessarily $|\mathcal{S}|$ distinct pure phase.

Remark

1) If there exists a finite family of fundamental bonds then $N_n \leq a\, q^n$; moreover in general we will be able to show that $s_n \leq b n^\nu$; if this is the case then it follows that the series

$$\sum_n e^{-2\beta J n} N_n s_n \leq \sum_n e^{-(2\beta J - \ln q)n} a b n^\nu$$

converges for $2\beta J > \ln q$ and is arbitrary small for β sufficiently large, therefore for such systems $\omega^+[\sigma_x] \neq 0$ at low temperature.

2) For lattice defined by $\mathcal{L} = \mathbb{Z}^\nu$, with $\mathbb{Z}^\nu$-invariant interactions, explicit bounds for N_n and s_n were recently given in ref. [6].

3) We remark that $\Gamma^{\mathcal{L}} = \Gamma \cap \mathcal{Q}_{\mathcal{L}}(\mathcal{B})$ (condition c.2) if and only if $\mathcal{K}$ coincides with the closure of $\mathcal{K}_{\mathcal{L}}$; therefore if there exists an action of $\mathbb{Z}^\nu$ on the system we recover the condition that $|\mathcal{K}|$ must be of the order of $2^{|\Lambda|}$, condition we had discuss in sec. 2.8.

4. 4. Application

It is clear that the above results applied to the 2-dimensional Ising model with n.n interaction give exactly the standard Peierls argument ; in particular the definition of "contour" and "interior" given in sec. 4.3 in terms of the group structure, coincides with the graphical terminology introduced in Peierls argument (see Fig. 4 a).

The main theorem cannot be applied to the 1 dimensional Ising model with 2-body interactions between nearest neighbour, since condition c.2 is not satisfied ; the same remark also holds for the system defined by the square lattice with 4-body forces on each unit square ; in fact, we know that there exists no phase transitions for these systems. The same condition c.2 is also not satisfied for the model (Example 1, sec. 2.7) defined by the square lattice and 4-body forces.

We shall now illustrate the preceeding discussion with the triangular model with 3-body forces $J>0$ described in sec. 2.7 (Example 2). The infinite system has 4 ground states which are the 4 elements of $\mathcal{S}$. Let us consider a finite system Λ with boundary condition $y=\phi$; i.e. all spins outside of Λ have the fixed value + 1. To follow the arguments of sec. 4.3.2 we introduce the HT-LT dual transformation defined by the identity mapping $d: B \longmapsto B^*=B$. In this case, the isomorphism $X \mapsto \beta^*_X$ between the configuration space $\mathcal{P}(\Lambda)$ and closed graphs is simply given by

$$\beta^*_X = \gamma(X) = \{B ; \sigma_B(X) = -1\}$$

and the interior of β^*_X is exactly the set of sites y where $\sigma_y(X) = -1$. A typical configuration and its associated closed graphs is shown on fig. 4 b; it is immediately seen that the closed graphs are in general "thick" boundary (+) which separates pure phases or ground state.

(+) In fact, the elements $\beta^* \in \mathcal{K}^*$ are called "hypergraphs" in standard graph theory. [51]

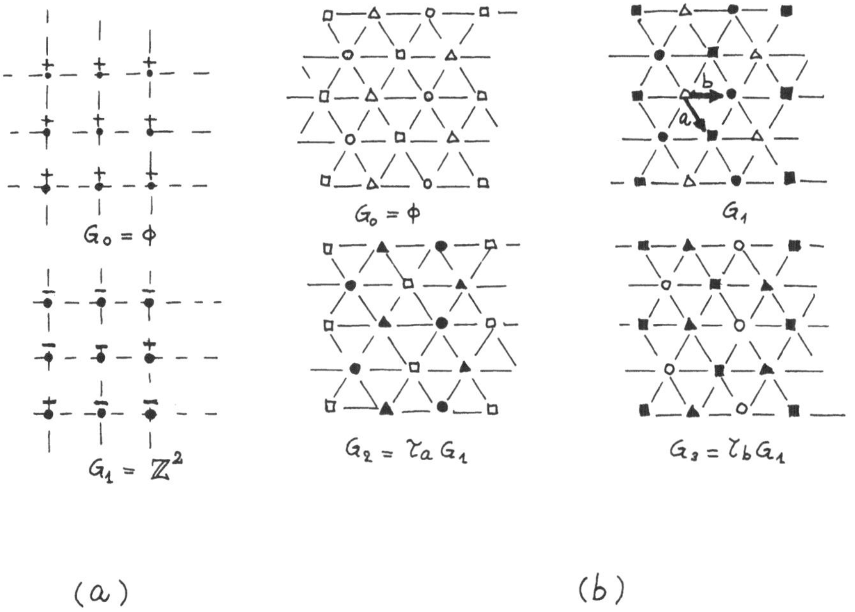

Fig. 3 The family of ground states for a) the Ising model with $J > 0$ and b) the triangular model with 3-body interactions $J > 0$. (The heavy dots represent the sites with $s=-1$)

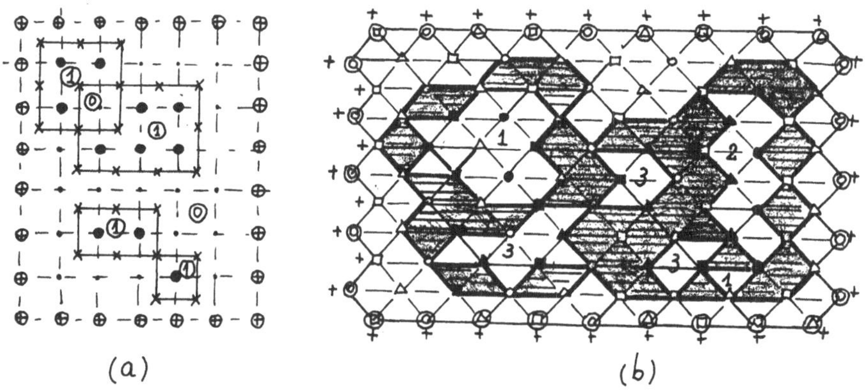

Fig. 4 Closed contours associated with a typical configuration for a finite system with boundary condition $G = \phi$, the spins outside of Λ are represented by circles (the heavy dots represent the sites with spins -1)

a) $\{ \times\!\!-\!\!\times \}$ set of B^* defining the closed graph β^*_X .

b) $\{ \blacktriangle \}$ set of B^* defining the closed graph β^*_X .

To obtain explicit bounds we then transform this problem in the following way [39] :

For all connected closed graph $\tilde{\beta}^*$ with $\ell(\tilde{\beta}^*)=\ell$ we associate a multigraph with length 3ℓ on the triangular lattice by considering the following mapping:

$\forall B^* \in \beta^* \; : \; B^* \longrightarrow (b_1, b_2, b_3)$ where b_i are the three sides of the "triangle" B^* . In this way, each point of the multigraph so obtained has an even coordination number and thus there exists an Euler cycle on the multigraph [51] . We then easily obtain $N_\ell < (6)^{3\ell}$. On the other hand $\mathrm{Int}\,\beta^* \leq \frac{\ell^2}{24}$ (the equality is realized for some contours β^* with $\mathrm{Int}\,\beta^* \subset \mathbb{E}_i$, $\mathbb{E}_i$ one of the 3 hexagonal lattices).
From Eq. 4.4 and the main theorem we then have :

$$\langle \sigma_x \rangle_\phi \geq 1 - 2 \sum_{\ell=6,8,\ldots}^{\infty} e^{-2\beta J \cdot \ell} \, 6^{3\ell} \cdot \frac{\ell^2}{24}$$

The series converge for $e^{-2\beta J} \cdot 6^3 < 1$ i.e $e^{-2K_0} < 6^{-3}$
A lower bound for T_c is thus given by : $e^{-2\beta J} \cdot 6^3 = 0{,}67$.

In conclusion for $T < T_0$ then there exists at least 4 distinct extremal equilibrium states : ω^+_β which is $\mathbb{Z}^\nu$-invariant and ω^i_β , i = 1,2,3 which are not $\mathbb{Z}^\nu$-invariant ; thus for $T<T_0$ there coexist two pure phases ω^+_β and $\omega'_\beta = \frac{1}{3} \cdot (\omega^1_\beta + \omega^2_\beta + \omega^3_\beta)$ which can be interpreted respectively as a fluid and a solid phase. (see also sec. 4.6).

4. 5. Invariant States and States with Free Boundary Conditions

It is expected that for ferromagnetic $\mathbb{Z}^{\nu}$-invariant systems with $|\mathcal{S}| < \infty$, the invariant state $\omega_\beta^I = |\mathcal{S}|^{-1} \sum_{s\in\mathcal{S}} \tau_s' \omega_\beta^+$ will coincide with the state ω_β^{op} defined by means of the free (or open) boundary conditions. This conjecture is of interest since it would imply there exists a unique invariant equilibrium state (see ch.7). This result is of course true at high temperature where it is known that the equilibrium state is unique ; we shall now show that for essentially ferromagnetic systems this same conjecture also holds for $T < T_0'$ when T_0' is larger then the T_0 obtained for the existence of a phase transition (Main theorem of last section).

To derive this property we use again the group structure to generalize the proof given by A. Martin-Löf [38] for the 2-dimensional Ising model with nearest neighbour interactions. To obtain such a generalisation, we first have to introduce the concepts of "inside" and " outside " of a closed graph.

Let $\{\mathcal{L},\mathcal{B}\}$ be a lattice system such that $\mathcal{B}$ satisfies conditions c1 - c3 of sec. 4.3.2 and let $\{\mathcal{L}^*,\mathcal{B}^*\}$ be any LT-HT dual; as we have seen there is an isomorphism $X \mapsto \beta_X^* = \gamma(X)$ between the group $\mathcal{P}(\Lambda)$ and the group of closed graphs $\mathcal{K}_{\Lambda^*}^* \subset \mathcal{P}(\mathcal{B}_\Lambda^*)$ where $\mathcal{B}_\Lambda = \{ B \in \mathcal{B} ; B \cap \Lambda \neq \phi \}$. With β^* a closed graph in $\mathcal{K}^*_{\Lambda^*}$ we introduce the following subset of $\mathcal{L}$:

$$\operatorname{Int}\beta^* = \gamma^{-1} D^{-1} \beta^* \qquad \text{i.e.} \qquad \operatorname{Int}\beta_X^* = X$$

<u>Outside β^*</u> $= \{ y \in \mathcal{L} ;\ \forall B \ni y,\ B \cap \mathrm{Int}\,\beta^* = \phi \}$

i.e. subset of sites which are not connected with the interior of β^*

<u>Boundary of β^*</u> $= \{ y \in \mathcal{L}$; y connected with Int β^* and either y connected with outside or $\exists B \ni y$ such that $\sigma_B(\mathrm{Int}\,\beta^*) = -1 \}$

<u>Inside β^*</u> $= \mathcal{L} \ / \ \{$ outside $\beta^* \cup$ boundary $\beta^* \} =$
$= \{ y \in \mathcal{L} ; y$ connected with Int β^*, y not connected with the outside and $\sigma_B(\mathrm{Int}\,\beta^*) = +1$ $\forall B \ni y \}$.

<u>Property</u> :

β^* induces a partition of $\mathcal{L}$ into "Inside", "Outside", and "Boundary" ; moreover Int β^* and Inside β^* are not connected to Outside β^*.

These concepts are illustrated on Fig. 5

<u>Theorem</u>

Let $\{\mathcal{L}, \mathcal{B}, K\}$ be any essentially ferromagnetic system with an action of $\mathbb{Z}^\nu$ on $\mathcal{L}$ and $\mathcal{B}$ satisfying the conditions c1-c3 and let $\{\mathcal{L}^*, \mathcal{B}^*\}$ be a LT-HT dual for $\{\mathcal{L}, \mathcal{B}\}$. If furthermore $\mathcal{B}$ satisfies the condition c4-c5 below and if $\sum_n e^{-2Kn} N_n \mathcal{S}_n < \infty$ then $\omega_\beta^I = \omega_\beta^{op}$. In particular in this case $\omega_\beta^+ = \omega_\beta^{op}$ on the symmetric algebra.

c.4 For any connected subset $\Lambda \subset \mathcal{L}$ and any $X \subset \mathcal{L}$ the condition $\sigma_B(X) = 1$ for all B in $\mathcal{B}_\Lambda$ implies that there exists $S \in \mathcal{S}$ such that $X \cap \bar{\Lambda} \equiv S \cap \bar{\Lambda}$ where $\bar{\Lambda} = \bigcup_{B \in \mathcal{B}_\Lambda} B$ $=$ set of sites in $\mathcal{L}$ connected to Λ.

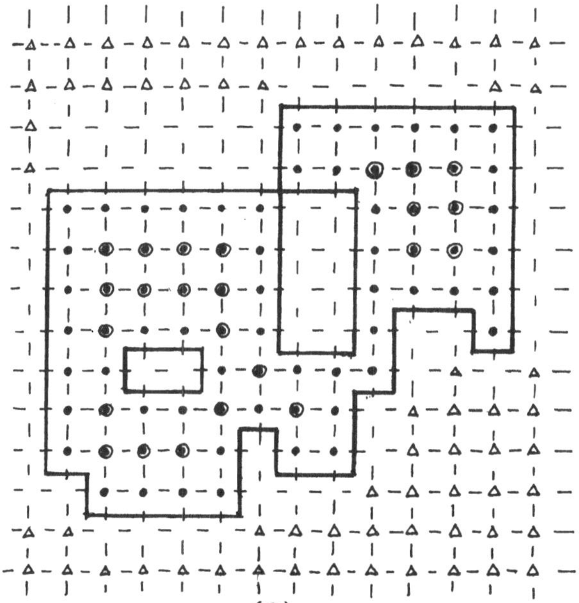

(a)

(b)

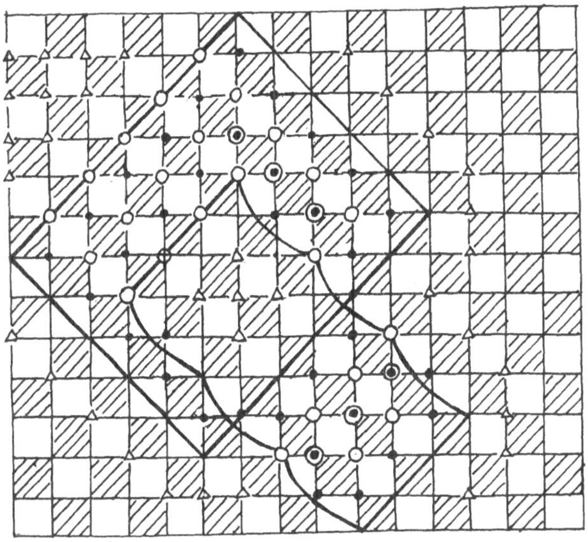

(c)

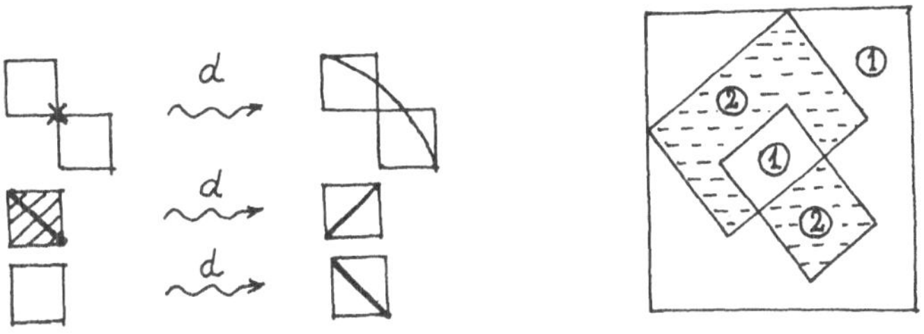

$\{\bullet\} = Int\ \beta^*$ $\qquad$ $\{\Delta\} = Outside\ \beta^*$

$\{\circ\} = Inside\ \beta^*$ $\qquad$ for a), b), c).

Fig. 5 Examples of Interior, Inside, Outside

a) Ising model c.1 - c.5 are satisfied

b) Triangular model c.1 - c.5 are satisfied

c) Model with 4 and 2 body forces and external field where c.1 - c.4 are satisfied but not c.5.

c.5 With β_1^*, β_2^* two connected, non intersecting, closed graphs on the dual then

either i) $\mathrm{Int}\,\beta_i^* \subset$ Outside β_j^* which we denote by $\beta_i \vee \beta_j^*$: "β_i^* outside β_j^*"

or ii) (a) $\mathrm{Int}\,\beta_i^* \subset$ Inside β_j^* and (b) $\alpha^*[\beta_i^*] \subset \alpha^*[\beta_j^*]$,

where $\alpha^*[\beta^*] = \bigcup_{x \in Int\beta^*} \gamma(x)^* = \{B \in \mathcal{B}\ ;\ B \cap Int\,\beta \neq \phi\}^*$

which we denote by $\beta_i^* < \beta_j^*$: "β_i^* inside β_j^*".

Remarks

1.- The condition (i) is equivalent to $\alpha^*[\beta_i^*] \cap \alpha^*[\beta_j^*] = \phi$

2.- The conditions c1-c4 seem to be necessary conditions; c5 has been essentially introduced to simplify the following proof. For systems such that c5 is not satisfied, we could introduce "contours" defined as interacting closed graphs for which the proof could be adapted. (Fig. 5 (c) would be then one contour).

3.- If for all $\mathcal{B} \ni y$ where $y \in$ Inside β^*, we have $\mathcal{B} \cap$ Int $\beta^* \neq \phi$ then (ii-a) implies (ii-b)

Proof

From the preceeding discussion it is sufficient to consider ferromagnetic systems with + boundary conditions. From GKS inequality $\omega_\beta^{op}[\sigma_{\bar{B}}] \leq \omega_\beta^+[\sigma_{\bar{B}}] \; \forall \bar{B} \in \bar{\mathcal{B}}$,
it follows that we just need to show that $\omega_\beta^{op}[\sigma_{\bar{B}}] \geq \omega_\beta^+[\sigma_{\bar{B}}] \; \forall \bar{B} \in \bar{\mathcal{B}}$.
With X any configuration of the finite system $\Lambda \supset \bar{B}$ with free boundary conditions we associate the closed graph $\beta_X^* = \gamma(X)^*$ on the LT-HT dual $\{\mathcal{L}^*, \mathcal{B}^*\}$. Let $\beta_X^* = \prod_i \beta_i^*$ be the decomposition into non-intesecting closed graphs, we say that β_i^* is "open" if there exists some $B \in \mathcal{B}$, $B \not\subset \Lambda$, such that $B \cap \text{Int}\,\beta_i^* \neq \phi$; moreover we say that "X does not separate $A \subset \Lambda$ " if for any β_i^* open, either A is contained in a connected subset of Inside β_i^*, or A is contained in a connected subset of outside β_i^*. With any X which does not separate A, we write :

$$\beta_X^* = \prod_i \bar{\beta}_i^* \prod_j \bar{\beta}_j^* \prod_k \beta_k^*$$

where $\{\bar{\beta}_i^*\}$ are the open graphs with $A \subset$ Inside β_i^*
$\{\bar{\beta}_j^*\}$ are the open graphs with $A \subset$ Outside β_j^*
$\{\beta_k^*\}$ are the graphs which are not open.

and introduce $\Lambda_1 = \bigcap_i \text{Inside}\,\bar{\beta}_i^* \bigcap_j \text{Outside}\,\bar{\beta}_j^*$

Finally with any X which does not separate A, we define the outer set $\Lambda' \subset \Lambda$ as the largest connected set containing A which is a subset of Λ_1 .

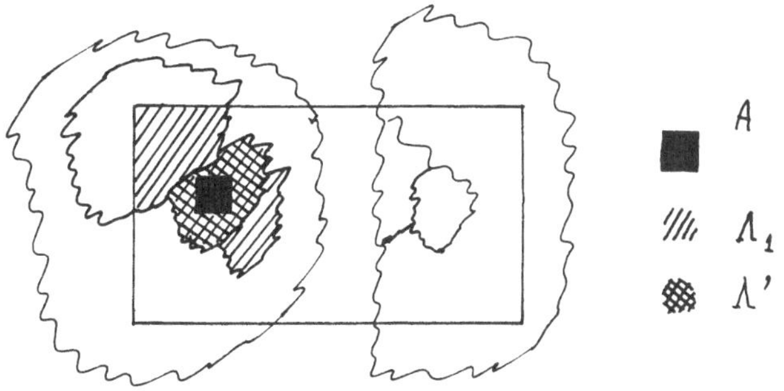

Fig. 6

Since $\sigma_B\,(Int\,(\prod_i \bar\beta_i^* \prod_j \bar\beta_j^*)) = \prod_i \sigma_B\,(Int\,\bar\beta_i^*)\prod_j \sigma_B\,(Int\,\bar\beta_j^*) = +1$ for all $B \in \mathcal{B}_{\Lambda'}$ (because of the definitions), we conclude with condition c.4 that if **A** is connected then there exists S in $\mathcal{S}$ such that $S \cap \overline{\Lambda'} \equiv Int\,(\prod_i \bar\beta_i^* \prod_j \bar\beta_j^*) \cap \overline{\Lambda'}$.

To go on with the proof, we need the following result :

Lemma 1

With $\beta^* = \prod_i \bar\beta_i^* \prod_j \bar\beta_j^* \prod_k \beta_k^*$ and $\bar\beta^* = \prod_i \bar\beta_i^* \prod_j \bar\beta_j^*$ then $\sigma_x\,(Int\,\beta^*) = \sigma_x\,(Int\,\bar\beta^*)$ for any $x \in \partial\Lambda' = \overline{\Lambda'}/\Lambda'$.

In other words for a given set of open contours the values of the spin at any site x connected to Λ' but not in Λ' is independent of the contours β_k^* which are not open.

Proof

If $\beta_1^* < \beta_2^*$ and if β_1^* is open then by definition β_2^* is also open ; therefore, using the condition c.5, for any β_k^* which is not open and any $\bar\beta^*$ open we have either $\beta_k^* < \bar\beta^*$ or $Int\,\beta_k^* \subset$ Outside $\bar\beta^*$ and thus $Int\,\beta_k^* \subset$ inside $\bar\beta^* \cup$ outside $\bar\beta^*$.

With $x \in \partial\Lambda' = \overline{\Lambda'}/\Lambda'$, $\exists$ some $\bar\beta^*$ open such that $x \in$ Boundary $\bar\beta^*$; therefore any closed graph β_k^* which is not open is such that $Int\,\beta_k^* \not\ni x$, which conclude the proof.

With this lemma and condition c.4 we conclude that for any connected **A** and any family of open contours which do not separate **A**, there exists **S** in $\mathcal{S}$ such that all spins at the boundary $\partial\Lambda' = \bar{\Lambda}' / \Lambda'$ are in the configuration defined by **S**. It then follows that for any $\bar{B} \subset A$, $\bar{B} \in \bar{\mathcal{B}}$, the conditionnal average of $\sigma_{\bar{B}}$ given the outer set is Λ' is identical with $\omega^{S}_{\Lambda'}[\sigma_{\bar{B}}]$; but from Theorem 1, sec. 4.2 and GKS inequality it follows that $\omega^{S}_{\Lambda'}[\sigma_{\bar{B}}] = \omega^{+}_{\Lambda'}[\sigma_{\bar{B}}] \geq \omega^{+}[\sigma_{\bar{B}}]$. With $\mathcal{E}_A$ the set of all configurations **X** of Λ which do not separate **A** we have thus obtained

$$\omega^{op}_{\Lambda, \mathcal{E}_A}[\sigma_{\bar{B}}] \geq \omega^{+}[\sigma_{\bar{B}}] \quad \forall \bar{B} \in \bar{\mathcal{B}} \quad \bar{B} \subset \Lambda$$

Lemma 2

The probability $P^{f}_{\Lambda, \mathcal{E}_A}$ tends uniformly to **1** when $d(A, \partial\Lambda) \geq |\Lambda|^{1/\nu}$ as $|\Lambda| \to \infty$ if $\sum_n e^{-2Kn} N_n \delta_n < \infty$.

Indeed if **X** separates **A**, it follows that there exists some $y \in A$ and some open graph $\bar{\beta}^{*}$ such that $y \in$ boundary $\bar{\beta}$; condition c.5 implies then $\gamma(y)^{*} \cap \bar{\beta}^{*} \neq \phi$ and the end of the proof is indentical with the one given in ref. [38]

The proof of the theorem is thus completed since

$$\omega^{op}_{\Lambda}[\sigma_{\bar{B}}] = P^{f}_{\Lambda, \mathcal{E}_A} \, \omega^{op}_{\Lambda, \mathcal{E}_A}[\sigma_{\bar{B}}] + (1 - P^{f}_{\Lambda, \mathcal{E}_A}) \, \omega^{op}_{\Lambda, \bar{\mathcal{E}}_A}[\sigma_{\bar{B}}]$$

combined with the above results yield

$$\omega^{op}[\sigma_{\bar{B}}] = \omega^{op}_{\mathcal{E}_{\Lambda}}[\sigma_{\bar{B}}] \geq \omega^{+}[\sigma_{\bar{B}}]$$

4.6. Phase Transitions without Breakdown of the Internal Symmetry Group

The existence of a phase transition with breakdown of the translation group was obtained by R.L. Dobrushin [53] using the anti-ferromagnetic Ising model with an external field (see also J. Ginibre [54]), later O.J. Heilman gave for certain lattice systems an extension of Peierls argument to establish the existence of phase transitions with breakdown of the euclidean group [55] ; recently S.A. Pirogov and Y.G. Sinaï [56] gave a very interesting method to prove the existence of phase transitions which are not associated with a breakdown of the symmetry group. They showed that for $T < T_0$ there exists chemical potentials $\mu_i = \mu_i(T)$ at which several phases coexist. This work of Pirogov and Sinaï has been useful to study explicit models, for example to prove the existence of phase transitions for lattice gas with hard cores [57,58] .

In this section, we shall restrict ourselves to a discussion of phase transitions associated with spontaneous breakdown of the translation group, i.e. using the group structure we generalise Dobrushin's idea to arbitrary lattice systems. We consider a system for which the argument of sec. 4.3.2 have established the existence of a phase transition and we investigate the possible phase transition when this same model is placed in an external field. For systems with an external field the internal symmetry group contains only the identity and thus there can be no phase transition associated with a breakdown of $\mathcal{S}$.

Let $\{ \mathcal{L} = \mathbb{Z}^\nu, \mathcal{B}, K \}$ be a $\mathbb{Z}^\nu$-invariant, essentially ferromagnetic, lattice system (sec. 4.2) satisfying the conditions c1-c5 of sec. 4.3.1. and 4. 5 . We recall that c1-c2 imply the existing of LT-HT dual ; c-3 yields for any $\Lambda \subset \mathbb{Z}^\nu$ and any boundary condition Y an isomorphism between $\mathcal{P}(\Lambda)$ and closed graphs $\mathcal{K}^*_{\Lambda^*}$ on the dual ; c.4 implies that for any connected closed graph the inside of the graph decomposes into connected subsets associated with one of the ground states. Those four conditions seem to be necessary; on the other hand, c.5 is rather a simplifying condition. From c.5 it follows that any closed graph β^* in $\mathcal{K}^*_f$ can be uniquely decomposed into components such that :

$$\beta^* = \bigcup_{j=1}^{q} \{ c_j^* \bigcup_r \beta_{jr}^* \} \tag{4.12}$$

where the c_j^* are the "<u>outer components</u>" (connected closed graph which is not "inside" any other components in the sense of sec. 4. 5), Int β_{jr}^* is contained in some connected subset of the inside of c_j^* , and β_{jr}^* is outside β_{ks}^* for any (k,s). Therefore there exists S_{jr} in $\mathcal{S}$ such that <u>"The ground state $S_{jr} Y$ surrounds β_{jr}^* " i.e.</u>

$$S_{jr} \cap \left(\bigcup_{B \in Int \beta_{jr}^*} B \right) = Int\, c_j^* \cap \left(\bigcup_{B \in Int \beta_{jr}^*} B \right) \tag{4.13}$$

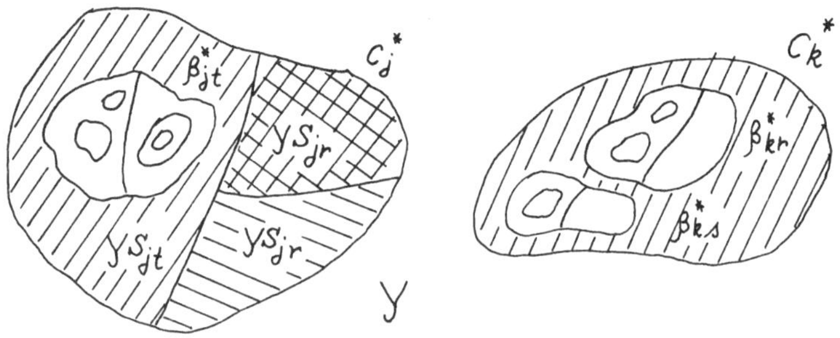

Fig. 7 Decomposition of β^* into outer components and components inside the outer components. (The contours represent "Thick boundary" between ground states).

We then consider this same model in the presence of an external field h; this new model is defined by $\{\mathbb{Z}^\nu, \mathcal{B}', K'\}$ where $\mathcal{B}' = \mathcal{B} \cup \mathbb{Z}^\nu$ and $K'(B) = K(B)$ for $B \in \mathcal{B}$, $K'(x) = h$ for $x \in \mathbb{Z}^\nu$ i.e.

$$H'_{\Lambda,Y} = H_{\Lambda,Y} + h \sum_{x \in \Lambda} \sigma_x$$

Using theorem 2 of sec. 4.3.1 we have :

$$H'_{\Lambda,Y}[\beta^*] = H'_{\Lambda,Y}[\phi^*] + 2 \sum_{B \in D^{-1}\beta^*} J(B)\sigma_B(Y) + 2h \sum_{x \in Int\beta^*} \sigma_x(Y)$$

ϕ^* = empty set in $\mathcal{P}(\mathcal{B}^*)$.

which gives since $\{\mathcal{L}, \mathcal{B}, K\}$ is essentially ferromagnetic

$$H'_{\Lambda,Y}[\beta^*] = H'_{\Lambda,Y}[\phi^*] + 2 \sum_{B \in D^{-1}\beta^*} |J(B)| + 2h \sum_{x \in Int\beta^*} \sigma_x(Y) \qquad (4.14)$$

for any ground state Y of $\{\mathcal{L}, \mathcal{B}, K\}$

Let Y be a ground state of $\{\mathbb{Z}^\nu, \mathcal{B}, K\}$ and $\omega^Y_{(\Lambda,K')}$ be the Gibbs state of the new model $\{\mathbb{Z}^\nu, \mathcal{B}', K'\}$ with boundary condition Y. As we have discussed in sec. 4.3. it will be sufficient to obtain a bound of the form

$$Prob'_{\Lambda,Y}[s_x = -\sigma_x(Y)] \leq \eta < 1/2$$

to conclude at the existence of Gibbs states which are not $\mathbb{Z}^\nu$-invariant, if Y is not $\mathbb{Z}^\nu$-invariant.

Moreover a slight modification of Eq. 4.11 yield

$$\mathrm{Prob}'_{\Lambda,Y}[s_x = -\sigma_x(Y)] \leq \sum_{\substack{C_o^*,\text{connected}, \in \mathcal{K}^*_{\Lambda^*} \\ \mathrm{Int}\, C_o^* \ni x}} \mathrm{Prob}'_{\Lambda,Y}[C_o^*]$$

$$\mathrm{Prob}'_{\Lambda,Y}[C_o^*] = \frac{\sum^{\blacktriangledown} e^{-\beta H'_{\Lambda,Y}[\beta^*]}}{\sum_{\beta^* \in \mathcal{K}^*_{\Lambda^*}} e^{-\beta H'_{\Lambda,Y}[\beta^*]}} \tag{4.15}$$

where the sum $\sum^{\blacktriangledown}$ at the numerator is over all $\beta^* \in \mathcal{K}^*_{\Lambda^*}$ containing C_o^* as outer component .

It follows from Eq. 4.14 together with the condition c.4 - c.5 (Eq. 4.12, 4.13) that for any β^* containing C_o^* as outer component then

$$\beta^* = \left(C_o^* \bigcup_{r=1}^{n} \beta^*_{or}\right) \cup \tilde{\beta}^* \tag{4.16}$$

$$\beta^*_{or} < C_o^* \qquad \text{and} \qquad C_o^* \vee \tilde{\beta}^*$$

and $$H'_{\Lambda,Y}[\beta^*] - H'_{\Lambda,Y}[\phi^*] = 2\sum_{B\in D^{-1}\beta^*} |J(B)| + 2h \sum_{x\in \mathrm{Int}\,\tilde{\beta}^*} \sigma_x(Y) + 2h \sum_{x\in \mathrm{Int}\, C_o^*} \sigma_x(Y) + 2h \sum_{r=1}^{n} \Big\{ \sum_{x\in \mathrm{Int}\,\beta^*_{or}} \sigma_x(Y\cdot S_r)\Big\}$$

where S_r is the element in $\mathcal{S}$ such that $S_r Y$ surrounds β^*_{or}

<u>Theorem</u>

Let $\{\mathbb{Z}^\nu, \mathcal{B}, K\}$ be an essentially ferromagnetic $\mathbb{Z}^\nu$-invariant system satisfying the conditions c.1 - c.5 and such that Peierls argument has shown the existence of a phase transition. (Main theorem of sec. 4.3.2) ; this same model placed in a constant magnetic field h has at low temperature a phase transition associated with a breakdown of the translation group if there exists a ground state Y of $\{\mathbb{Z}^\nu, \mathcal{B}, K\}$ which is not $\mathbb{Z}^\nu$-invariant and such that the following condition c.6 is satisfied.

c.6 With C_0^* a connected closed graph in $\mathcal{K}_f^*$ and β_r^*, $r=1,\dots,n$ a family of closed graphs inside C_0^*, which are outside each other and surrounded by $S_r Y$, then for any S in $\mathcal{S}$ there exists a translation $T_S \in \mathbb{Z}^\nu$, such that $T_S = \mathbb{1}$ if $S \cdot Y$ is not a translate of Y and $T_S Y = SY$ otherwise, with the properties

1) $\mathrm{Int}\ (T_{S_r}^{-1} \beta_r^*) \cap \text{Outside } C_0^* = \phi$
   $\mathrm{Int}\ (T_{S_r}^{-1} \beta_r^*) \subset \text{Outside } (T_{S_t}^{-1} \beta_t^*)$ for all (r,t)

2) $h\left[\sum_{x\in \mathrm{Int}\, C_0^*} \sigma_x(Y) + \sum_{r=1}^{n}\left\{\sum_{x\in \mathrm{Int}\,\beta_r^*}[\sigma_x(S_r Y) - \sigma_x(T_{S_r} Y)]\right\}\right] > \alpha h |C_0^*|$

   with $J(B) + \alpha h > 0$ for all $B \in \mathcal{B}$

Proof

With $\beta^* = C_0^* \bigcup_{r=1}^{n} \beta_{0r}^* \cup \tilde{\beta}^*$, (Eq. 4.16) we define $T\beta^*$ to be the closed graph given by :

$$T\beta^* = \tilde{\beta}^* \prod_{r=1}^{n} T_{S_r}^{-1} \beta_{0r}^*$$

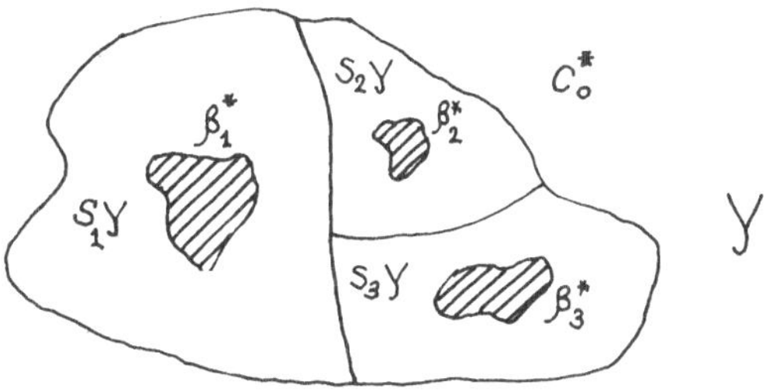

Fig. 8

where T_S is the translation introduced in c.6 (1).

We thus obtain

$$H'_{\Lambda,Y}[\beta^*] - H'_{\Lambda,Y}[T\beta^*] = 2\sum_{B\in D^{-1}C_o^*} + 2h\sum_{x\in Int\, C_o^*}\sigma_x(Y) + $$
$$+ 2h\sum_{r=1}^{n}\Big\{\sum_{x\in Int\,\beta^*_{or}}[\sigma_x(S_r Y) - \sigma_x(T_{S_r}Y)]\Big\}$$

and from the condition c.6 (2) we have :

$$H'_{\Lambda,Y}[\beta^*] - H'_{\Lambda,Y}[T\beta^*] \geq 2(|J| + \alpha h)|C_o^*|$$

where $|J| = \max_{B\in\mathcal{B}} |J(B)|$

Since distinct elements β^* which appears in the sum $\sum^*$ Eq. 4.15 are associated with distinct element $T\beta^*$ in the denominator we conclude that :

$$Prob'_{\Lambda,Y}[C_o^*] \leq e^{[-2\beta(|J|+\alpha h)|C_o^*|]}$$

and thus there exists a phase transition with breakdown of the translation group for $\beta > \beta_o$ where β_o is obtained by

$$\sum_{n=0}^{\infty} e^{-2\beta_o(|J|+\alpha h)\cdot n} N_n s_n = \frac{1}{2}\,.$$

Applications :

1- Ferromagnetic Ising model : all ground states are $\mathbb{Z}^\nu$-invariant and there exists a unique state for $h \neq 0$ [59]

2- Anti-ferromagnetic Ising model :
There are two ground states Y_1 and $Y_2 = T_a Y_1 = \mathbb{Z}^\nu Y_1$; the condition c.6 (1) is trivially satisfied; the condition c.6 (2) becomes :

$$\hbar\left[\sum_{x\in Int\, C_o^*} \sigma_x(Y)\right] > \hbar\cdot\alpha\cdot|C_o^*|$$

which is satisfied since :

$$\sum_{x\in Int\, C_o^*} \sigma_x(Y_1) + \;=\; \sum_{x\notin Y_1}\left(\delta_{Int\, C_o^*,x} - \delta_{Int\, C_o^*, T_a x}\right)$$

$$\geq -\sum_{\substack{x\in Y_1\\ x\notin Int\, C_o^*\\ T_a x\in Int\, C_o^*}} \geq -\frac{|C_o^*|}{2}$$

Furthermore, this last inequality can be improved to $\frac{|C_o^*|}{4}$ with a good choice of the translation T_a

In conclusion we recover the result that there exists a phase transition at low temperature if $|J| - \frac{|\hbar|}{4} > 0$ [53,54].

3- Triangular model with 3-body forces $J>0$.

There are 4 ground states given by : $Y_0=\emptyset, Y_1, Y_2=T_aY_1, Y_3=T_bY_1$, the condition c.6 (1) is trivially satisfied ; let us then choose Y_1 as boundary condition ; it is immediately clear that c.6 (2) cannot be satisfied if $\hbar>0$ and we expect that there will be a unique equilibrium state in this case ; for $\hbar<0$ the condition c.6 (2) yields.

$$-|\hbar|\left\{\sum_{x\in\mathbb{Z}^\nu}\sigma_x(Y_1)\cdot\delta_{x,Int\, C_o^*} + \sum_{\substack{r \text{ such that}\\ \beta_r^* \text{ sourrounded}\\ \text{by } Y_o=\phi}}\left\{\sum_{x\in Int\,\beta_r^*}[1-\sigma_x(Y_1)]\right\}\right\}\geq$$

$$\geq -|\hbar|\left\{\sum_{x\notin Y_1}\left[\delta_{Int\, C_o^*,x} - \delta_{Int\, C_o^*, T_a x} - \delta_{Int\, C_o^*, T_b x}\right] + \right.$$

$$\left. + \sum_{\substack{x\notin Y_1\\ x\notin Int\, C_o^*\\ T_a x\in Int\, C_o^*\\ \text{or } T_b x\in Int\, C_o^*}} 2\right\} \geq -|\hbar|\sum_{\substack{x\notin Y_1\\ x\in Int\, C_o^*\\ T_a x\notin Int\, C_o^*\\ T_b x\notin Int\, C_o^*}} \geq -|\hbar||C_o^*|$$

Therefore if $h < 0$ and $|h| < J$ there exists 3 coexisting equilibrium states at low temperature which are not translationnally invariant. In conclusion, we expect the following phase diagramme for this model, phase diagramme which is typical of a solid-liquid phase transition.

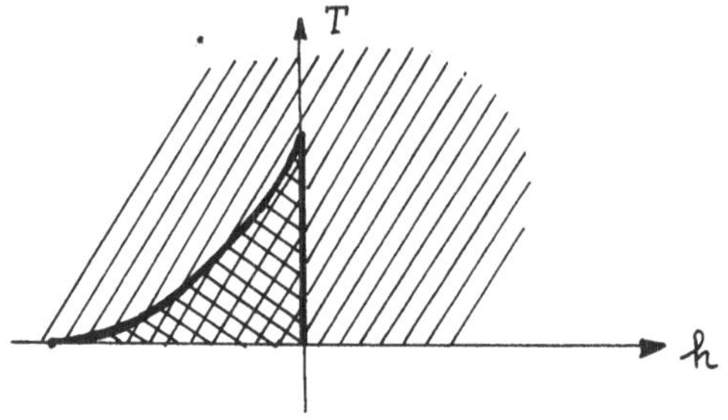

Fig. 9 Expected phase diagramme for the triangular model with external field

a) //// : 1 pure phase which is an extremal equilibrium state thus corresponds to the "fluid phase".

b) #### : 1 pure phase which decomposes into 3 extremal equilibrium states which are not translationnally invariant and thus correspond to "solid phase".

c) ——— : coexistence between solid phase and fluid phase, there exists two pure phases which decompose into 4 extremal equilibrium state, 1 invariant under translation and 3 which are not.

CHAPTER 5. - SERIES AND CLUSTER EXPANSION
AN APPLICATION OF UNIVERSALITY HYPOTHESIS
A " GENERALIZED " DROPLET-MODEL

5.1 Introduction

For lattice systems, the hamiltonian is a bounded operator and we can expand the partition function in powers of $\beta = (kT)^{-1}$. However, more information can be gained if we consider the expansions in terms of the groups Γ and $\mathcal{K}$ (chapter 1); in particular the computation of a few coefficients yields numerical estimates for the critical indices. Several methods, such as "ratio method", "Padé approximant", ... , have been developed to investigate such expansions and they yield in general satisfactory approximations for the critical temperature as well as for the critical indices. (For a general review of series expansions methods, see ref. [60])

In this chapter, we apply the simplest of these methods, the ratio method [36] to discuss critical properties of the triangular model with 3-body forces of sec.2.7 ; as we shall see, this model has a critical behaviour which is different from the usual Ising model. Therefore, this model, just as the Baxter model [23], shows that the critical behaviour might be strongly dependent on the nature of the short range interactions. This analysis is then completed with a few remarks concerning the "scaling" and "universality" hypothesis :

we discuss the relationship between the critical behaviour of this model and the Baxter model for which the values of the critical parameters have been given by Barber and Baxter [61]. Using their results, we conjecture the critical indices for the triangular model and compare them with recent numerical results.

In the last part of this chapter, we then consider a "generalized droplet-model", in connection with Shermann's theorem on path for the 2-dimensional Ising model in zero field. This generalized droplet model has two parameters (J,h) (J analogous to the n.n. 2-body interactions and h analogous to the external magnetic field in the Ising model). We first show that up to some order, it is possible to find a function I(c) on 1-cycle such that the first few coefficients of the series (specific heat, magnetization and zero field susceptibility) coincide with those of the corresponding series for the Ising model. We then briefly investigate the asymptotic character of these series and conclude with some remarks regarding the relationship between this model, the droplet model, and the Ising model.

5.2 Ratio method [36]

Let F(x) be a function known only through its power series expansion :

$$F(x) = \sum_n a_n x^n \qquad \text{with } a_n \geqslant 0$$

then F(x) has its first singularity on the real axis and its position determines the radius of convergence of the series. If we have physical reasons to believe that F(x) becomes infinite for some critical value $x_c > 0$, and if we identify this critical value with the above singularity, then the problem is reduced to estimate the radius of convergence of the series given its coefficient. For this purpose, we use the general formula :

$$\mu = x_c^{-1} = \lim_{n \to \infty} a_n^{1/n}$$

and for a smoothly varying sequence we expect that :

$$\mu_n = \frac{a_n}{a_{n-1}} = \mu\left(1 + \frac{c}{n} + \mathcal{O}\left(\frac{1}{n^2}\right)\right)$$

so that the ratio should vary linearly with $^1/n$.

We shall now apply this method to the triangular model with 3-body interactions and positive boundary conditions (Example 4, sec.2.7) ; let us consider this model in the presence of an external field h and let us use the expansion in terms of the group Γ. With $x = e^{-2K}$ and $z = e^{-2h}$ the LT. expansion is given by :

$$Z_\Lambda(x,z) = e^{\sum\limits_{B\in\mathcal{B}} K(B)} \sum_{\beta\in\Gamma} w(\beta) = e^{-|\Lambda| \ln (x z^{1/2})} \sum_{\beta\in\Gamma} w(\beta)$$

Moreover in terms of closed graphs $\beta \in \mathcal{K}_0$ on $\{\bar{\Lambda}, K, h=0\}_{op}$, $\bar{\Lambda} = \bigcup\limits_{B\in\mathcal{B}_\Lambda} B$ we have :

$$\sum_{\beta\in\Gamma} w(\beta) = \sum_{\beta\in\mathcal{K}_0} x^{|\beta|} z^{|\mathrm{Int}\,\beta|} = \exp\left[\sum_{m=1}^{\infty} \alpha_m(z)\cdot x^{2m+4}\right]$$

i.e.

$$1 + \sum_{n\geq 1} x^{2n+4} c_n(z) = \exp\left[\sum_{m=1}^{\infty} \alpha_m(z)\cdot x^{2m+4}\right] \qquad (5.1.)$$

where : $c_n(z) = \sum\limits_{\substack{\beta\in\mathcal{K}_0 \\ |\beta| = 2n+4}} z^{|\mathrm{Int}\,\beta|}$

We want to obtain the coefficients α_m up to m = 6.

Comparing the coefficients up to order 16 in x in Eq. 5.1 we find

$$c_1 = \alpha_1 \qquad c_2 = \alpha_2 \qquad c_3 = \alpha_3$$

$$c_4 = \alpha_4 + \frac{\alpha_1^2}{2} \qquad c_5 = \alpha_5 + \alpha_1 \alpha_2 \qquad c_6 = \alpha_6 + \alpha_2 \alpha_3 + \frac{\alpha_2^2}{2}$$

To compute $c_n(z)$, we must consider all closed graphs of length $|\beta| = 2n + 4$ and one can show that $|\mathrm{Int}\beta| \leqslant \frac{|\beta|^2}{24}$

It is immediately seen that

$$c_1(z) = z \cdot |\Lambda| \qquad c_2(z) = 3|\Lambda| z^2 \qquad c_3(z) = 11 \cdot |\Lambda| \cdot z^3;$$

we shall now give the computation of $\alpha_4 = c_4 - \frac{\alpha_1^2}{2} = c_4 - \frac{|\Lambda|^2 z^2}{2}$;

the evaluation of c_5 and c_6 follows the same line.

We first remark that $|\beta| = 12$ implies $|\mathrm{Int}\,\beta| \leqslant 6$, moreover if $|\mathrm{Int}\beta| = 2$ the closed graph is disconnected and these graphs yield a contribution $\frac{|\Lambda|(|\Lambda|-7)}{2} \cdot z^2$; if $|\mathrm{Int}\beta| = 1,3$ or 5 there is no closed graphs with $|\beta| = 12$; if $|\mathrm{Int}\beta| = 4$ or 6 the only closed graphs are connected and represented together with their respective contribution by the following :
(the heavy dots represent $\mathrm{Int}\beta$)

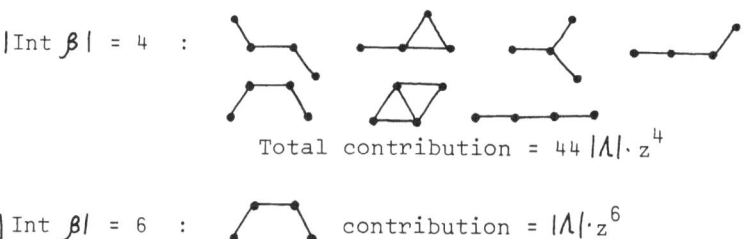

$|\mathrm{Int}\,\beta| = 4$:

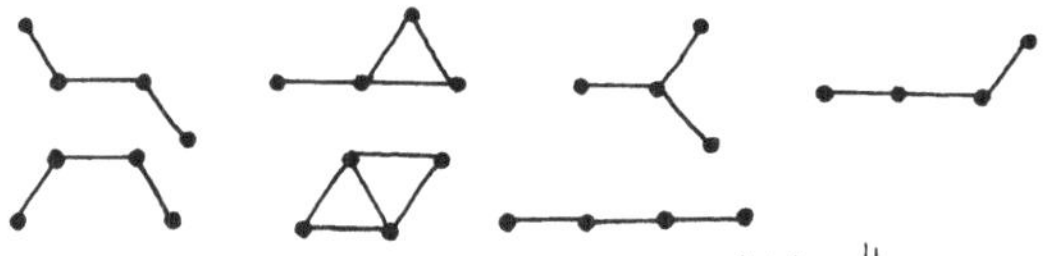

Total contribution = $44\,|\Lambda| \cdot z^4$

$|\mathrm{Int}\,\beta| = 6$:

contribution = $|\Lambda| \cdot z^6$

We thus have :

$$\alpha_4 = c_4 - \frac{\alpha_1^2}{2} = \frac{|\Lambda|(|\Lambda|-7)}{2} Z^2 + 44|\Lambda| Z^4 + |\Lambda| Z^6 - \frac{|\Lambda|^2 Z^2}{2} =$$

$$= |\Lambda| Z^6 + 44 |\Lambda| Z^4 - \frac{7}{2}|\Lambda| Z^2 .$$

The free energy is given up to order 16 in $X = e^{-2\beta J}$ by :

$$-\beta f = -\ln XZ^{1/2} + ZX^6 + 3Z^2X^8 + 11Z^3 X^{10} +$$

$$+ (Z^6+44Z^4 - \frac{7}{2} Z^2) X^{12} + (12Z^7 + 186Z^5 - 30Z^3) X^{14} +$$

$$+ (3Z^{10} + 99Z^8 + 813Z^6 - \frac{405Z^4}{2}) X^{16} + \ldots .$$

From this we obtain (assuming differentiability) :

"Spontaneous Magnetization" :

$$m = \frac{1}{|\Lambda|} \frac{\partial \ln Z}{\partial h}\Bigg)_{h=0} = 1 - 2 X^6 - 12 X^8 - 66X^{10} - 350X^{12} - 1848X^{14} - 9780X^{16}$$

"Zero Field Susceptibility" :

$$\chi = \frac{\partial m}{\partial h}\Bigg)_{h=0} = 4X^6 + 48X^8 + 396X^{10} + 2904X^{12} + 19872X^{14} + 130656X^{16}$$

Specific Heat : ($Z = 1$)

$$\frac{c_v}{K^2 k} = 4X \frac{\partial}{\partial X}\left[X \frac{\partial}{\partial X}(-\beta f)\right] = 144X^6 + 768X^8 + 4400X^{10} + 23904X^{12} + 131712X^{14} + 729600X^{16}$$

The coefficient ratios $\mu_n = \frac{a_n}{a_{n-1}}$ are given in Table 5.1.

n	3	4	5	6	7	8
C_v	x	5.33	5.723	5.433	5.510	5.6248
χ	x	12	8.25	7.33	6.843	6.637
m	2	6	5.5	5.3	5.28	5.35

Table 5.1.

By analogy with the known results for Ising models with 2-body forces, it is expected that, within this simple method, the best results for these ratio, if extrapolated at the "critical" point $X_c = e^{-2\beta J_c} = \sqrt{2}-1$ (we assume that the critical point is unique) are obtained for the specific heat. Proceeding in this way, extrapolation of the last two ratios at $X_c = \sqrt{2}-1$ gives us approximate values of the α indices

We get : $\alpha = 0.617614$

$\alpha = 0.7206$

The average values give : $\alpha \simeq 1{,}33/2 = 2/3$. This analysis, even if approximate and crude can be expressed by the following numerical "results" :

Result 1 :The triangular model with n.n. 3-body interactions has a different critical behaviour than the usual Ising model.

Result 2 : The system shows a relatively high specific heat anomaly, i.e. $\alpha \sim 0.6-0.7$.

On the other hand, using Padé approximant, Griffith and Wood [62] have obtained the following numerical results for this model :

$$\frac{3}{5} \leq \alpha' \leq 4/5 \qquad 5/4 \leq \gamma' \leq 7/5 \qquad 0.07 \leq \beta' \leq 0.071 \qquad (5.2)$$

and recently the exact solution for the free energy has been obtained by R.J. Baxter and F.Y. Wu [63] which yields $\alpha = 2/3$ as we had conjectured.

5.3. Remark about Universality and the Critical Behaviour of the Triangular Model

In this section, we want to argue that the critical behaviour of the triangular model with 3-body interactions can be obtained from the critical indices of the Baxter model and from a careful application of the scaling ideas. The starting point relies upon the following observations :

1. From "universality", it is expected that all phase transitions can be grouped into a small number of classes depending upon the dimensionality and the symmetry of the system ; moreover all phase transitions within a given class have the same critical behaviour. [66]

2. The Baxter model with parameter (J_1, J_2, λ) is equivalent to a model with pure 3-body interactions on a square lattice (see Ch.6) ; furthermore it is also equivalent to the 8- vertex model.

3. The triangular model which we are considering is equivalent to a Baxter model on an hexagonal lattice. (It is also equivalent to a 32-vertex model [64]).

To see this equivalence, we note that a partial trace on one of the sublattice $\mathbb{Z}^2/\mathbb{E}_i$ (see example 2, sec.2.7 and Ch.6 for partial trace method) yields an hexagonal lattice with two and 4-body interactions i.e. $-\beta' H' = \text{const.} +$

$$+ \sum_{\text{cells}} \frac{1}{16} \cdot \ln\left(\operatorname{Cosh} 6K / \operatorname{Cosh} 2K\right) \cdot \left(\sum \sigma_{x_i} \sigma_{x_j} + 1\right) \cdot \left(\sum \sigma_{x_k} \sigma_{x_\ell} + 1\right) \qquad (5.3)$$

where the indices (ij) resp. (kl) refer to points on the remaining two sublattices in each hexagonal cell. This model (5.3) can thus be considered as a Baxter model on an hexagonal lattice, i.e. a model consisting of two triangular lattice interacting by means of an energy-energy coupling, as explained by Kadanoff and Wegner [65] for the usual Baxter model; now the scaling idea is valid for the class of Baxter model on the square lattice with coupling (J_1, J_2, λ). We expect furthermore that the general anisotropic case of the model (5.3) bares the same relationship to the Baxter model (J_1, J_2, λ) as the hexagonal Ising model for the usual Ising model on the square lattice (in the sense of critical behaviour). In this latter case however, we know that the critical behaviour of both model is the same ; we are thus led to conclude that the triangular model has the same critical behaviour as the usual Baxter model (J_1, J_2, λ).

In this context, some recent results on the eight-vertex model [61] suggest that the relations between critical exponents

$$\alpha + 2\beta + \gamma = 2 \qquad \delta = 1 + \frac{\gamma}{\beta} \qquad (5.4)$$

is valid for the eight-vertex model ; moreover the indices η and δ , respectively the exponent for the decay of the correlation function and the critical isotherm, appear to have a value, independent of the coupling parameter (J_1, J_2, λ), equal to those of the Ising model, i.e. $\delta = 15$, $\eta = \frac{1}{4}$. From these results and the above discussion, we conjecture that the relations Eq. 5.4 remains valid for the triangular model ; since $\alpha = {}^2/_3$ and using the universality hypothesis, combined with Eq. 5.4, we have :

$$\alpha = {}^2/_3 \;, \quad \delta = 15, \quad \eta = \tfrac{1}{4}, \quad \beta = \tfrac{1}{12} = 0{,}0833, \quad \gamma = \tfrac{7}{6} = 1{,}166 \qquad (5.5)$$

results which are not in agreement with those obtained by Padé approximant Eq. 5.2. Moreover improved numerical values of the critical indices have been obtained recently on the basis of the exact solutions of Baxter and Wu which yields [67]

$$\beta = 0{,}080 \pm 0{,}005 \qquad \gamma = 1{,}15 \pm 0{,}15$$

In conclusion, the values we have obtained from the universality arguments are not too far from those recent numerical values. Let us also note that a new "ansatz" for the spontaneous order of this model has been introduced in Ref. [68] ,by means of elliptic function. Again, the first few terms of the expansion in power of $x = e^{-2K}$ agree with those we have given above.

It is also interesting to note that the Baxter model with $J_1=0$, $J_2=\lambda=J$ has the same critical point $\sinh 2\beta J = 1$ and the same critical indices Eq.(5.5) as the triangular model [69] ; indeed since the model is self-dual the critical temperature is given by $\tanh K_c = e^{-2K_c} = \sqrt{2}-1$ and the α index given by the solution of the equation [61,65]

$$\sin \frac{\pi\alpha}{4(1-\frac{\alpha}{2})} = \text{th}\, 2\lambda_c = \frac{\sqrt{2}}{2} \qquad \text{i.e. } \alpha = \frac{2}{3}$$

Moreover evaluation of the first few terms of the series expansions yield the same coefficient as for the triangular model. We thus expect that these two models are equivalent in the sense that there should exists a transformation (duality, partial trace) mapping one on the other ; in fact such a transformation has appeared recently as far as the free energy is concerned [70] but the problem of the equivalence of state is still open.

In conclusion, the triangular model gives another example of an exactly soluble model with different critical behaviour than the Ising model.

5.4. A Generalized Droplet-Model

In this section, we consider the 2-dimensional Ising Model with external magnetic field and show that up to some order in the coupling constants $x = e^{-2\beta J}$ and $z = e^{-2\beta h}$ the expansions for the free energy density, susceptibility, specific heat and spontaneous magnetization, can be obtained by a "generalized" Feynmann-Kac conjecture for the case $h \neq 0$ More precisely, in zero field, the free energy of the 2-dimensional Ising Model is given by :

$$Z = e^{-\beta|\Lambda|f} = e^{-|\Lambda| \ln x + \sum_{\{C\}} \frac{W(C)}{\mu(C)}(x)} \qquad (5.6)$$

where $x = e^{-2\beta J}$, $\{C\}$ denotes the set of all 1-cycle which can be drawn on the finite square lattice Λ [71] and $W(C)$ is a function defined over all 1-cycle, i.e. $W(C) = (-1)^{N_C} \cdot x^{|C|}$ with N_C the number of self-crossing of C, $|C|$ the 1-cycle length and $\mu(C)$ its multiplicity. In fact, the known result (the Kac and Ward's determinant) obtained by the introduction of a 4x4 matrix propagator in the fourier space of the trajectory, give for the low-temperature expansion, (or the high-temperature expansion), because of the self-dual property, the solution in closed form :

$$\lim_{|\Lambda| \to \infty} \frac{1}{|\Lambda|} \cdot \sum_{\{C\}} \frac{W(C)}{\mu(C)}(x) = \frac{1}{8\pi^2} \cdot \iint_0^{2\pi} dk_1 dk_2 \ln f(x, k_1, k_2)$$

$$f(x, k_1, k_2) = [(1+x^2)^2 - 2x(1-x^2)(\cos k_1 + \cos k_2)] .$$

An interesting combinatorial question related to these known facts is then the following : is it possible to find an extension of the Feynmann Kac-Ward's conjecture, which holds when $h \neq 0$?

That is, is there an analogous of formula (5.6) for $h \neq 0$? Since the problem for $h \neq 0$, in graphical language, has to do with the difficulty of global properties (e.g. interior of contours), it is clear that one must look at the eventual possibility to define a function I(c) on the set of important cycles, such that the counting procedure correctly holds.[72]

We first consider the ferromagnetic Ising model with + boundary conditions, coupling constant J and an homogeneous magnetic field h on Λ : fig. 1 (Model 1).

Fig. 1.

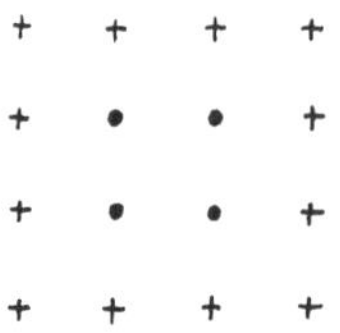

• point of Λ

Model 1

Applying a HT-LT duality to the model 2, defined on Figure 2, we just obtain model 1

Model 2

Fig. 2.

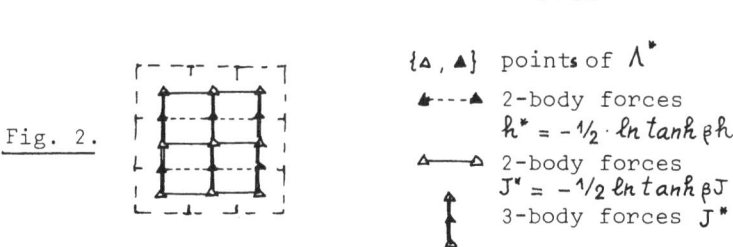

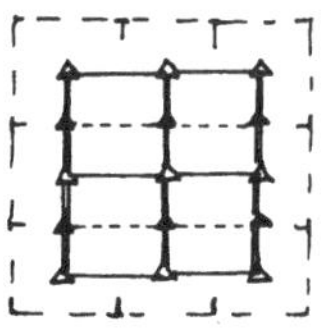

{△, ▲} points of Λ^*

▲---▲ 2-body forces $h^* = -1/2 \cdot \ln \tanh \beta h$

△—△ 2-body forces $J^* = -1/2 \ln \tanh \beta J$

3-body forces J^*

The high temperature expansion of model 2 coincides with the low temperature expansion of model 1. If $h \neq 0$ the factor (the weight) for a graph $\beta \in \mathcal{K}$ of model 2, is given by $w_\beta = e^{-2\beta J \cdot \ell} \cdot e^{-2\beta h S}$ where ℓ is the length of the corresponding contour for $h = 0$ and S is the interior of the graph (Number of interactions h^*). We then consider the following function (ansatz) :

$$Z'_{\Lambda,J,h} = e^{-|\Lambda| \ln(x z^{1/2}) + \sum_{\{C\}} \frac{W'(C)}{\mu(C)}(x,z)} \qquad (5.7)$$

where :
$$W'_C(x,z) = (-1)^{N_C} \cdot z^{I(C)} x^{|C|}$$

$\sum_{\{C\}}$ means as for $h=0$, the sum over all 1-cycle on Λ and I(c) is the interior of the cycle C defined in the following way :

Definition 1:

Given a 1-cycle C, and let $(P_1, P_2, \cdots P_n)$, be a decomposition of C into simple polygon(*) such that $\cup P_i = C$, then $I(C) = \sum_K |Int\, P_K|$ where $Int\,(P_K)$ is the interior of the polygon P_K and is well defined.

Result :

Up to order 12 in x and simultaneously to order 3 in z, all thermodynamic functions, $\beta f(J,h)$, $m(h=0)$, $\chi(h=0)$ and $C_V(h=0)$ coincide with that of the Ising model.

Lemma :

Let $C = \bigcup_K P_K$ and $\ell(C)$ be the length of C, then

$$I(C) = \sum_K I(P_K) \leq \sum_K \frac{\ell^2(P_K)}{16} \leq \frac{\ell^2(C)}{16}$$

(*) By definition, a simple polygon C is a 1-cycle which cannot be decomposed into two 1-cycle such that $C = C_1 \cup C_2$.

This property will be used in the sequel.

The detailed analysis has been given in [72] . We consider here an example , namely the explicit computation of the contribution to the order 10 in x , which has the form

$$x^{10} f(z) \qquad \text{i.e.} \sum_{\substack{C \\ |C|=10}} \frac{(-1)^{N_C}}{\mu(C)} x^{10} \cdot z^{I(C)}, \quad |C| = \ell(C)$$

To do this, we consider all 1-cycle such that $|C| = 10$ and order them with increasing "interior"; notice that by the Lemma $I \leq 6$ and $\mu \leq 2$.

We have nine types of "important" cycles :

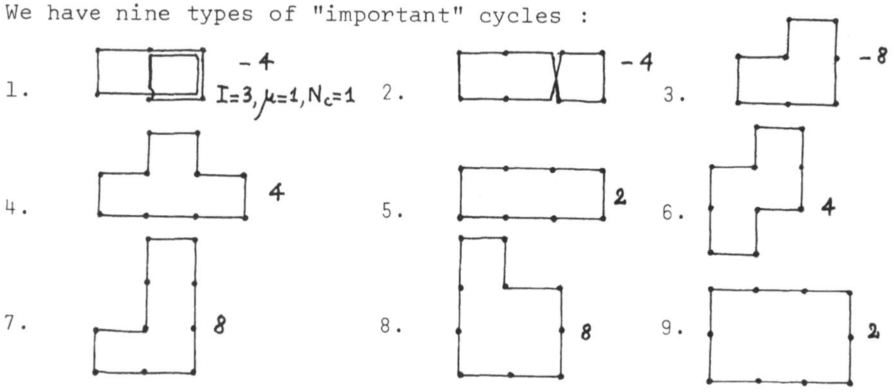

The other types of cycle are not "important" in the sense that their partial contribution vanishes (the strong cancellation of the series).

They are :

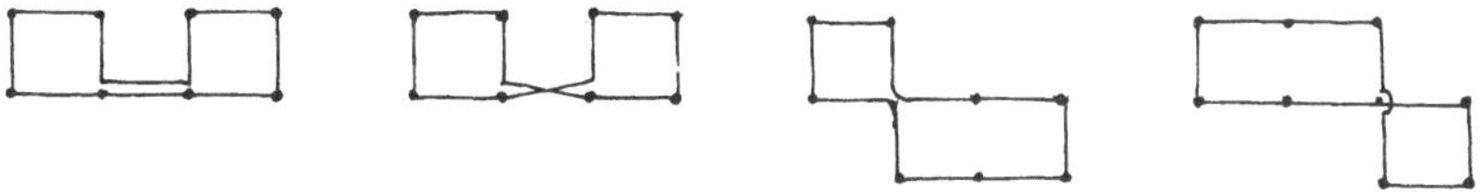

Thus the total contribution to this order is given by :

$$f(Z)\cdot X^{10} = (-16Z^3 + 18Z^4 + 8Z^5 + 2Z^6)\cdot X^{10}$$

In this way, the free energy density $f(X,Z)$ of the Ising model with $h \neq 0$ $(h \geq 0)$ can be obtained exactly to the order 12 in X and 3 in Z, by choosing a well defined decomposition of C (Def.1). The other quantities of interest are given by

$$M = \lim_{h\to 0^+} \frac{\partial \ln Z'}{\partial h} = |\Lambda|\cdot m$$

$$m = 1 - 2\sum_C \frac{W'(C)}{\mu(C)} I(C) \quad \text{where}$$

$$\lim_{|\Lambda|\to\infty} \sum_C f(C)\,|\Lambda|^{-1} = \sum_{C_x} f(C_x), \; x \in \mathbb{Z}^2 \qquad (*)$$

1. Magnetization

$$m(X,Z) \cong 1 - 2X^4 Z - 2X^6(4Z^2) - 2X^8(-5Z^2 + 18Z^3 + 4Z^4) - $$
$$- 2X^{10}(-48Z^3 + 72Z^4 + 40Z^5 + 12Z^6) - 2X^{12}(9Z^9 + 48Z^8 +$$
$$+ 154Z^7 + 240Z^6 + 215Z^5 - 340Z^4 + 31Z^3) - \ldots\ldots \qquad (5.8)$$

In particular to the order 12 in $X = e^{-2\beta J}$, we obtain for m, the spontaneous magnetization of the pure phase,

$$m \cong 1 - 2X^4 - 8X^6 - 34X^8 - 152X^{10} - 714X^{12} - \ldots\ldots$$

and this coincides with the expansion, for $T < T_c$ of the exact Lee-Yang's result, for the spontaneous magnetization m_0 announced by Onsager. Notice that the equality $m_0 = \frac{\partial f}{\partial h}\Big|_{h\to 0^+}$ has also been established only recently by G. Benettin, G. Gallavotti, G.Jona Lasinio and A.L. Stella [73].

(*) C_x denotes a cycle passing through the site x.

Remark : The total contribution of all cycles, such that $I \leq 3$ is contained in the truncated series 5.8 , thus for J=0 we obtain to the order z^3 :

$$m(0,h) \cong 1 - 2z + 2z^2 - 2z^3(18 - 48 + 31) \cong 1 - 2z + 2z^2 - 2z^3 + \dots$$

and this coincides (to this order) with the exact result for the Ising model ; in fact, for J = 0 we have :

$$m(0,h) = \tanh h = \frac{1-z}{1+z} = 1 - \frac{2z}{1+z} \cong 1 - 2z + 2z^2 - 2z^3 + \cdots$$

2. Zero Field Susceptibility

From(5.7) we have : $\chi(x,0) = 4 \cdot \sum_{C} \frac{W(C) I^2(C)}{\mu(C)}$

and to the order 12 in x :

$$\chi(x,0) = 4x^4 + 32x^6 + 240x^8 + 1664x^{10} + 11164x^{12} + \dots\dots$$

which also coincides with the exact expansion recently found with standard refined techniques [74].

From this analysis, we define a "generalized droplet Model" (G.D. Model) by the following :

Definition 2: The free energy of the generalized droplet model is given by formula

$$e^{-\beta f \cdot |\Lambda|} = e^{-|\Lambda| \ln(x z^{1/2}) + \sum_{C} \frac{W(C)}{\mu(C)} z^{I(C)}} \qquad (5.9)$$

$W(C) = (-1)^{N_C} \cdot x^{|C|}$.

and where for $|C|>12$, I(c) is a function as expressed by the definition **1.**

Remark : In the generalized droplet model, the free energy density appears as a sum of Boltzmann factors containing the two parameters $\ell(C), I(C)$ and has a simple geometrical meaning, as in the usual droplet model [75] : $\ell(C)$ is reminiscent of the line (multipolygon) of separation between the two "phases" in each configuration (surface effect) while I(c) is reminiscent of the part occupied in one phase (volume effect).

Property : If $h=0$ the free energy of the G.D. Model coincides with that of the Ising model in zero field. (by definition).

Moreover, the thermodynamic quantities of the G.D. Model are obtained by differentiation of the free energy. We now consider them in more details :

1. Magnetization m

$$m = 1 - 2\sum_{C} \frac{W(C)}{\mu(C)} I(C) z^{I(C)} \qquad (5.10)$$

for $z=1$ $(h \to 0^+)$ we obtain $m_s = 1 - 2\sum_{C} \frac{W(C)}{\mu(C)} I(C)$

A lower bound is then obtained, i.e. :

$$m_s = 1 - 2\sum_{C} \frac{W(C)\, I(C)}{\mu(C)} > 1 - 2\alpha \quad \text{if}$$

$$\sum_{C} \frac{W(C)}{\mu(C)} I(C) < \alpha \qquad \text{for some } \alpha > 0$$

Proposition 1 If $x = e^{-2\beta J} < 1/3$, $m_s > 0$.

Notice that the bound $e^{-2\beta J/T_0} = 1/3$ coincides with that given by Peierls argument for the Ising model in zero field [41] .

Proof :

$$W(C)>0 \text{ or } W(C)<0 \quad , \text{ i.e. } W(C) = (-1)^{N_C} x^{|C|}$$

Using the definition of I(c), and the lemma, we obtain :

$$\sum_C \frac{W(C)}{\mu(C)} I(C) \leq \sum_C \frac{|W_C|}{\mu(C)} I(C) \leq \sum_C \frac{x^{|C|} \cdot |C|^2}{16} \leq \sum_{\ell=4}^{\infty} (3x)^{\ell} \cdot 3^{\ell} 4^{-2} \leq (48)^{-1} \cdot \sum_{\ell=4}^{\infty} (3x)^{\ell} \cdot 4$$

Since from Euler's theorem there is at most $|\Lambda| \cdot 3^{\ell-2} \cdot 4$ cycles such that $\ell(C) = \ell$ on $\Lambda \subset \mathbb{Z}^2$; for $x < 1/3$ the series converge and the proposition is proved.

2. Zero Field Susceptibility χ

$$\chi = 4 \sum_C \frac{W(C) I^2(C)}{\mu(C)} z^{I(C)} \tag{5.11}$$

for $z = 1$ $\chi_0 = 4 \sum_C \frac{W(C) \cdot I^2(C)}{\mu(C)}$ an upper bound can analogously be obtained as for m_s ; moreover, we have :

Proposition 2

If $x < 1/3$ the zero field susceptibility is finite ; it can only diverge for $x > 1/3$.

Proof : Using Lemma 1, we conclude in fact :

$$\chi_0 < 4 \sum_{\ell=4}^{\infty} (3x)^{\ell} \frac{\ell^4}{(16)^2} < \frac{1}{64} \sum_{\ell=4}^{\infty} (3x)^{\ell} \cdot \ell^4 \qquad \text{and}$$

the series converges for $X < 1/3$.

Remark : The bounds we have obtained are the consequence of simply removing the factor $(-1)^{N_c}$ in W(c), and corresponds to neglect the strong cancellation of the series as for h=0, in giving a positive weight to all cycles C.

3. Relation with the Droplet Model

As pointed out in [36] , the droplet model is not a well defined model with a specific Hamiltonian (the same for our GD Model) but it is expected that in the low temperature region, it gives us an idea of some important features of phase transition.

In fact [75] , the droplet model corresponding to the Ising model is based on the two following hypothesis :

1. The contribution of the configuration where a droplet (a contour) is inside another droplet is small.

2. The droplets can superpose one another (effect of excluded volume). It turns out that these hypothesis are reasonable in the low temperature region only.

With 1,2, one then obtain :

$$m \cong 1 - 2 \sum_{\gamma \ni x} X^{\ell(\gamma)} z^{S(\gamma)}$$

where ℓ , S are respectively the length and the inside of the contour γ .

To make a connection with this model, we consider :

$$m = 1 - 2 \sum_{C} \frac{W(C)}{\mu(C)} I(C) z^{I(C)}$$

If we restrict the sum to cycles C such that $N_C = 0$ and $\mu(C)=1$ we obtain :

$$m \cong 1 - 2 \sum_{C, I(C) \ni x} x^{|C|} z^{S(C)} = 1 - 2 \sum_{\gamma \ni x} x^{\ell(\gamma)} z^{S(\gamma)}$$

which is presicely the expression of the magnetization for the droplet model.

It must be noted that the $N_C = 0$ and $\mu_C = 1$ approximations are consistent with the hypothesis (1,2) of the droplet model, since N_C and μ_C small means essentially that the droplets (cycles) are simply connected regions of the plane.

4. Some Analytic Properties of the G.D.M. in zero field

We now consider some analytic properties and first investigate the convergence radius for m and χ and their relations with the critical point of the Ising model.

Consider for example, the zero field susceptibility :

$$\chi = 4 \sum_{C} \frac{W(C)}{\mu(C)} I^2(C) = 4 \sum_{\ell} a_\ell x^\ell$$

Notice that up to $\ell = 12$, $a_\ell > 0$; we then expect that this series has a meaning below some limiting point called the critical point of the G.D.M.

From Hadamard's theorem, we have :

$$\rho = \lim_{\ell \to \infty} (a_\ell)^{-1/\ell} \text{, thus } \quad a_\ell = \sum_{\substack{c \\ \ell(c)=\ell}} W(c, J=0, h=0) \frac{I^2(c)}{\mu(c)}$$

where $W(C, z=1, x=1) = (-1)^{N_c}$; let $\{c ; \ell(c)=\ell\} = \{c_1, \dots, c_{|\alpha|}\} = (\alpha)_\ell$

then $a_\ell = \sum_{c_i \in (\alpha)_\ell} x_i f_i$, $x_i = (-1)^{N_{c_i}}$, $f(c_i) = \frac{I^2(c_i)}{\mu(c_i)} = f_i$

and from lemma : $0 \leq f_i(c_i) \leq \left(\frac{\ell}{4}\right)^4 \quad \forall c_i \in (\alpha)_\ell$

we then define :

$$a_\ell = \sum_{c_i \in (\alpha)_\ell} x_i f_i = \sum_{c_i \in (\alpha)_\ell} x_i \cdot \bar{f}(\bar{c})$$

Assuming that I(c) as expressed by the definition 1 can be chosen such that $\bar{f}(\bar{c}) = e^{g(\ell)}$ with $\lim_{\ell \to \infty} \frac{g(\ell)}{\ell} = 0$ we then obtain :

$$\rho = \lim_{\ell \to \infty} \left| \sum_{c_i \in (\alpha)_\ell} x_i \right|^{-1/\ell}$$

In terms of the low-temperature 4x4 matrix propagator $M_{\alpha\beta}(\vec{k})$ (+) $(x = e^{-2\beta J} = 1)$ we then obtain :

$$\rho = \lim_{\ell \to \infty} \left(\left| \sum_{\vec{k}=(k_1,k_2)} \mathrm{Trace}\, M^\ell(\vec{k}) \right| \right)^{-1/\ell} \cdot e^{-\frac{g(\ell)}{\ell}} =$$
$$= \max_{\vec{k}, i} |\lambda_i(\vec{k})|^{-1} , \quad \lambda_i(\vec{k}) , \; i \in (1,2,3,4)$$

The eigenvalues of the propagator : $M \xi = \lambda \xi$ are solutions of $\mathrm{Det}(M - \lambda I) = 0$ and $\max |\lambda(\vec{k})| = 1 + \sqrt{2}$ then $\rho = \frac{1}{\sqrt{2}+1} = \sqrt{2} - 1$.

Thus, <u>under assumption that</u> $\lim_{\ell \to \infty} \frac{g(\ell)}{\ell} = 0$, the convergence radius for χ coincides with the critical point of the Ising model $x_c = e^{-2\beta J_c} = \sqrt{2} - 1 = tg \frac{\pi}{8}$

(+) See foot note page 121

Notice that, in the case we give a positive weight to all cycles, then $\max_{\vec{k},i} |\lambda_i(\vec{k})|$ is given by the solution of the equation :

$$(\lambda^2-1)\cdot(3+\lambda^2-4\lambda)=0 \quad , \quad \max_i |\lambda_i| = 3$$

In this case, the convergence radius is given by $\rho = 3^{-1}$ and is precisely the condition which comes out from Peierls argument. In fact, the latter equation is given by $\mathrm{Det}(M-\lambda \cdot I)=0$ where the propagator does not contain the factors $e^{i\pi/4}$, $e^{-i\pi/4}$.

However , we have not succeded in proving that $\lim_{\ell\to\infty} g(\ell)/\ell = 0$

5.5 Analytic Properties of the Generalized Droplet Model (in $z = e^{-2\beta h}$)

To conclude this section we add some remarks in the form of a non rigourous discussion about the unsolved problem associated with the analytic behaviour in the z variable at a first order phase transition; this question is also of interest in connection with metastable states[19,37]. Recently several studies have appeared on this question for the planar Ising Model in order to gain some insight into the nature of the singularity which occurs in diluted Ising ferromagnet [91 , 92, 93].

For the usual Ising Model with external field the problem is whether the magnetisation $m(J,h)$ defined for $h>0$ can be analytically continued into the region $h<0$; for liquid-gaz systems it is whether the density ρ (p) defined for $p > p_o$ can be analytically continued into the region $p < p_o$ (see fig. 3). Such an analytic continuation, if it exists, would presumably define the magnetisation (or density) of a metastable state. In this connection let us recall that it was rigourously shown by Lanford and Ruelle [48] that there can not exist metastable states defined as "Translationally Invariant Equilibrium States (sec. 7.1.) with a free energy larger then the Gibbs States"; on the other hand it was also shown in this same reference that if the free energy is not analytic in a neighbourhood of a point $z = z_o$, then it is impossible to analytically continue the state in the neighbourhood of this singularity; these results go in the same direction as Fisher's [76] in the way of excluding metastable states. However it could well be that Eq. 7. 6 is not sufficiently general to define metastable states; it could also well be that the first few correlation functions have an analytic continuation but not all the correlation functions.

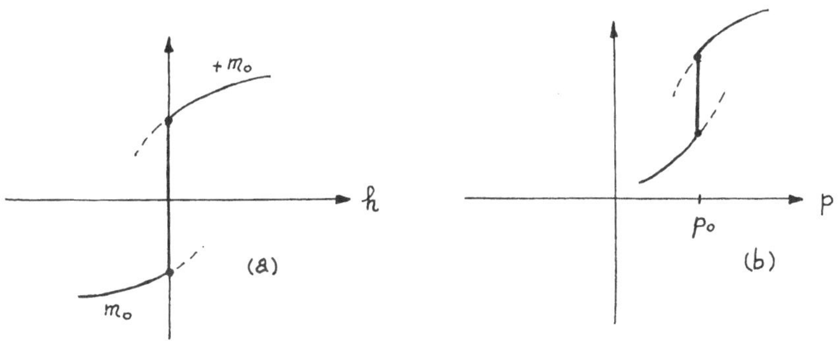

Fig. 3 a) magnetisation $m(h)$ for the Ising Model
b) density $\rho(p)$ for liquid-gaz systems

To discuss this problem we follow Fisher's analysis of the usual droplet model [76] and apply it to our "generalised droplet model"; as we have already seen the radius of convergence of the series for χ in the x-variable (for h = o) coincides with the critical point of the Ising model (provided only that the mean interior of all cycles of length ℓ is asymptotically equal to $e^{g(\ell)}$ where $\lim_{\ell\to\infty} g(\ell)/\ell = 0$) and the same conclusion also holds for the magnetisation. We shall now show that in the h variable the magnetisation m (J, h) has an essential singularity at $h = 0$.

The free energy of our G.D. model is given by Eq. 5.7.

$$-\beta f(x,z) = \ln x + \frac{1}{2}\ln z + \sum_{n=1}^{\infty} a_n(x) z^n \quad .$$

$x = e^{-2\beta J}$, $z = e^{-2\beta h}$,

$$\text{where :} \quad a_n = \sum_{\substack{C_o \\ \mathrm{Int}\, C_o = n}} \frac{W_{C_o}(x)}{\mu(C_o)} \quad ; \quad W_{C_o}(x) = (-1)^{N_c} \cdot |x|^{|C|} \tag{5.12}$$

Property :

i) For fixed $x = e^{-2\beta J}$, the radius of convergence of the series Eq. 5.10 for the magnetisation is given by $z_o = e^{-2\beta h} = 1$ i.e $h = 0$ for all x .

ii) Under the assumption that for large n all a_n have the same sign then $z_o = 1$ (i.e. $h = 0$) is an essential singularity of m (x, z).

To establish this property we consider the expansion Eq. 5.10 i.e.

$$g(x,z) = \frac{1}{2}[1 - m] = \sum_{n=1}^{\infty} b_n(x) z^n \; ; \; b_n(x) = \sum_{\substack{C_o \\ I(C_o) = n}} \frac{W_{C_o}(x) \cdot I(C_o)}{\mu(C_o)}$$

for which the radius of convergence is again obtained by :

$$z_o = \lim_{n\to\infty} |b_n|^{-\frac{1}{n}}$$

It follows from our previous lemma that $I(C) < \ell^2(C)/16$ and from definition 1 that $I(c) \geq 4\ell$; we thus have asymptotically

$$z_0 = \lim_{n\to\infty} \frac{1}{\left|\sum\limits_{\substack{C_0 \\ I(C_0)=n}} \frac{W(C_0)}{\mu(C_0)} I(C_0)\right|^{\frac{1}{n}}} = \lim_{n\to\infty} \left| n \cdot \sum_{\substack{C_0 \\ 4\sqrt{n} \leq \ell(C_0) \leq \frac{n}{4}}} \frac{W(C_0)}{\mu(C_0)} \right|^{-\frac{1}{n}}$$

and repeating the previous evaluation we obtain :

$$z_0 \cong \lim_{n\to\infty} \left| \sum_{\vec{k}} \mathrm{Trace} \sum_{\ell=4\sqrt{n}}^{n/4} M^{\ell} \right|^{-\frac{1}{n}} = \lim_{n\to\infty} \left| \sum_{\vec{k}} \mathrm{Trace}\, M^{4\cdot\sqrt{n}} \frac{(M^{n/4-4\sqrt{n}} - 1)}{M^2 - 1} \right|^{-\frac{1}{n}}$$

Since $\|M\| < 1$

$$z_0 = \lim_{n\to\infty} \left| \max_{\vec{k},i} \lambda_i(\vec{k},x) \right|^{-\frac{1}{n}\cdot 4\cdot\sqrt{n}} = 1 \tag{5.13}$$

Moreover with the assumption that all a_n have asymptotically the same sign we conclude that $z_0 = 1$ is a singularity of the analytic function $m(x,z)$ (*).

To establish the second part of the property, we have show that all derivative are finite at $z_0 = 1$.

with
$$g^{(n)}(x,z_0) = \frac{\partial^n}{\partial h^n} g(x,z)\Big|_{z_0} = \frac{\partial^{n-1}}{\partial h^{n-1}} \cdot z\frac{\partial g}{\partial z}\Big|_{z_0} (-2\beta) =$$
$$= (-2\beta)\frac{\partial^{n-1}}{\partial h^{n-1}} \left[\sum \frac{W(C_0)}{\mu(C_0)} z^{I(C)} I^2(C_0) \right] =$$
$$= (-2\beta)^n \sum_{C_0} \frac{W(C_0)}{\mu(C_0)} I(C_0)^{n+1} = \sum_{\ell} b_{\ell}^{(n)} x^{\ell}$$

(*) Let us notice that we could also write Eq. 5.13 as $\lim\limits_{n\to\infty} \left| \max\limits_{\vec{k},i} \lambda_i(\vec{k},x) \right|^{\frac{-4}{n^{1-\sigma}}}$ with $\sigma = \frac{1}{2}$ which coincide with the assumption introduced by Fisher in the droplet model.

where :

$$b_\ell^{(n)} = (-2\beta)^n \sum_{\substack{C_o \\ I(C_o)=\ell}} \frac{(-1)^{N_{C_o}}}{\mu(C_o)} \cdot I(C_o)^{n+1}$$

Computing the radius of convergence for the series $g^n(x, z_o=1)$ (in the x variable) we have

$$|b_\ell^{(n)}|^{\frac{1}{\ell}} \longrightarrow |\sum_{\vec{k}} \mathrm{Trace}\, M^\ell I(\vec{c})^{n+1}|^{\frac{1}{\ell}}$$

and $I^{n+1}(\bar{c}) \sim e^{g(\ell)(n+1)} \sim \ell^{2(n+1)}$

which gives :

$$|b_\ell^n|^{-\frac{1}{\ell}} \longrightarrow \sqrt{2}-1$$

Therefore the series for $g^{(n)}(x, z_o=1)$ converges for all n if $x < \sqrt{2}-1$.

To compare our G.D. model with the usual droplet model of Fisher [76] or the modified droplet model of Reatto [94], we recall that the coefficients $a_n(x)$ for the free energy are given in the usual model by :

$$a_n(x) = q_o n^{-\tau} \cdot x^{n^\sigma}$$

while in the modified droplet model they are given by :

$$a_n(x) = q \cdot n^{-\tau} \cdot x^{n^\sigma} \cdot f(n,x) \; , \; f(n,x) = (B x^{bn^\sigma} + 1)^{-1}$$

A first difference lies in the fact that the first few coefficients of the G.D. model Eq. 5.12 can alternate in sign while this is not the case for the other two droplet models (however we have assumed that for large n the coefficient a_n have also the same sign in our G.D. model). Another difference lies in the fact that, because of the multiplicity of the cycles, the perimeters $\ell(c)$ of our model can be proportionnal to the square root of the area as well as proportionnal to the area; on the other hand for the other droplets models the perimeter of a connection with this difference we shall note that recent investigations on dilute ferromagnet [93] indicate that it is not sufficient to consider droplet of conventionnal size ($\ell \sim \sqrt{\text{area}}$) but it is necessary to consider also contribution due to ramified domains ($\ell \sim$ area).

(+) The low-temperature 4 x 4 matrix propagator in Fourier space of trajectory is given by :

$$M(\vec{k},x) = \begin{pmatrix} xe^{-ik_1} & xe^{-ik_2-\frac{i\pi}{4}} & 0 & xe^{ik_2+\frac{i\pi}{4}} \\ xe^{-ik_1+\frac{i\pi}{4}} & xe^{-ik_2} & xe^{ik_1-\frac{i\pi}{4}} & 0 \\ 0 & xe^{-ik_2+\frac{i\pi}{4}} & xe^{ik_1} & xe^{ik_2-\frac{i\pi}{4}} \\ xe^{-ik_1-\frac{i\pi}{4}} & 0 & xe^{ik_1+\frac{i\pi}{4}} & xe^{ik_2} \end{pmatrix}$$

$x = e^{-2\beta J}$, $k_i \in [0, 2\pi]$.

See for example :

P.N. Burgoyne J. Math. Phys. 4, 1320 (1963)

M. Glasser Am. J. of Phys. 38, 1033 (1970)

CHAPTER 6 - THE PARTIAL TRACE TRANSFORMATION. EQUATION FOR THE CORRELATION FUNCTIONS AND REPRESENTATION OF THE SYMMETRY GROUP

In this chapter, we use the local property of the group structure to introduce another transformation called partial trace. This transformation relates the partition function and the correlation functions of a general spin lattice system defined on $\Lambda \subset \mathbb{Z}^{\nu}$ to the corresponding functions for a system defined on a sublattice Λ/Λ_0 . This transformation is then applied to derive equilibrium equations for the correlation functions, equations which reflects in particular the symmetry properties of the lattice. They will be investigated in connection with phase transition and symmetry breakdown of the state.

6.1. Partial Trace Transformation

It is a transformation of the model and includes in particular the star-triangle and decoration transformation. It is a reduction of the phase space, which can be derived in a simple way, using the group structure associated with the lattice.

To study phase transition and critical phenomena, much interest is given by a knowledge of some internal symmetry of the system. As for the duality transformation, given a model, one would like to obtain another model whose physical properties are identical to the original ones. A well-known example is the hexagonal lattice in zero field, which is not self-dual; moreover internal symmetry, "like self-duality" can be easily obtained by combining the duality and partial trace transformations. The same procedure is applied to the correlation functions in the next Section and allow us to find a set of linear equations for the correlation functions.

The partial trace operation is defined as follows [78] :

Given a general spin lattice system (Λ, K) with $\Lambda \subset \mathbb{Z}^\nu$ finite, and let $\Lambda_0 \subset \Lambda$ be a sublattice such that the subsystems on Λ_0 do not interact "directly" with each others, i.e. $K(X) = 0$ if $|X \cap \Lambda_0| \geq 2$. Then the partition function and the correlation functions of the system (Λ, K) are related to those of a system defined on Λ / Λ_0 . The support $\mathcal{B}'$ of the interaction on Λ / Λ_0 is obtained directly from the group structure associated with the lattice (Λ, K) . The elements adapted to this reduction are the generators $\gamma(x)$, $x \in \Lambda_0$, $\Gamma_0 \subset \Gamma$. We thus introduce the following :

<u>Definition</u> : $\mathcal{D}_{x_0}$ is the subset of $\mathcal{P}(\Lambda / \Lambda_0)$ defined by :

$$\mathcal{D}_{x_0} = \{ x_0 \cdot B \; ; \; B \ni x_0 , K(B) \neq 0 \} = \{ D \subset \Lambda / \Lambda_0 \; ; \; K(x_0 \cdot D) \neq 0 \}$$

$$\Delta_{x_0} = \bigcup_{D \in \mathcal{D}_{x_0}} D$$

Notice that $x_0 \cdot \Delta_{x_0}$ is precisely the range of the interaction at x_0 .

<u>Definition</u> : $\bar{\mathcal{D}}_{x_0} = \{ \bar{D} = \prod_i D_i \; ; \; D_i \in \mathcal{D}_{x_0} \}$ is the subgroup of $\mathcal{P}(\Delta_{x_0})$ generated by $\mathcal{D}_{x_0}$.

$$\bar{\mathcal{D}}_{x_0}^{\,even} = \{ \bar{D} = \prod_{i=1}^{n} D_i \; ; \; D_i \in \mathcal{D}_{x_0} , \; n \text{ even} \}$$

$$\bar{\mathcal{D}}_{x_0}^{\,odd} = \{ \bar{D} = \prod_{i=1}^{n} D_i \; ; \; D_i \in \mathcal{D}_{x_0} , \; n \text{ odd} \}$$

Theorem 1

Let (Λ, K) be a general spin lattice system such that $\exists \Lambda_0 \subset \Lambda$ with $K(R) = 0$ if $|R \cap \Lambda_0| \geq 2$. Then

1. $Z(\Lambda, K) = 2^{|\Lambda_0|} Z(\Lambda/\Lambda_0 ; K')$

2. $K'(X') = K(X') + k(X') \quad \forall X' \subset \Lambda/\Lambda^0$, and $k(X') = \sum_{x_0 \in \Lambda_0} k(x_0; X')$

   $k(x_0; X')$ is given by Eq. (6.2) or (6.3).

3. The support of $k(x_0; \cdot)$ is a subset of $\overline{\mathcal{D}}^{\text{even}}_{x_0}$, i.e. $k(x_0; X') \neq 0$ only if X' can be written as $X' = \prod_i^n D_i$ $D_i \in \mathcal{D}_{x_0}$ n even. In particular if $|D|$ is even (or odd) $\forall D \in \mathcal{D}_{x_0}$ then $k(x_0; X') = 0$ for all X' such that $|X'|$ is odd.

To prove this theorem, we need the following lemma :

Lemma :

With f any function defined on $\mathcal{P}(\Lambda)$ having the following property : $\exists \Delta_0 \subset \Lambda$ with $f(X) = f(X \cap \Delta_0) \quad \forall X \subset \Lambda$ (6.1.) then f admits the following Fourier decomposition :

$$f = \sum_{R \subset \Delta_0} \tilde{f}(R) \sigma_R \qquad \text{where} \quad \tilde{f}(R) = 2^{-|\Delta_0|} \sum_{S \subset \Delta_0} f(S) \sigma_S(R)$$

Proof : Since any function defined on $\mathcal{P}(\Lambda)$ admits a Fourier decomposition given by :

$$f = \sum_{X \subset \Lambda} \tilde{f}(X) \sigma_X \qquad \text{with} \quad \tilde{f}(X) = 2^{-|\Lambda|} \sum_{Y \subset \Lambda} f(Y) \sigma_Y(X)$$

We have, using condition (6.1.) :

$$\tilde{f}(X) = 2^{-|\Lambda|} \sum_{S \subset \Delta_0} \sum_{Y \subset \Lambda/\Delta_0} f(S) \sigma_S(X) \sigma_Y(X) \quad ;$$

this completes the proof since

$$\sum_{Y \subset \Lambda/\Delta_0} \sigma_Y(X) = \begin{cases} 2^{|\Lambda| - |\Delta_0|} & \text{if } X \subset \Delta_0 \\ 0 & \text{otherwise} \end{cases}$$

i.e.

$$\tilde{f} = 2^{-|\Delta_o|} \sum_{S \subset \Delta_o} f(S)\, \sigma_S \qquad \text{and} \qquad f = \sum_{R \subset \Delta_o} \tilde{f}(R)\, \sigma_R$$

Proof of Theorem 1

From the definition :

$$Z(\Lambda, K) = \sum_{X' \subset \Lambda/\Lambda_o} e^{\sum_{R' \subset \Lambda/\Lambda_o} K(R')\sigma_{R'}(X')} \prod_{x_o \in \Lambda_o} 2 \cosh\Big(\sum_{D \subset \Lambda/\Lambda_o} K(x_o \cdot D)\, \sigma_D(X') \Big)$$

Since $\forall D \in \mathcal{D}_{x_o}$, $\sigma_D(X' \cap \Delta_{x_o}) = \sigma_D(X')$ $\forall X' \subset \Lambda/\Lambda_o$, the function $f(X') = \cosh\Big(\sum_{D \in \mathcal{D}_{x_o}} K(x_o D) \cdot \sigma_D(X')\Big)$ satisfies the condition of the above lemma; thus we obtain :

$$\cosh\Big(\sum_{D \in \mathcal{D}_{x_o}} K(x_o \cdot D)\, \sigma_D(X')\Big) = e^{\sum_{R \subset \Delta_{x_o}} k(x_o; R)\, \sigma_R(X')}$$

where

$$k(x_o; R) = \begin{cases} 0, \text{if } R \not\subset \Delta_{x_o} \\ 2^{-|\Delta_{x_o}|} \sum_{S \subset \Delta_{x_o}} \sigma_S(R)\, \ln \cosh\Big(\sum_{D \in \mathcal{D}_{x_o}} K(x_o D)\, \sigma_D(S)\Big) & \text{if } R \subset \Delta_{x_o} \end{cases} \tag{6.2}$$

and (1), (2) are proved.

To prove 3, as well as to obtain a useful method to compute the coefficient $k(x_o; R)$, we further use the group property of the lattice; with $\mathcal{D}_{x_o}$ we define the corresponding symmetric group $\mathcal{S}_{x_o} \subset \mathcal{P}(\Delta_{x_o})$ by :

$$\mathcal{S}_{x_o} = \{ S \subset \Delta_{x_o} \mid \sigma_D(S) = +1 \ \ \forall D \in \mathcal{D}_{x_o} \}$$

Then, from Chapter 2, we have :

$\mathcal{P}(\Delta_{x_o}) = \mathcal{T}_{x_o} \cdot \mathcal{S}_{x_o}$ where $\mathcal{T}_{x_o} \subset \mathcal{P}(\Delta_{x_o})$ is a subgroup of $\mathcal{P}(\Delta_{x_o})$ isomorphic to $\overline{\mathcal{D}}_{x_o}$. From (6.2) and the definitions, it follows that $\forall R \subset \Delta_{x_o}$:

$$k(x_0;R) = 2^{-|\Delta_{x_0}|} \sum_{T \in \mathcal{T}_{x_0}} \sum_{S \in \mathcal{S}_{x_0}} \sigma_S(R)\sigma_T(R)\,\ln(\cosh \sum_{D \in \mathcal{D}_{x_0}} K(x_0 D)\sigma_D(T))$$

and using the result (see Chapter 2)

$$\sum_{S \subset \mathcal{S}_{x_0}} \sigma_S(R) = \begin{cases} |\mathcal{S}_{x_0}| & \text{if } R \in \overline{\mathcal{D}}_{x_0} \\ 0 & \text{if } R \notin \overline{\mathcal{D}}_{x_0} \end{cases}$$

we obtain :

$$\begin{cases} k(x_0;R) = 0 & \text{if } R \notin \overline{\mathcal{D}}_{x_0} \\ k(x_0;\prod_i D_i) = |\mathcal{S}_{x_0}| \cdot 2^{-|\Delta_{x_0}|} \ln\left\{ \prod_{T \in \mathcal{T}_{x_0}} [\cosh(\sum_{D \in \mathcal{D}_{x_0}} K(x_0;D)\sigma_D(T))]^{\sigma_T(\prod_i D_i)} \right\} \end{cases}$$

To have a simple method to compute $k(x_0;\prod_i D_i)$, we use the isomorphism between the subgroups $\mathcal{T}_{x_0}$ and $\overline{\mathcal{D}}_{x_0}$ of $\mathcal{P}(\Delta_{x_0})$. Let $\mathcal{D}_0 = (d_1,\cdots d_n)$, $d_i \in \mathcal{D}_{x_0}$ be a set of generators for the subgroup $\overline{\mathcal{D}}_{x_0}$ then :

$$\begin{cases} k(x_0;R) = 0 & \text{if } R \notin \overline{\mathcal{D}}_{x_0} \\ k(x_0;\bar{D}) = 2^{-|\mathcal{D}_0|} \ln \prod_{\delta \subset \mathcal{D}_0} [\cosh \sum_{D \in \mathcal{D}_{x_0}} K(x_0 D)\sigma_\delta(D)]^{\sigma_\delta(\bar{D})} \end{cases} \tag{6.3}$$

where $\forall\, \bar{D} = \prod_{i \in I} d_i$ and $\delta = \{d_j\}_{j \in J} \subset \mathcal{D}_0$; $\sigma_\delta(\bar{D}) = (-1)^{|I \cap J|}$.

From(6.3)it follows, that if $\forall D \in \mathcal{D}_{x_0}$, D is an odd product of generators $d_i \in \mathcal{D}_0$ then $k(x_0;\bar{D}) \neq 0$ only if $\bar{D}$ is a product of an even number of generators d_i and in those cases (3) is established. In all other cases (3) only says that support of $k(x_0;\cdot)$ is a subset of $\overline{\mathcal{D}}_{x_0}$ and this has been established: Indeed, if $\exists\; D \in \mathcal{D}_{x_0}$ with $D = \prod_{i=1}^{n} d_i$, n even, then every $\bar{D} \in \overline{\mathcal{D}}_{x_0}$ can be expressed as an even product of $D_i \in \mathcal{D}_{x_0}$ since $\forall\, \bar{D} = \prod_{j \in J} d_j$ with $|J|$ odd, we have :

$\bar{D} = \prod_{j \in J} d_j \; D \prod_{i \in I} d_i$.

Thus $\bar{D}$ can be written as the product of an even number of $D_i \in \mathcal{D}_{x_0}$ and this completes the proof of Theorem 1.

Theorem 2

Let (Λ, K) be a general spin lattice system as in Theorem 1

1) For any $X_1 \subset \Lambda/\Lambda_0$ and $X_0 \subset \Lambda_0$ we have :

$$<\sigma_{X_1 \cdot X_0}>_{(\Lambda,K)} = <\prod_{x_0 \in X_0} \left(\sum_{R \subset \Lambda/\Lambda_0} a(x_0;R)\sigma_{X_1 \cdot R}\right)>_{(\Lambda/\Lambda_0, K')}$$

where $a(x_0; R)$ is given by Eq. (6.5) or (6.6)

2) The support of $a(x_0;\cdot)$ is a subset of $\bar{\mathcal{D}}^{odd}_{x_0}$ i.e $a(x_0;R)\neq 0$ only if R can be expressed as an odd product of $D_i \in \mathcal{D}_{x_0}$.

In particular if $|D|$ is odd (resp. even) $\forall D \in \mathcal{D}_{x_0}$, then $a(x_0; R) = 0$ for all R such that $|R|$ is even (resp.odd).

Proof :

From definition and the condition $K(R)=0$ if $|R \cap \Lambda_0| \geq 2$ we have :

$$\begin{aligned} <\sigma_{X_1 \cdot X_0}>_{(\Lambda,K)} &= Z^{-1} \sum_{Y_1 \subset \Lambda/\Lambda_0} \sum_{Y_0 \subset \Lambda_0} \sigma_{X_1 \cdot X_0}(Y_1 \cdot Y_0) \cdot \\ &\cdot e^{\sum_{S' \subset \Lambda/\Lambda_0} \left(K(S') + \sum_{x_0 \in \Lambda_0} K(x_0 S')\sigma_{x_0}(Y_0)\sigma_{S'}(Y_1)\right)} = \\ &= Z^{-1} \sum_{Y_1 \subset \Lambda/\Lambda_0} \sigma_{X_1}(Y_1)\, e^{\sum_{S' \subset \Lambda/\Lambda_0} K(S')\sigma_{S'}(Y_1)} \cdot \prod_{x_0 \in \Lambda_0} 2\cosh\left(\sum_{D \in \mathcal{D}_{x_0}} K(x_0 D)\sigma_D(Y_1)\right) \cdot \\ &\cdot \prod_{x_0 \in X_0} \tanh\left(\sum_{D \in \mathcal{D}_{x_0}} K(x_0 D)\sigma_D(Y_1)\right) \end{aligned} \qquad (6.4)$$

and using Theorem 1 :

$$<\sigma_{X_1 X_0}>_{(\Lambda,K)} = <\sigma_{X_1} \prod_{x_0 \in X_0} \tanh\left(\sum_{D \in \mathcal{D}_{x_0}} K(x_0 D)\sigma_D\right)>_{(\Lambda/\Lambda_0, K')}$$

but $\tanh\left(\sum_{D \in \mathcal{D}_{x_0}} K(x_0 D)\sigma_D(Y_1)\right) = \sum_{R \subset \Delta_{x_0}} a(x_0; R)\sigma_R(Y_1)$

where
$$\begin{cases} a(x_o;R) = 0 & \text{if } R \not\subset \Delta_{x_o} \\ a(x_o;R) = 2^{-|\Delta_{x_o}|} \sum\limits_{S \subset \Delta_{x_o}} \sigma_S(R) \tanh\left(\sum\limits_{D \in \mathcal{D}_{x_o}} K(x_o D)\, \sigma_D(S)\right) \end{cases} \tag{6.5}$$

Using the group structure, as before, we have :

$$\begin{cases} a(x_o;R) = 0 & \text{if } R \notin \bar{\mathcal{D}}_{x_o} \\ a(x_o;\bar{D}) = 2^{-|\mathcal{D}_o|} \sum\limits_{\delta \subset \mathcal{D}_o} \sigma_\delta(\bar{D}) \tanh\left(\sum\limits_{D \in \mathcal{D}_{x_o}} K(x_o D)\, \sigma_\delta(D)\right) \end{cases} \tag{6.6}$$

where $\mathcal{D}_o = (d_1, d_2, \cdots d_n) \subset \mathcal{D}_{x_o}$ is a set of generators for $\bar{\mathcal{D}}_{x_o}$ and $\forall \bar{D} \in \bar{\mathcal{D}}_{x_o}$, $\sigma_\delta(\bar{D})$ is defined as before. To prove (2), we remark that as for (3) of Theorem 1, it follows from (6.6) that if $\forall D \in \mathcal{D}_{x_o}$ is an odd product of generators d_i, then $a(x_o;\bar{D})$ only if $\bar{D}$ is an odd product of generators d_i. Moreover if $\exists D \in \mathcal{D}_{x_o}$ which is an even product of generators d_i, then each $\bar{D} \in \bar{\mathcal{D}}_{x_o}$ can be written as an odd product of $D_i \in \mathcal{D}_{x_o}$ and this concludes the proof of Theorem 2. As we shall see in the next Section, this will be useful to discuss symmetry breaking.

6.2 Equation for the Correlation Functions

For lattice systems, the correlation functions give a definition of the equilibrium state for finite as well as infinite system (Chapter 1,3,4) ; a central point is to investigate and prove their existence in the thermodynamic limit (Chapter 4). Some of these properties will be investigated by means of a set of equations relating the correlation functions involving a given spin and those interacting with its "neighbours".

Several sets of linear equations exist and are generally investigated with the standard method of Banach space and Neumann series [1].

To derive this set of equations, first for the correlation functions of finite systems, we apply the partial trace transformation, which is very well adapted to this purpose. These equations can then be used to define and discuss the state of finite as well as infinite systems. The interest of such equations is then analyzed and discussed in relation to phase transition and symmetry breaking of the state.[78]

From Chapters 1 and 4, the Gibbs states are invariant for finite systems under the symmetry group $\mathcal{S}$; from a result derived in [79] , the Gibbs state for $\Lambda \subset \mathbb{Z}^{\nu}$ finite, is then the unique physical solution of this set of equations (6.8) i.e. the only physical solution of these sets yields the Gibbs state and is obtained by iteration of (6.8). On the other hand, in the infinite volume limit $\Lambda \to \mathbb{Z}^{\nu}$, there may exist states which are no longer invariant under $\mathcal{S}$ and the problem is then to investigate the set of different solutions which can appear in this limit.

Not all solutions within this set define necessarily a state of the system; some conditions for such a solution to describe an equilibrium state of the system are further required and will be precisely specified in the next Section.

Let us first derive this set of equations for finite system $\Lambda \subset \mathbb{Z}^{\nu}$; this is given by the content of the following theorem :

Theorem 3

For any subset X of the finite system $\{\Lambda, K\}$ where K is a complex function such that $K(B) \neq \pm i\pi$ for all $B \in \mathcal{B}$, and any $x_0 \notin X$

$$< \sigma_{x_0 \cdot X} >_{(\Lambda,K)} = \sum_{Y \subset \Delta_{x_0}} a(x_0 ; Y) < \sigma_{X \cdot Y} >_{(\Lambda,K)} \qquad (6.7)$$

where $x_0 \cdot \Delta_{x_0}$ is the range of the interactions at x_0 and $a(x_0 ; Y)$ is given by Theorem 2.

Proof : The proof of this theorem follows easily from Theorem 2 for the particular choice $\Lambda_0 = \{x_0\}$.

To study these equations, we define a vector space V , of functions f defined on $\mathcal{P}_f(\mathbb{Z}^\nu)$ and define a linear operator $\mathcal{A}$ on V by :

$$\mathcal{A} f(\phi) = 0$$

$$\mathcal{A} f(X) = \sum_{Y \subset \Delta_{x_1}} a(x_1 ; Y) f(x_1 \cdot Y \cdot X) \qquad X \neq \phi$$

where x_1 is the first point of X in some lexicographic order; let $\mathbb{1} \in V$ be defined as $\mathbb{1}(X) = \delta_{X;\phi}$; the equations for the correlation functions (6.7) , for each $\Lambda \subset \mathbb{Z}^\nu$, can then be expressed as a linear equation on V given by :

$$\sigma_\Lambda = \mathbb{1} + \chi_\Lambda \mathcal{A} \sigma_\Lambda \qquad (6.8)$$

where χ_Λ is the characteristic function for $\Lambda \subset \mathbb{Z}^\nu$ and the vector $\sigma_\Lambda \in V$ is defined by $\sigma_\Lambda(X) = 0$ if $X \not\subset \Lambda$ and $\sigma_\Lambda(X) = < \sigma_X >_{(\Lambda,K)}$ if $X \subset \Lambda$ which is in particular bounded by 1 in absolute value.

Notice that these equations have a structure similar to the usual Kirkwood-Salzburg equations [1] and one can use them, in the same manner, to prove existence, unicity and analyticity properties of the solution in the case of finite as well as infinite systems. For the infinite system, the equations are then defined by Eq.(6.8), with $\Lambda = \mathbb{Z}^\nu$, i.e.

$$\sigma = \mathbb{1} + \mathcal{A}\sigma \tag{6.9}$$

With $\sigma(X) = \omega[\sigma_X]$, the above equation can then be written as :

$$\omega[\sigma_{x_0 \cdot X}] = \omega[\sigma_X \cdot \tanh(\sum_{B \in \mathcal{Y}(x_0)} K(B)\sigma_{B \cdot x_0})] \tag{6.10}$$

Notice that we restrict ourselves to the case of finite range interaction.
We can then define the equilibrium states of the infinite system by means of the solution of the set of equations satisfying some conditions.

Definition

Any solution σ of Eq.(6.9) defines an equilibrium state for the interaction K , if

$$\mu_\Lambda(Y) = 2^{-|\Lambda|} \sum_{X \subset \Lambda} \langle \sigma_X \rangle \cdot \sigma_X(Y) = (\mathcal{F}_\Lambda^{-1}\sigma)(Y) \geq 0$$

for all finite $\Lambda \subset \mathbb{Z}^\nu$ and $Y \subset \Lambda$. This result was established recently in [80,81] where it was shown that the equilibrium state equations for the probability measure μ , for the lattice correlation functions ρ and for the spin correlation functions σ , are essentially equivalent.

In this respect, in [81], the following interesting result has been obtained.

Theorem 4

Any bounded solution of (6.9) defines a solution $\mu_\Lambda = \mathcal{F}_\Lambda^{-1}\sigma$ of the equilibrium equations Eq.(7.6) satisfying the normalization and compatibility conditions; a necessary condition for a solution σ to define a state is given by the boundedness condition $|\sigma(X)| \leq 1$ (however, it must be remarked that it is not known under what conditions such a condition could be sufficient).

Property 1 [79]

$\forall \Lambda \subset \mathbb{Z}^{\nu}$ finite, every state σ_{Λ}, solution of the linear equation (6.8) is invariant under the internal symmetry group, i.e. $I_S \sigma = \sigma$.

Property 2

For an infinite system, $\exists\ T_0$ such that for $T > T_0$ there exists a unique bounded solution σ of Eq. (6.9). In fact for $T > T_0, \|\mathcal{A}\| < 1$ (Sect. 6.4). Moreover this state is invariant under the internal symmetry group $\mathcal{S}$, i.e. $I_S \sigma = \sigma$.

Property 3

The equations for the correlation function (6.9) may admit non physical solutions even above "criticality". See example 4 in Sect. 6.4.

Remark 1

The equations for the correlation functions, Eq. (6.7) have been given implicitly, early by Fisher for the case of 2-body interactions only [77,82]; the intermediate result (Eq. 6.4) was obtained for general systems by Suzuki [83], the generalisation to arbitrary spin $\frac{1}{2}$ lattice system was given by C. Gruber and D. Merlini. [78]

Remark 2

The structure of Eq. (6.7) is reminiscent of the generalized Griffith's inequality : $< \sigma_{x_0 \cdot X} > \leq \sum_{R \subset \Delta_{x_0}} \tau(x_0; R) < \sigma_{X \cdot R} >$

Notice that in the above inequality, we have a sum on $R = D_i$, $D_i \in \mathcal{D}_{x_0}$, while in the equations (6.7) we have a sum on those R which can be expressed as an odd product of $D_i \in \mathcal{D}_{X_0}$; moreover Eq.(6.7) is not restricted to positive interactions.

Remark 3

As mentioned at the beginning of this chapter, several linear sets of linear equations have been derived for the correlation functions of classical spin ½ lattice systems. The linear equation we have considered differs from those equations, insofar they have a non zero contribution from those $<\sigma_{X \cdot R}>$ with $X \cap R \neq \phi$. Moreover, in studying the equation (6.8) with the help of the group structure of the lattice, it will be possible to include the action of the symmetry group. This was not reflected in the set of equations previously obtained and will be done in the next section.

6.3. Representation of the Symmetry and Translation Group ($\mathcal{S}$ resp. $\mathcal{T}$)

We now discuss the question of symmetry breakdown with respect to the internal symmetry group $\mathcal{S}$ and the translation group $\mathcal{T} = \mathbb{Z}^\nu$ of the lattice. The first point is to define how these groups act on the linear space $\mathcal{V}$ [78].

Symmetry Group $\mathcal{S}$ (we restrict ourselves to the case $|\mathcal{S}| < \infty$)

Let $\bar{\mathcal{B}} \subset \mathcal{P}_f(\mathbb{Z}^\nu)$ be the subgroup generated by $\mathcal{B}$ (Chapter 2) . Then $\mathcal{P}_f(\mathbb{Z}^\nu)/_{\bar{\mathcal{B}}} \cong \mathcal{S}$ (Chapter 2) : $\mathcal{P}_f(\mathbb{Z}^\nu) = \mathcal{R} \cdot \bar{\mathcal{B}}$, where $\mathcal{R} = \{ R_j \}$, $j = 1, 2, \ldots |\mathcal{S}|$, $\mathcal{R}$ is a subgroup of $\mathcal{P}_f(\mathbb{Z}^\nu)$ isomorphic to $\mathcal{S}$.

We then consider the decomposition of $\mathcal{V}$, i.e. $\mathcal{V} = \bigoplus_{j=1}^{|\mathcal{S}|} \mathcal{V}_j$ which is defined by :

$$\forall j \in [1, 2, \ldots |\mathcal{S}|] : \quad \mathcal{V}_j = \{ f \in \mathcal{V} ; \ supp\, f \subset R_j \cdot \bar{\mathcal{B}} \}$$

Let f_j be the component of f in V_j . The symmetry group $\mathcal{S}$ acts on V as :

$$(\tau'_S f)(X) = \sigma_S(X) f(X)$$

From the above definition and properties, it follows that $V_j \subset V$ is the subspace of functions f which is invariant under the subgroup $\mathcal{S}_j$ of $\mathcal{S}$ defined as $\mathcal{S}_j = \{ S \in \mathcal{S} ; \sigma_{R_j}(S) = +1 \}$ with the choice $R_1 = \phi$, V_1 is then the subspace of functions invariant under the internal symmetry group $\mathcal{S}$.

The internal symmetry group $\mathcal{S}$ can then be used to discuss some properties of the linear equation(6.7), in particular we have the following result

Theorem 5

1) The equations for the correlation functions decouple into $|\mathcal{S}|$ linear equations defined on each V_j by :

$$\sigma_\Lambda^1 = \mathbb{1} + \chi_\Lambda \mathcal{A}^1 \sigma_\Lambda^1 \qquad j = 1 \qquad (a) \qquad (6.11)$$

$$\sigma_\Lambda^j = \chi_\Lambda \mathcal{A}^j \sigma_\Lambda^j \qquad j \neq 1 \qquad (b)$$

where $\mathcal{A}^j$ is the restriction of $\mathcal{A}$ to V_j and Λ is either finite or infinite.

2) with f any solution of Eq. (6.11) then $\tau'_S f$ is also a solution for all $S \in \mathcal{S}$.

Proof :

It follows from Theorem 3 ; indeed it was shown that $a(x; Y) \neq 0$ only if Y can be written as an odd product of $D_i \in \mathcal{D}_{x_1}$; from this, it follows that $\forall f^j \in V_j$ we have :

$$(\mathcal{A} f^j)(X) = \sum_{\substack{Y \subset \Delta_{x_1} \\ Y = \prod_{i=1}^{2n+1} D_i}} a(x_1; Y) f^j(X \cdot \prod_{i=1}^{2n+1} (x_1 \cdot D_i))$$

Since $x_1, D_i \in \mathcal{B}$ we have $\mathcal{A} f^j \in \mathcal{V}_j$ and $\mathcal{A}$ is a linear operator leaving $\mathcal{V}_j$ invariant. From the previous discussion, it follows that any σ_Λ (resp. σ) defined as solution of Eq. (6.8) resp. (6.9) is invariant under $\mathcal{S}_\sigma$ the subgroup being defined by :

$$\mathcal{S}_\sigma = \bigcap_{j \in J} \mathcal{S}_j$$

where J is the subset of indices such that $\sigma^j \neq 0 \quad \forall j \in J$.

Discussion

The study of phase transition associated with a symmetry breakdown of the internal symmetry group, within this approach, is reduced to the analysis of the spectrum of the operators $\mathcal{A}^j$, $j = 1, 2, \ldots, |\mathcal{S}|$.

In fact, if at some temperature T , some of the operators $\mathcal{A}^j$, $j \neq 1$ have eigenvectors with eigenvalues +1, then it may exist states of the infinite systems which are only invariant under a subgroup $\mathcal{S}_\sigma$ of $\mathcal{S}$. In this case there exists a critical temperature T_c and a symmetry breakdown of the system. We notice that eigenvectors of $\mathcal{A}^j$ are defined up to a multiplicative constant; if an eigenvector σ^j with eigenvalue +1 is associated with a state then $\lambda \sigma^j$ with $\lambda \in [0, 1]$ will also define a state.

From general results on equilibrium states [1] , the extremal equilibrium states can then be defined as those physical solutions of Eq. (6.11) having the cluster property. The equations (6.7) yield also upper bounds for the critical temperature, improving those obtained from other Kirkwood - Salzburg types of equations. In particular, for ferromagnetic systems bounds can be calculated with Eq. 6.7 improving also those obtained recently by means of refined Griffith's inequality [87]. This analysis will be done in Section (6.4) devoted to applications.

Translation Group

More difficult is the problem to introduce the translation group $\mathcal{T}$ of the lattice in analogous manner as for the symmetry group $\mathcal{S}$:

This group acts on $\mathcal{V}$ as :

$$(\tau_a' f)(X) = f(X+a)$$

Assuming that the interaction J is translationally invariant, it then follows that the kernel $\mathcal{K}$ associated with the set of linear equations is translationally invariant, and then with any solution f of 6.11 (with $\Lambda = \mathbb{Z}^{\nu}$), $\tau_a' f$ is also solution.

Therefore, if the equation 6.11 has a unique solution, it defines a translation invariant equilibrium state (and also invariant under the full symmetry group).

We shall now come back on the uniqueness of the invariant equilibrium state in Chapters 7 and 9.

6.4. Application

1. Internal Symmetry of the Eight Vertex Model and the Ashkin-Teller Model

As a first application we investigate the internal symmetry of the eight-vertex model and the Ashkin-Teller model on a square lattice Λ [84]. These models are not self-dual in the sense of Chapter 2 Section 2 ; moreover, symmetry property can easily be obtained by application of a duality transformation followed by a partial trace operation.

A. The Eight-Vertex Model [23]

Let J_1 , J_2 , λ be the three coupling parameters of the model; J_1 , J_2 the 2-body interactions along the diagonals and λ the four body interactions within a unit square. By a HT-LT [Chapter 2] duality, we obtain a dual model (Λ^*, K^*) with magnetic field and 3-body interactions as in Figure 6.1.

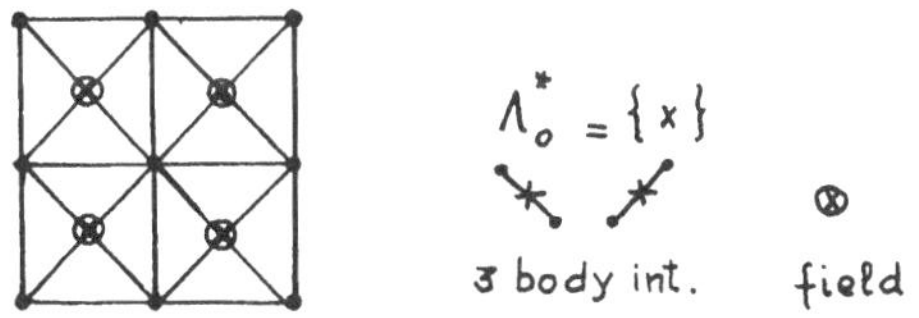

Fig. 6.1 : The dual model (Λ^*, K^*)

The sublattice $\Lambda_o^* \subset \Lambda^*$ (center of each unit square) is such that it satisfies the condition of Theorem 1 sec. 6.1

From this theorem, it follows that, since $\mathcal{D}_0$ is generated by the 2 point subsets along the diagonals, the support of $k(x_0;\cdot)$ is given by two and four point subsets. We obtain at once the original model (up to boundary terms) and the internal symmetry is found. After computation, this symmetry is expressed by the following relation for the free energy density :

$$F_{Bax}(\beta J_1, \beta J_2, \beta \lambda) = g + F_{Bax}(\beta' J_2', \beta' J_1', \beta' \lambda') \qquad (6.12)$$

where $\beta' J_2' = \frac{1}{4} \ln \frac{ac}{bd}$ $\quad \beta' J_1' = \frac{1}{4} \ln \frac{ad}{bc}$ $\quad$ and $\beta' \lambda' = \frac{1}{4} \ln \frac{ab}{cd}$

with $\quad g = \frac{1}{2} \ln \operatorname{sh} 2\beta J_1 \operatorname{sh} 2\beta J_2 \operatorname{sh} 2\beta\lambda + \frac{1}{4} \ln abcd$

$$a = 2 \cosh(\beta^*(J_1^* + J_2^* + \lambda^*))$$
$$b = 2 \cosh(\beta^*(-J_1^* - J_2^* + \lambda^*))$$
$$c = 2 \cosh(\beta^*(-J_1^* + J_2^* + \lambda^*))$$
$$d = 2 \cosh(\beta^*(J_1^* - J_2^* + \lambda^*))$$

$$\beta^* J_1^* = -\frac{1}{2} \ln \operatorname{th} \beta J_1 \;, \quad \beta^* J_2^* = -\frac{1}{2} \ln \operatorname{th} \beta J_2$$

Assuming the uniqueness of the critical point, then this must be given by the fixed point of (6.12) , yielding :

$$g = 0 \quad \beta J_1 = \beta' J_2' \quad \beta J_2 = \beta' J_1' \quad \beta\lambda = \beta'\lambda'$$

and we obtain the equation :

$$e^{-2\beta\lambda} \cosh \beta(J_1 - J_2) = \sinh \beta(J_1 + J_2) \qquad (6.13)$$

This equation coincides with the one obtained by Baxter in the exact solution of the eight-vertex model and the only singularity which appears, is that given by (6.13); in particular, for $J_1 = 0$, $\lambda = J_2$, the critical point coincides with that of the usual Ising Model $e^{-2\beta\lambda} = th\,\beta\lambda$, i.e. $Sh\,2\beta\lambda = 1$ [Chapter 2] . Moreover, we can conclude that the model (Λ^*, K^*) with magnetic field and 3-body interactions has a phase transition for any value of the interactions λ^* , J_1^* , J_2^* and continuous critical indice α depending on the values of the critical field λ^*[85].

B. <u>The Ashkin-Teller Model</u>

This model was introduced by Ashkin and Teller [84] ; they considered a four-component system on the square lattice Λ , such that each point $x \in \Lambda$ can be occupied by either one of the four kinds of atoms denoted (A, B, C, D) .

If we represent the configuration of $\{A, B, C, D\}$ by the four states of two spin $\frac{1}{2}$ particles ar each point of $\Lambda : (S_x, T_x)$ $S_x = \pm 1$, $T_x = \pm 1$, $\forall x \in \Lambda$ by the choice $A = (+,+)$, $B = (+,-)$, $C = (-,+)$, $D = (-,-)$ and assume the following interaction constants :

A - A	A - B	A - C	A - D
B - B	C - D	B - D	B - C
C - C			
D - D			
$-J_0$	$-J_1$	$-J_2$	$-J_3$

We then obtain a parametrization of the model in the spin language, given by the Hamiltonian :

$$-\beta H = \sum_{B \in \mathcal{B}} K(B)\sigma_B + C = C + \sum_{SS'} K_1 \sigma_{SS'} + \sum_{TT'} K_2 \sigma_{TT'} + \\ + \chi \sum_{SS'TT'} \sigma_{SS'TT'}$$

where the parameters are given by [86] :

$$K_1(SS') = K_1 = \beta(J_0 + J_1 - J_2 - J_3)/4 \qquad SS' \text{ nearest neighbours}$$

$$K_2(T,T') = K_2 = \beta(J_0 - J_1 + J_2 - J_3)/4 \qquad TT' \text{ nearest neighbours}$$

$$\chi(SS'TT') = \chi = \beta(J_0 - J_1 - J_2 + J_3)/4$$

$$C = 2|\Lambda|\beta \sum_{i=0}^{3} J_i \ .$$

We represent the 4-body interactions χ in Figure 6.2.

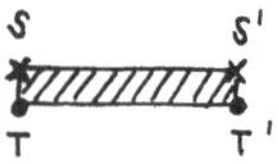

Fig. 6.2 : 4-body interactions χ

As for Baxter's model, we can apply a duality transformation (HT-LT) followed by a partial trace. By the first transformation, we obtain a model (Λ^*, K^*) on a square lattice Λ^* containing 3-body forces and magnetic field. For simplicity, we represent this model on Fig. 6.3 together with the duality mapping d.

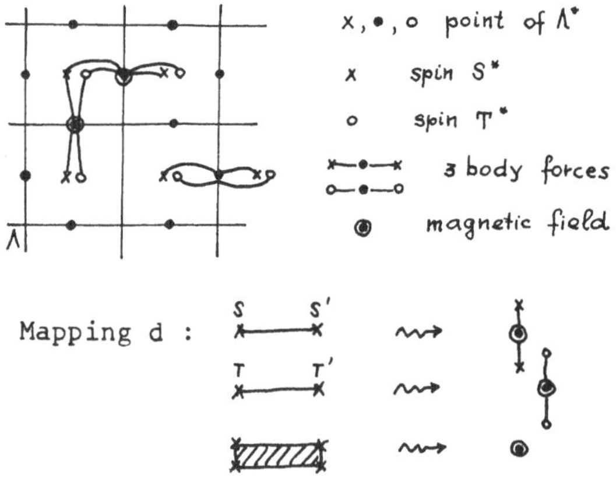

Fig. 6.3 : Mapping d for the Ashkin – Teller Model

As for model A., $\exists\ \Lambda_0^* \subset \Lambda$ such that the points of Λ_0^* satisfy the condition of Theorem 1. $\mathcal{D}_{x_0}$ is given by the even subsets $S^*S^{*'}$, $T^*T^{*'}$, $S^*S^{*'}T^*T^{*'}$. Application of this theorem yields also in this case, the original model; and a symmetry property is thus found.

The relationship between the free energy density is given by :

$$F_{A.T.}(K_1, K_2, \chi) = g + F_{A.T.}(K_1^{**}, K_2^{**}, \chi^{**})$$

$$g = \frac{1}{4} \ln abcd + \frac{1}{2} \ln \operatorname{Sh} 2K_1 \operatorname{Sh} 2K_2 \operatorname{Sh} 2\chi$$

$$\begin{aligned} a &= 2 \cosh (K_1^* + K_2^* + \chi^*) & K_1^* &= \frac{1}{4} \ln \frac{ac}{bd} \\ b &= 2 \cosh (-K_1^* - K_2^* + \chi^*) & K_2^* &= \frac{1}{4} \ln \frac{ad}{bc} \\ c &= 2 \cosh (K_1^* - K_2^* + \chi^*) & \chi^* &= \frac{1}{4} \ln \frac{ab}{cd} \\ d &= 2 \cosh (-K_1^* + K_2^* + \chi^*) \end{aligned} \tag{6.14}$$

Assuming the uniqueness postulat, we obtain, in this case :

$g = 0 \quad \chi = \chi^{**} \quad K_1^{**} = K_1^* \quad K_2^{**} = K_2^*$ yielding the same equation as for Model A : $e^{-2\chi} \cosh (K_1 - K_2) = \sinh (K_1 + K_2)$.

C. Relation between the Two Models

Since the two models have "similar symmetry", it is interesting to find a relationship between them. This relation is dictated by the fact that if we make a rotation of $\pi/2$ on each bond SS' of the Teller model, we obtain precisely the structure of the Baxter's model and vice-versa (Figure 6.4).

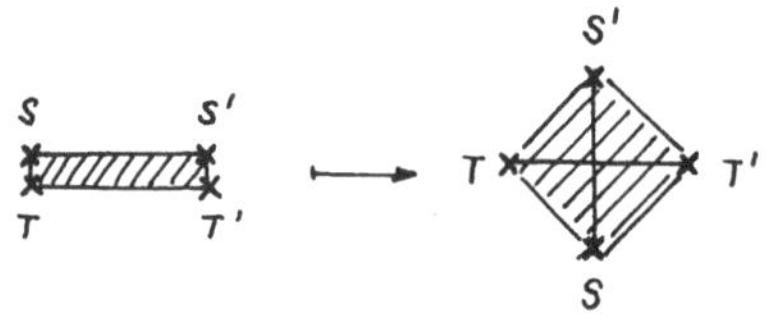

Fig. 6 .4

This operation is a partial duality transformation on one of the two superposed square sublattices, for example $\{S\}$. From Chapter 2 , we obtain at once :

$$Z_{A-T}(K_1,K_2,\chi) = Z_{Bax}(K_1^*,K_2^*,\chi^*)\cdot F$$

where $$F = [\,\mathrm{Sh}\,2(K_2+\chi)\,\mathrm{Sh}(2)(K_2-\chi)]^{|\Lambda|/2} \qquad (6.15)$$

$$K_1^* = K_1 + \frac{1}{4}\ln\frac{\mathrm{Sh}\,2(K_2+\chi)}{\mathrm{Sh}\,2(K_2-\chi)}$$

$$K_2^* = -\frac{1}{4}\ln \mathrm{th}(K_2+\chi)\cdot\mathrm{th}(K_2-\chi)$$

$$\chi^* = -\frac{1}{4}\ln\frac{\mathrm{th}(K_2+\chi)}{\mathrm{th}(K_2-\chi)}$$

Thus we have a parametrization of the Ashkin-Teller model (Λ,K) by a Baxter's model (Λ^*,K^*) with $|\Lambda^*|=|\Lambda|$ and with K^* given above.

As it has been pointed out first by F. Wegner [86] the existence of a similar symmetry for the two models and their relationship, does not imply necessarily that the model B has a unique critical point, as model A. Moreover, the critical indices α for the A-T model cannot be obtained from the known values of that found for the Baxter's model [86] . Besides the two different structures of the corresponding symmetry group, which has been partially discussed in chapter 4, a first difference between the two models arises in considering the case $\chi=0$ (we only consider the ferromagnetic case $K_i \geqslant 0$, $\chi \geqslant 0$). In fact for model A, the two sublattices are, as for model B, uncorrelated but have the same critical point; in case B, the two Ising models have different critical points except for $K_1 = K_2$ and one expects that the same holds for $\chi \neq 0$.

We now give some evidence for this possibility; our point is the following : to have a real parametrization of model B by model A (6.15) we must have : $K_2 > \chi$, by permutation $K_1 > \chi$. The unique critical point of model A is given by $e^{-2\chi^*} \cosh(K_1^* - K_2^*) = \sinh(K_1^* + K_2^*)$, yielding for model A the equation $K_1 = K_2$ as for $\chi=0$. We thus conclude that the Ashkin-Teller's model has a unique critical point if $K_1 = K_2 > \chi$. If this condition is not satisfied, then one just expects the possibility of two phase transitions and the two critical temperatures are related by the relation obtained above, from duality. Moreover, one can verify that this symmetry relation can also be obtained without partial trace operation, namely, by application of two partial duality transformations, the first one on the S -sublattice, the second one on the T sublattice.

2. Thermodynamic Equivalence between the Eight-Vertex Model and a Model with 3-Body Forces

In this example, it is intended to give a thermodynamic equivalence of the Baxter's model with an anisotropic model with 3-body interactions, like the triangular lattice with 3-body forces we have considered in Chapters 2,4. Consider a square lattice Λ with pure, diluted 3-body interactions, as in Figure 6.5 .

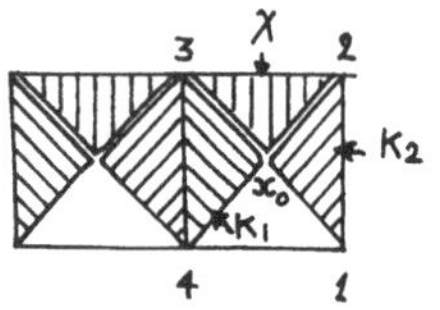

Fig. 8.5 : The diluted model

We can apply Theorem 1 on $\Lambda_0 = \{x_0\}$, $\forall x_0$ center of a unit square. At x_0 we have three 3-body interactions

$B_i = (x_0, x_K, x_\ell) = x_0 D_i$, $i = 1,2,3$.

We have $\quad \mathscr{D}_{x_0} = \{(12),(23),(34)\} \qquad \Delta_{x_0} = \{1,2,3,4\}$

$$\overline{\mathscr{D}}_{x_0} = \{(12),(23),(34),(13),(24),(14),(1234),\phi\}$$

From Theorem 1, we conclude at once that a partial trace on the site x_0 will introduce 2-body and 4-body forces between the point of Δ_{x_0} . Moreover, since $\overline{\mathscr{D}}_{x_0}$ is generated by the element of the set $\mathscr{D}_0 = \{d_1 = (12),\ d_2 = (23),\ d_3 = (34)\}$ and since every $D \in \mathscr{D}_{x_0}$ is the product of an odd number of generators, it follows from Theorem 1 that the only non-zero supplementary interactions are $k(x_0; X')$ with $X' = \phi$, $X' = d_1 d_2$, $X' = d_2 d_3$, $X' = d_1 d_3$. The explicit values of $k(x_0; X')$ are then obtained from Theorem 1, with the help of the character table of the subgroup $\overline{\mathscr{D}}_{x_0}$, yielding :

$$\begin{aligned}
k(x_0; X') = 2^{-3} \ln \Big\{ & (\cosh(K_1+K_2+\lambda))^{1+\sigma_{d_1 d_2 d_3}(X')} \cdot \\
& \cdot (\cosh(-K_1+K_2+\lambda))^{\sigma_{d_1}(X')+\sigma_{d_2 d_3}(X')} \cdot \\
& \cdot (\cosh(K_1-K_2+\lambda))^{\sigma_{d_2}(X')+\sigma_{d_1 d_3}(X')} \cdot \\
& \cdot (\cosh(K_1+K_2-\lambda))^{\sigma_{d_3}(X')+\sigma_{d_1 d_2}(X')} \Big\}
\end{aligned}$$

Table 1 : Character table $\sigma_\delta(\bar{D})$

$\bar{\mathcal{D}}_{x_0}=\{\delta\subset\mathcal{D}_0\}$ / $\bar{\mathcal{D}}_{x_0}=\{\bar{D}\}$	ϕ	d_1	d_2	d_3	d_1d_2	d_1d_3	d_2d_3	$d_1d_2d_3$
ϕ	1	1	1	1	1	1	1	1
d_1	1	-1	1	1	-1	-1	1	-1
d_2	1	1	-1	1	-1	1	-1	-1
d_3	1	1	1	-1	1	-1	-1	-1
d_1d_2	1	-1	-1	1	1	-1	-1	1
d_1d_3	1	-1	1	-1	-1	1	-1	1
d_2d_3	1	1	-1	-1	-1	-1	1	1
$d_1d_2d_3$	1	-1	-1	-1	1	1	1	-1

Therefore $k(x_0; X')$ is immediately obtained and the property of the subgroup $\bar{\mathcal{D}}_{x_0}$ yields at once the values of the interactions for the model Λ/Λ_0 . This gives us a parametrization of the Baxter's model in terms of a model with pure 3 spin interactions [sec. 5.3] . On the basis of the established results for the eight-vertex model [23] , this example also illustrate how an anisotropic character of a pure 3-body forces gives rise to a continuous changing in some of the critical indices (see also [24] for a similar equivalence of the Baxter model with another model on $\mathbb{Z}^2$ with pure anisotropic 3-body forces).

3. Upper and Lower Bounds for the Critical Temperature of Ferromagnetic Systems

In this example, we derive upper and lower bounds for the critical temperature of ferromagnetic systems, using standard methods to analyze the kernel associated to the set of equations (6.7) . These bounds improve those obtained from others Kirkwood-Salzburg type of equations From eq. (6.7), it follows that the condition :

$$\max_{x \in \mathbb{Z}^\nu} \sum_{R \subset \Delta_x} |a(x;R)| = 1 \tag{6.16}$$

yields an upper bound T_a for the critical temperature. In fact, T_a is a bound for the radius of convergence of the iterative process, i.e. the solution given by the series expansion :

$$\sigma = \sum_{k=0}^{\infty} [\mathcal{A}]^k \sigma_0 \qquad \text{where } \sigma_0 = \mathbb{1}$$

Notice that the best bounds obtained from previous linear equations are those given by Greenberg. In the case of ferromagnetic systems with translationally invariant 2 -body forces, bounds can also be obtained from Eq. 6.11 , improving those recently got by Thompson [87] using refined Griffith's inequality. In fact, it follows from Eq.(6.11) that if $\sum_{R \subset \Delta_x} a^+(x;R) < 1$, where $a^+ = a$ if $a > 0$ and $a^+ = 0$ if $a < 0$ then $\sigma(X)$ with $|X|$ odd is zero for any translationally invariant state σ such that $\sigma \geqslant 0$. The condition :

$$\max_{x \in \mathbb{Z}^\nu} \sum_{R \subset \Delta_x} a^+(x;R) = 1 \tag{6.17}$$

gives us an upper bound T_0 for the critical temperature of ferromagnetic systems. For $T > T_0$, σ is invariant under the full internal symmetry group $\mathcal{S}$ and the correlation functions with $|X|$ odd vanished. This condition is obtained as follows : $(\sigma \geq 0)$

Let $\tilde{\sigma} = \max_{\substack{X \subset \mathbb{Z}^\nu \\ |X| \text{ odd}}} \sigma(X)$ then

$$\tilde{\sigma} \leq \sum_{R \subset \Delta_X} a^+(x;R)\,\tilde{\sigma} < \tilde{\sigma}$$

For illustration, we explicit the computation in the case of the two-dimensional Ising Model : $K = \beta J$, with K a translationally invariant 2-body forces.

From Theorem 2, we obtain :

$$a(x_0;i) = \frac{1}{16}(2\,\mathrm{th}\,4K + 4\,\mathrm{th}\,2K)$$

$$a(x_0;ijk) = \frac{1}{16}(2\,\mathrm{th}\,4K + 4\,\mathrm{th}\,2K)$$

Thus the condition 1 gives us :

$$\max_{x \in \mathbb{Z}^\nu} \sum_{R \subset \Delta_x} a(x;R) = \frac{1}{16}(32\,\mathrm{th}\,2K) < 1 \Rightarrow \mathrm{th}\,2K < \frac{1}{2}$$

with $x = \mathrm{th}\,\beta J \Rightarrow x^2 - 4x + 1 > 0$, $x < 2-\sqrt{3} \sim 0{,}27$.

On the other hand, using condition 2 (ferromagnetic system) we have :

$$\max_{x \in \mathbb{Z}^\nu} \sum_{R \subset \Delta_x} a^+(x;R) = \frac{1}{4}(2\,\mathrm{th}\,4K + 4\,\mathrm{th}\,2K) \qquad \text{with } x = \mathrm{th}\,\beta J$$

as before $x^3 - x^2 + 2x - 1 < 0$

$$x = \mathrm{th}\,\beta J \lesssim 0{,}32$$

Clearly, in all cases, we obtain upper bounds using the equations we got from the properties of the local structure, associated with the lattice. As remarked, the solution obtained by the iterative process is constructed from the unperturbed "solution" $\sigma = 1$ in the high temperature region $T \sim \infty$.

In the one-dimensional Ising case,

$$a(x_0; i) = \frac{1}{2} \operatorname{th} 2K \quad \text{thus :}$$

$$\max_{x \in \mathbb{Z}^\nu} \sum_{R \subset \Delta x_0} a(x_0; R) = \operatorname{th} 2K < 1 \ , \ T > 0$$ which is the "critical temperature" of the one-dimensional case. Moreover $T = 0$ is a singular point, since there exists non-physical solutions which are obtained around $T = 0$. (See next example). It is clear that such bounds only depend on the local structure of the interaction set $\mathfrak{D}_{x_0}$ and thus the bound for the diamond is the same as for the two-dimensional Ising Model. The results are listed below for some 2- and 3-dimensional models and compared with other previous results :

Table 2 : Upper bounds for $\operatorname{th} \beta J$ as given by Greenberg, Thompson, Eq. (6.16) and Eq. (6.17) [89,87]

Lattice	Greenberg	Thompson	Eq 6.16	Eq 6.17	Exact
1-dimensional			1	1	1
Square	0,14	0,28	0,27	0,32	0,414
Honeycomb	0,18	0,38	0,39	0,44	0,58
Triangular 3-body forces	-	-			0,414
Diamond	0,14	0,28	0,27	0,32	0,35
Simple cubic	0,09	0,18	0,16	0,19	0,22

4. One-Dimensional Ising Model; Existence of Non-Physical Solutions for the Set of Equations

We consider an infinite one-dimensional Ising Model with 2-body interactions $K = \beta J$ between nearest neighbours. We treat this example explicitly to show what kind of situation occurs in studying the solution of the infinite set of equations (6.11) and show the existence of non-physical solutions.

Since $\mathcal{S} = (S_1, S_2) = (\phi, \mathbb{Z}^1)$, the vector space $\mathcal{V}$ of function on $\mathcal{P}(\mathbb{Z}^1)$ decouples into $\mathcal{V} = \mathcal{V}^{\text{even}} \oplus \mathcal{V}^{\text{odd}}$ where $\mathcal{V}^{\text{even}}$ (resp. odd) is the subspace of functions defined on subsets X with $|X|$ even (resp. odd); we have :

$$\tau'_{S_1} f^{\text{even}} = f^{\text{even}} \qquad \tau'_{S_2} f^{\text{even}} = f^{\text{even}}$$
$$\tau'_{S_2} f^{\text{odd}} = f^{\text{odd}} \qquad \tau'_{S_2} f^{\text{odd}} = -f^{\text{odd}}$$

The translation group $\mathbb{Z}^1$ acts on $\mathcal{V}$ in the usual manner, i.e. $(T_a f)(X) = f(X+a)$, where $X+a$ denotes the translate of X by a. The equations then decouples into equations for $\mathcal{V}^{\text{even}}$ and $\mathcal{V}^{\text{odd}}$:

$$\sigma(\phi) = 1$$
$$\sigma(x_1 \cdots x_{2n}) = \alpha\,(\sigma(x_1 - 1, x_2 \cdots x_{2n}) + \sigma(x_1 + 1, x_2 \cdots x_{2n})) \tag{6.18}$$

and
$$\sigma(x_1 \cdots x_{2n+1}) = \alpha\,(\sigma(x_1 - 1, x_2 \cdots x_{2n+1}) + \sigma(x_1 + 1, x_2 \cdots x_{2n+1}))$$

with $\alpha = \frac{1}{2}\operatorname{th} 2K$. Let us order the points in increasing order, i.e. $x_1 < x_2 < \cdots x_m$. The solution σ^{even} of Eq. (6.18) is then given by the Ansatz $\sigma(x_1 \cdots x_{2n}) = A^{x_{2n} - x_{2n-1} \cdots + x_2 - x_1}$. The equations are then fulfilled if α is solution of the quadratic equation $A = \alpha(A^2+1)$ yielding the two values :

$$A = (2\alpha)^{-1}\,[\,1 \pm (1 - 4\alpha^2)^{1/2}\,] = [\tanh K]^{\mp 1} \tag{6.19}$$

1. Physical Solution : It is obtained by the choice $A = \operatorname{th} K$, which corresponds to the boundedness condition $|\sigma| < 1$. $|2\alpha| = |\operatorname{th} 2K| < 1$. σ^{odd} has no solution; thus for $T \neq 0$, $\exists$ only one solution with $\sigma^{\text{odd}} = 0$ which is invariant under the

full symmetry group (symmetric - translationally invariant state ω).

$2\alpha = 1$ i.e. $J>0$, $T=0$, then $\sigma^{odd} = \sigma$ with σ arbitrary. The boundedness condition implies $\alpha \in (-1,+1)$. Moreover, the extremal states σ_e^{odd} (which possess the cluster property) are given by $\sigma_+^{odd} = +1$, $\sigma_-^{odd} = -1$. Thus there exists two extremal states invariant, under $\mathbb{Z}^1$, but invariant only under a subgroup of the symmetry group $\mathcal{S}$.

$2\alpha = -1$ i.e. $J<0$, $T=0$ $\sigma(x_1 \cdots x_{2n+1}) = (-1)^{x_1}\sigma$, with σ arbitrary. Again, the boundedness condition imposes $\sigma \in [-1,+1]$; the extremal states are given in this case by $\sigma_+(x_1 \cdots x_{2n+1}) = (-1)^{x_1}$ and $\sigma_-(x_1 \cdots x_{2n+1}) = -(-1)^{x_1}$. In this case, we also have two extremal states; however, they are invariant only under a subgroup of the translation group and under a subgroup of internal symmetry group. Since the only singular point is given by $T=0$ we expect that the solution $\neq 0$ in the whole range of temperature, i.e. σ^{even}, is obtained by iteration from $\sigma^{even} = \sum_k \mathcal{A}^k \sigma_0$. In fact, the radius of convergence is given by $th\,2K = 1$; for example for the two points correlation function, it yields :

$$\langle \sigma_x \sigma_{x+1} \rangle = \sum (2\alpha)^{k+1} (2k-1)!!/(2k+1)!!$$

the series obtained by expansion of the square root in the solution for A , Eq. (6.19).

2. Non-Physical Solution

It is obtained by the choice : $A = (th\,K)^{-1}$. Some properties are relevant for this solution of the set of equations.

Property 1 : The non-physical solution σ does not possess the boundedness condition since $|\sigma(X)| > 1 \quad \forall X \subset \mathbb{Z}^\nu$.

Property 2 : This solution is such that the correlation functions are increasing functions with the temperature.

Remark : The existence of such a solution (which turns out to be unphysical) can be noted without explicit computation by simply analysing the structure of the kernel of the equation we have derived. In fact, it can be directly checked that the equilibrium equations (6.11) in zero field are, as it was remarked recently in [88] for the Kirkwood-Salzburg type of equations, invariant under the transformation $K \to K + i\frac{\pi}{2}$; this is because only $\tanh 2K$ enters into the kernel and this in any dimension; therefore with σ_K any solution of Eq. 6.11 associated with the interaction K , then the analytic continuation in the variable K to the value $K' = K + i\frac{\pi}{2}$, i.e. $\sigma_{K+i\frac{\pi}{2}}$ will give another solution of the same equation. In particular for the one-dimensional case, the kernel is given by : $\alpha = \frac{1}{2}\tanh 2K$; let $K \to K' = K + i\frac{\pi}{2}$, then $\alpha \to \alpha' = \frac{1}{2}\tanh 2K' = \alpha$.

It should be noticed that in the one-dimensional case, this solution can be obtained by a suitable limit of finite systems with some non physical (in fact complex) boundary conditions, namely that, putting two imaginary magnetic fields of strength $h = i\frac{\pi}{2}$ on the two points boundary of the open chain. As it was noted in previous section, the iterative solution (physical) was obtained from the unperturbed one at $T = \infty$ $(\sigma = 1, \tanh K = 0)$. The unphysical solution is precisely singular around $T = \infty$ and this explains why they cannot be obtained by the perturbative method, as for the physical solution.

From theorem 4 however, these non physical solution yield local distribution μ_Λ which satisfy the equilibrium equation [80] as well as normalization and compatibility condition.

CHAPTER 7 - INVARIANT EQUILIBRIUM STATES AND DUALITY TRANSFORMATION FOR INFINITE SYSTEMS

In this chapter we shall present general properties of "Invariant Equilibrium States" as defined in Sec. 4.1. and derive certain consequences of duality which are of interest to prove the unicity of the invariant equilibrium state. In particular the duality transformation yields necessary and sufficient conditions for equilibrium states to be transformed into equilibrium states of a dual system; furthermore it follows from this condition that the invariant equilibrium states of ferromagnetic systems is unique if $\omega_+ = \omega_{op}$. As another illustration of duality transformation we shall briefly discuss the problem of "surface tension".

In the following we restrict ourselves to the case of HT-LT duality transformations; analogous properties for HT-HT as well as LT-LT duality transformations will be derived in Part. II, Sec. 4.1. Moreover the proofs of the theorems presented in this chapter are given in ref. [32] and will not be reproduce here.

7.1. Symmetric Equilibrium States

Symmetric states were introduced in Sec. 1.3.2. as "states invariant under the internal symmetry group " and their properties were discussed in Sec. 9.3, 4.2, 4.5, 6.2, 6.3. In particular it was shown in Sec. 3.3. that any symmetric state can be uniquely specified by any of the following functions :

$$\sigma(\beta) = \omega[\sigma_\beta] \qquad \beta \in \mathcal{P}_f(\mathcal{B}, \quad \sigma_\beta = \prod_{B\in\beta} \sigma_B \tag{7.1}$$

$$\mu(\beta) = \omega[\mu_\beta] \qquad \beta \in \mathcal{P}_f(\mathcal{B}), \quad \mu_\beta = \prod_{B\in\beta} \mu_B \tag{7.2}$$

It could also be defined by means of the following functions [32]

$$\mu_\Lambda(X) = \omega[\mu_{\Lambda,X}^{sym}] \qquad \Lambda \in \mathcal{P}_f(\mathcal{L}); X\subset\Lambda \tag{7.3}$$

where
$$\mu_{\Lambda,X}^{sym} = 2^{-|\Lambda|} \sum_{\bar{B}\in \bar{\mathcal{B}}\cap\mathcal{P}(\Lambda)} \sigma_{\bar{B}}(X)\,\sigma_{\bar{B}}$$

Furthermore it was shown in this same section that the functions $\sigma = \sigma(\beta)$ and $\mu = \mu(\beta)$ are related by :

$$\sigma = D_{K_*}\mu \quad ; \quad \mu = D_K\sigma \quad \text{where} \quad e^{-2K_*(B)} = \tanh K(B)$$

In Sec. 4.3.4 it was then shown that ferromagnetic systems (satisfying certain conditions) have the property that the symmetric equilibrium state defined by means of the " + " boundary conditions coïncide with the equilibrium state defined by the "open" boundary conditions. Finally in Sec.6.4. we have shown that the equations for the symmetric equilibrium states decouple from the full set of equations.
In this section we present equivalent definitions of "symmetric equilibrium states" which are useful in connection with duality transformation.

Definition [48 a,b; 53 (i)]

A state ω is said to be an "Equilibrium state for the interaction K" or "Gibbs State" if the conditionnal probabilities $\mu_\Lambda(X;Y)$ for the finite portion Λ given that the outside is in the configuration Y is given by the following D.L.R. Equation (Dobrushin - Lanford - Ruelle) :

$$\mu_\Lambda(X;Y) = \frac{e^{-\beta H_{\Lambda,Y}(X)}}{Z_{(\Lambda,Y)}}$$

where $H_{\Lambda,Y}$ is given by Eq. 1.14 and $Z_{(\Lambda,Y)}$ is the corresponding partition function.

The equivalence between this definition of equilibrium state and the definition adopted in chapter 3 is given by the following result of O. Lanford and D. Ruelle [48,b] .

Proposition

1) The closed convex hull of states obtained as the l i m i t of finite volume Gibbs states with boundary condition Y (ch. 3) coïncide with the set Δ_K of all "Gibbs states" for the interaction K.

2) Δ_K is convex and compact; it is a Choquet simplex.

Theorem 1

The following statements are equivalent :

(1) ω is a symmetric equilibrium state for the interaction K.

(2) The function σ on $\mathcal{P}_f(\mathcal{B})$ is a solution of the following Eq. (7.4), defining a positive form :

$$\begin{cases} \sigma(\phi)=1 \qquad \sigma(\beta)=\sigma(\beta\cdot\varkappa) & \text{for all } \beta\in\mathcal{P}_f(\mathcal{B}),\ \varkappa\in\mathcal{K}_f \\ \omega[\sigma_\beta]=\omega[\sigma_\beta \tanh\sum_{B\in\gamma} K(B)\sigma_B] & \text{for all } \beta\in\mathcal{P}_f(\mathcal{B}) \text{ and} \\ \qquad\qquad \gamma\in\Gamma^f & \text{such that } |\beta\cap\gamma| \text{ is odd} \end{cases} \tag{7.4}$$

where ω is the linear form on the symmetric algebra $\mathfrak{A}_f$ (sec.3.3) defined by $\omega[\sigma_\beta]=\sigma(\beta)$; ω is positive if $\omega[\prod_{B\in\beta}\frac{1}{2}(1+\sigma_B(X)\sigma_B)]\geq 0$ for all $X\in\mathcal{P}_f(\mathcal{L})$ and $\beta\in\mathcal{P}_f(\mathcal{B})$.

3) The function μ on $\mathcal{P}_f(\mathcal{B})$ is a solution of the following Eq. (7.5), defining a positive form :

$$\begin{cases} \mu(\phi)=1 \qquad \mu(\beta)=\mu(\beta\cdot\gamma) & \text{for all } \beta\in\mathcal{P}_f(\mathcal{B}),\ \gamma\in\Gamma^f \\ (D_{K_*}\mu)(\beta)=(D_{K_*}\mu)(\beta\cdot\varkappa) & \text{for all } \beta\in\mathcal{P}_f(\mathcal{B})\ \varkappa\in\mathcal{K}_f \end{cases} \tag{7.5}$$

4) The functions $\mu_\Lambda(X)$ are solution of the following Eq. (7.6)

$$\mu_\Lambda(X) = e^{2\sum_{B\in\gamma(z)} K(B)\sigma_B(X)}\,\mu_\Lambda(X\cdot z) \quad \text{for all } \Lambda\in\mathcal{P}_f(\mathcal{L}),\ X\subset\Lambda \text{ and } z \text{ such that } \gamma(z)\subset\mathcal{P}(\Lambda) \tag{7.6}$$

satisfying the symmetry, normalization and compatibility conditions, i.e.

$$\mu_\Lambda(X)=\mu_\Lambda(S\cdot X) \qquad \text{for all } S\in\mathcal{S}$$

$$\sum_{X\subset\Lambda}\mu_\Lambda(X)=1$$

$$\sum_{Y\subset M/\Lambda}\mu_\Lambda(X\cup Y)=\mu_\Lambda(Y)$$

The proof of the equivalence of (1) and (4) was given in [80] ;

the equivalence of (2) and (4) in [81] and the equivalence (3) and (4) in [32] .

7.2. Symmetric Equilibrium States and Duality

As we have seen on several examples (e.g. p. 34), HT-LT duality transformations mapp " + " boundary conditions into "open" boundary conditions and "open" into " + ". In this section we shall see that this result is general for ferromagnetic systems satisfying the conditions C 1 and C 2 of Sec. 4.3 and yields very interesting consequences. (The proofs of the following theorems are based on Theorem 1 of the preceeding section).

Let $\{\mathcal{L}, K\}$ by any spin ½ lattice system satisfying the conditions C.1 and C.2 of Sec. 4.3. i.e.

C.1 $\mathcal{H}$ separates $\mathcal{B}$

C.2 $\Gamma^f \equiv \Gamma \cap \mathcal{P}_f(\mathcal{B})$

and let $\{\mathcal{L}^*, K^*\}$ be any HT-LT dual for $\{\mathcal{L}, K\}$(+) ; we then have the following result :

Theorem 2

With any equilibrium state ω of $\{\mathcal{L}, K\}$ we can associate a normalised linear form ω^* on $\mathfrak{O}^*_{f^*}$ defined by

$$\omega^*[\sigma_{\beta^*}] = \omega[\mu_{d^{-1}\beta^*}] \tag{7.7}$$

(+) We recall that it follows from C1-C2 that the duality mapping d is a bijection and that $\{\mathcal{L}^*, K^*\}$ is also a LT-HT dual for $\{\mathcal{L}, K\}$.

which is solution of the equilibrium equations Eq. (7.5).

Moreover the mapping $\omega \mapsto \omega^*$ is injectif and satisfies

$$\omega^*[\mu_{\beta^*}] = \omega[\sigma_{d^{-1}\beta^*}] \tag{7.8}$$

For the next theorems we shall need one more condition which is expected to hold for any $\mathbb{Z}^\nu$ - invariant system.

C.0. There exists a sequence of finite volumes $\Lambda_i \to \mathcal{L}$ such that for any Λ_i and X finite, the condition $\sigma_B(X)=+1$ for all $B \subset \mathcal{L}/\Lambda_i$ implies $X = Y \cdot S_f$ with $Y \subset \Lambda_i$ and $S_f \in \mathcal{S} \cap \mathcal{P}_f(\mathcal{L})$.

Theorem 3

For any ferromagnetic system $\{\mathcal{L}, K\}$ satisfying the condition C.0,C.1,C.2 and any HT-LT dual $\{\mathcal{L}^*, K^*\}$ we have :

$$(\omega_+)^* = \omega^*_{op} \quad \text{i.e.} \quad \langle \prod_{B\in\beta} \sigma_B \rangle_{\{\mathcal{L},K\},+} = \langle \prod_{B^*\in\beta^*} \mu_{B^*} \rangle_{\{\mathcal{L}^*,K^*\},op}$$

Moreover if the dual system $\{\mathcal{L}^*, K^*\}$ satisfies the same conditions then

$$(\omega_{op})^* = \omega^*_+ \quad \text{i.e.} \quad \langle \prod_{B\in\beta} \sigma_B \rangle_{\{\mathcal{L},K\},op} = \langle \prod_{B\in\beta^*} \mu_{B^*} \rangle_{\{\mathcal{L}^*,K^*\},+}$$

where ω^*_{op} and ω^*_+ denotes the Gibbs states of the dual system associated with "open" and " + " boundary conditions.

Theorem 4

With ω_I any $\mathbb{Z}^\nu$- invariant equilibrium state of the $\mathbb{Z}^\nu$- -invariant ferromagnetic system $\{\mathcal{L}, K\}$ satisfying C.0, C.1, C.2, then

ω_I^* is a symmetric, $\mathbb{Z}^\nu$-invariant, equilibrium state of $\{\mathcal{L}^*, K^*\}$ whenever the duality transformation commutes with translations.

Moreover if $\{\mathcal{L}^*, K^*\}$ satisfies the same conditions C.0, C.1, C.2, then

$$\omega_+[\sigma_\beta] \geq \omega_I[\sigma_\beta] \geq \omega_f[\sigma_\beta] \geq \prod_{B\in\beta} \tanh K(B)$$
$$1 \geq \omega_f[\mu_\beta] \geq \omega_I[\mu_\beta] \geq \omega_+[\mu_\beta] \qquad (7.9)$$

The proof of the first part of this theorem is a generalisation of the techniques introduced by Messager and Miracle-Sole in [97] and is given in [32].

Remarks :

1. The inequalities Eq. (7.9) will also hold for <u>any</u> equilibrium state ω (not nessarily $\mathbb{Z}^\nu$ - invariant) if ω^* is an equilibrium state of the dual system.

2. From theorem 1 and 2 it follows that to generalise theorem 3 and 4, i.e. to study under what condition the duality transformation yield by means of Eq. (7.7) an equilibrium state of the dual system, we just need conditions which will insure that ω^* is a positive form on $\mathfrak{A}^*_{\mathcal{L}^*}$; necessary and sufficient conditions were given in [32].

As consequences of the above theorems we have thus arrived at the following conclusions :

1. If $\omega_+ = \omega_{op}$ on $\mathfrak{A}_{\mathcal{L}}$ and if $\{\mathcal{L}, K\}$ satisfies the conditions C.0, C.1, C.2, there will exists a unique, invariant, symmetric equilibrium state; furthermore in this case ω_{op} is the only equilibrium state which yields a state by means of duality transformation.

   In this connection we recall that for $T < T_{(1)}$ (where $T_{(1)}$ is larger then the temperature obtained by means of

Peierl's argument) it was shown in Sec. 4.5. that $\omega_+ = \omega_{op}$ and thus <u>there exists in this temperature range a unique invariant equilibrium state</u>. This same conclusion also follows from theorem 4 together with the fact that there exists a unique equilibrium state at high temperature and the fact that the mapping $\omega \mapsto \omega^*$ is injective.

2. Let us note that it is expected that this last conclusion holds for all temperature. From theorem 4 it follows that it would be sufficient to show that $\omega_+ = \omega_{op}$ on $\mathfrak{A}_\mathcal{J}$ for all temperature. Such a result has been first proved for the Ising Model (in two dimension) with 2-body interaction between nearest neighbours only [97] ; is a consequence of the above theorem together with the result of J.L.Lebowitz and A. Martin-Löf showing the unicity of the equilibrium states above the critical temperature [95(a)] . Very recently J.L.Lebowitz was able to derive new inequalities from which follows that <u>for even interactions the invariant equilibrium state is unique in any dimension i.e. below the critical temperature only two phases can coexist [95(b)]</u> .

To conclude this general discussion on symmetric equilibrium states we mention that it is expected that the duality transformation together with Sherman's theorem on path [71] discussed in Ch. 5 may give a proof of the conjecture that for the 2-dimensional Ising Model all symmetric equilibrium states are necessarly $\mathbb{Z}^2$-invariant[145].

7.3. Duality Transformation and Surface Tension

The problem of surface tension for the Ising Model in the low temperature region has been extensively studied by Abraham, Gallavotti and Martin-Löf (see ref. [29] and the references quoted there). In this section we shall adopt one of the equivalent definitions which have been proposed in [29] and use duality transformation to prove the existence of the infinite volume surface tension at any temperature [32] .

Let $\{\mathbb{Z}^\nu, K\}$, be a lattice system; the idea is to decompose $\mathbb{Z}^\nu$ into $\mathbb{Z}_u^\nu \cup \mathbb{Z}_d^\nu$ where $\mathbb{Z}_u^\nu$ denotes those $x \in \mathbb{Z}^\nu$ such that $x_\nu > 0$. With $S \in \mathcal{S}$ we consider the finite system Λ with boundary conditions $Y = S \cap \mathbb{Z}_d^\nu$ where Λ is a parallelipiped with sides $(L_1, \dots, L_{\nu-1}, 2M)$ symmetric with respect to the plane $x_\nu = \frac{1}{2}$ (see fig. 1)

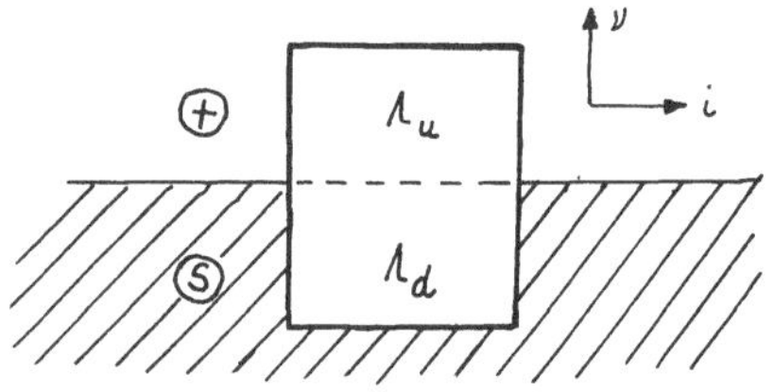

Fig. 1 : Finite system $\Lambda = \Lambda_d \cup \Lambda_u$ with boundary conditions $Y = S \cap \mathbb{Z}_d^\nu$

With $Z_\Lambda^{(+,S)}$ and $Z_\Lambda^{(+)}$ the partition functions corresponding respectively to "Y" and " + " boundary conditions, we have

Theorem 5

Let $\{\mathbb{Z}^\nu, K\}$ be a ferromagnetic system satisfying the conditions C.0, C.1, C.2; for any S in $\mathcal{S}$ the following limit

$$\tau_{(+,S)} = \lim_{L_1, \dots L_{\nu-1} \to \infty} \lim_{M \to \infty} \frac{1}{L_1 \cdots L_{\nu-1}} \cdot \log \frac{Z_\Lambda^{(+,S)}}{Z_\Lambda^{(+)}}$$

exists and is called the "Surface tension between the phase (+) and (S) ".

Moreover $\tau_{(+,S)}$ is non-positive and bounded below.

The proof of this theorem is based on theorem 3 and is given in [32] .

CHAPTER 8 - ASANO CONTRACTION AND GROUP STRUCTURE.

ANALYTICITY PROPERTIES OF THE FREE ENERGY

8.1. Introduction

In 1952 Yang and Lee approached the problem of phase transitions by means of the properties of the partition function $Z_\Lambda(z)$ which is for finite systems an entire function of the complex activity $z = e^{-2h}$ (h is the external field and the interactions between spins were considered to be fixed) [20]. As the thermodynamics of the problem enters through the function $\log Z_\Lambda(z)$, the zeroes of $Z_\Lambda(z)$ destroys the analylicity property of the thermodynamic functions and thus account for singularities, i.e. for phase transitions. Since the coefficients in the polynomial $Z_\Lambda(z)$ are real and positive (weighted Boltzmann factors), the zeroes cannot be on the positive axis for finite systems. The problem is thus to study what happens in the thermodynamic limit; the locus of the zeroes varies with the size of the system and could approach the positive real axis in this limit. This set of limiting points on the positive axis give the rise to a physical singularity, i.e. a manifestation of phase transition.

The method proposed by Yang and Lee was to consider the system in an external field h_x which is site dependent; the partition function becomes a polynomial in $|\Lambda|$ variables which is of order 1 with respect to each variable. However with the number of variables becoming infinite in the thermodynamic limit, the success of this approach was limited to a few models only (Ising Model, Monomer-Dimer [100] , Line graph [101] Ising for higher spin models [102]).

A new idea to study the zeroes of $Z_\Lambda(z)$ was introduced in [103] by T. Asano and renewed interest for this approach to phase transitions. This idea was to obtain $Z_\Lambda(z)$ by "contraction" of polynomials in few variables - also called "small polynomials" - and to relate properties of zeroes of the small polynomials to the zeroes of $Z_\Lambda(z)$; in fact those small polynomials were partition functions of small sub-systems. However, although the contraction process was defined for arbitrary systems, the relation between the zeroes of the small polynomials and those of the contracted polynomial was established only for small polynomials satisfying the so-called Yang - Lee Condition [104, 105].

Additional progress was then made by D. Ruelle [106] who gave general relations between the zeroes of a family of polynomials and the zeroes of the polynomial obtained by contraction. With this theorem it was thus possible to extract informations on the zeroes of $Z_\Lambda(z)$ from a knowledge of the zeroes of small sub-systems; moreover this prescription was valid for any lattice system. The remaining problem was to give a prescription to find the small sub-systems of interest. It should be noticed that in general the small polynomials were associated with the local structure and properties of the lattice; it appears thus natural to use the group structure to investigate this problem.

A first generalisation of the Asano - Ruelle technique was given by A. Messager and J.C. Trottin using an idea of S. Miracle Sole to discuss domain of zeroes in all complex variables $\{z_B = e^{-2K(B)}\}$ [107]; it was further extended by J. Slawny [108] and ourselves [109] . In particular Slawny gave a method to find the small sub-systems; he was also able to derive new results concerning the analyticity and uniqueness of the equilibrium states for ferromagnetic systems in the low temperature region.

In the next sections we give an extension of these methods based on the group structure we have developed; our results are valid for any polynomial in the variables $\{z_B\}_{B \in \mathcal{B}}$ which is associated with a subgroup $\mathcal{G}$ of $\mathcal{P}(\mathcal{B})$. Moreover the case where the polynomial is associated with a subset $\mathcal{E}$ of $\mathcal{P}(\mathcal{B})$ will be discussed in part II, Ch. 5, since it arises in connection with systems with constraints. The special cases $\mathcal{G} = \Gamma$ and $\mathcal{G} = \mathcal{K}$ correspond to the low (respectively high) temperature expansion and are up to now the only cases of interest; the results which we shall obtain concern "small" domain of analyticity for $|z| < 1$ with z the variable associated with the corresponding expansion.

To conclude this introduction we shall remark that in all those approaches, one obtains domains of the complex plane which are independent of Λ and free of zeroes of $Z_\Lambda(z)$ for all volumes Λ.

8.2. Asano Contraction for Polynomial Associated with a Group $\mathcal{G} \subset \mathcal{P}(\mathcal{B})$

In this section, we introduce the general definition of Asano Contraction, which associates with a family of "small polynomials" one "contracted polynomial". For completeness we then recall Ruelle's theorem which relates the zeroes of the small polynomials to those of the contracted polynomial. Finally, we give a method to find a family of small polynomials which yields by contraction the partition function of a given system $\{\Lambda, K\}$.

Let $\mathcal{B}$ be the set of bonds of the system as defined in Chapter 1 (*);

(*) In fact $\mathcal{B}$ could be any finite set, not necessarily related to any spin systems.

with each bond $B \in \mathcal{B}$ we associate a complex variable z_B and denote by $z_{\mathcal{B}}$ the family $\{z_B\}_{B \in \mathcal{B}}$ of complex variables.

With G any subgroup of $\mathcal{P}(\mathcal{B})$ we associate the following polynomial in the variables $z_{\mathcal{B}}$

$$M(z_{\mathcal{B}}) = \sum_{\beta \in G} z^{\beta} \tag{8.1}$$

where for all β in $\mathcal{P}(\mathcal{B})$

$$z^{\beta} = \prod_{B \in \beta} z_B \tag{8.2}$$

In particular it follows from Eq. (2.1) (resp. Eq. (2.2)) that for $G = \mathcal{H}$ (resp. $G = \Gamma$), $M(z_{\mathcal{B}})$ is up to a multiplicative factor the partition function of $\{\Lambda, K\}$ with $z_B = \tanh K(B)$ (resp. $z_B = e^{-2K(B)}$).

Definition : Asano Contraction

Let $\mathcal{B} = \bigcup_i \mathcal{B}_i$ be a finite covering of $\mathcal{B}$. The polynomial $P(z_{\mathcal{B}}) = \sum_{\beta \in \mathcal{B}} c_{\beta} z^{\beta}$ is the Asano contraction of the family of polynomials $\{P_i(z_{\mathcal{B}_i}) = \sum_{\beta_i \subset \mathcal{B}_i} c_{i,\beta_i} z^{\beta_i}\}$ if $c_{\beta} = \prod_i c_{i,\beta \cap \mathcal{B}_i}$; a variable z_B is said to undergo contraction if B belongs to more than one $\mathcal{B}_i$.

Theorem (D.Ruelle) [106]

Let $P(z_{\mathcal{B}})$ be the Asano contraction of $\{P_i(z_{\mathcal{B}_i})\}$; if $P_i(z_{\mathcal{B}_i}) \neq 0$ when $z_B \notin R_{i,B}$ $\forall B \in \mathcal{B}_i$, then $P(z_{\mathcal{B}}) \neq 0$ when $z_B \notin - \prod_i (-R_{i,B})$ $\forall B \in \mathcal{B}$, where $R_{i,B}$ $i = 1, 2, \ldots n$, $B \in \mathcal{B}_i$ are closed subsets of the complex plane which do not contain 0 if z_B undergo contraction.

The following generalization of a theorem due to J. Slawny [108] gives then all possible coverings of the set of bonds and the corresponding family of small polynomials which give $M(z_{\mathcal{B}})$ by contraction :

Theorem 1

For G any subgroup of $\mathcal{P}(\mathcal{B})$ and $\mathcal{B} = \bigcup_{i=1}^{n} \mathcal{B}_i$ a finite covering, let $G_i = \{\beta \cap \mathcal{B}_i \,;\, \beta \in G\}$ and $G_i^{\perp} = \{\beta \subset \mathcal{B}_i \,;\, \sigma_\beta(\beta_i) = +1 \ \forall \beta_i \in G_i\} \equiv G^{\perp} \cap \mathcal{P}(\mathcal{B}_i)$. Then $M(z_{\mathcal{B}}) = \sum_{\beta \in G} z^{\beta}$ is the Asano contraction of $\{M(\mathcal{B}_i) = \sum_{\beta_i \in G_i} z^{\beta_i}\}$ if and only if the subgroup of $\mathcal{P}(\mathcal{B})$ generated by $\bigcup_i G_i^{\perp}$ coincides with $G^{\perp}$, i.e. $[\bigcup_i G_i^{\perp}] = G$.

The proof is the same as the one given in [108] for the case $G = \Gamma$.

In the following, any covering of $\mathcal{B}$ satisfying the condition of theorem 1 will be said to satisfy the Asano Condition. This theorem yields the following method to find coverings satisfying the Asano Condition :

i) given G, compute $G^{\perp}$

ii) with $G_i^{\perp}$ $i = 1, \cdots m$ any family of subgroups of $G^{\perp}$ which generates $G^{\perp}$, choose

$$\mathcal{B}_i = \{ B \in \bigcup_{\beta \in G_i^{\perp}} \beta \} \qquad i = 1, \cdots, m$$

iii) then $\mathcal{B} = \bigcup_{i=1}^{m} \mathcal{B}_i \bigcup_{k} \mathcal{B}_k$ where

$\bigcup_k \mathcal{B}_k$ is any covering of $\mathcal{B} / \bigcup_{i=1}^{m} \mathcal{B}_i$

Remarks :

1) If $G = \mathcal{K}$, then $G^{\perp} = \Gamma$. Since Γ is generated by $\{\gamma_r\}_{r \in \Lambda}$ and since $\bigcup_{r \in \Lambda} \bigcup_{B \ni r} B = \mathcal{B}$. point iii) can be disregarded.

2) If $G = \Gamma$, then $G^{\perp} = \mathcal{K}$. The set of bonds defined by means of the subgroups of $\mathcal{K}_\Lambda$ does not always give a covering of $\mathcal{B}$ and point iii) is relevant. In the extreme case $\mathcal{K} = \{\phi\}$, we have $\Gamma = \mathcal{K}^{\perp} = \mathcal{P}(\mathcal{B})$ and the covering sets are determined by iii) only (in this case

$$M(z_{\mathcal{B}_\Lambda}) = \sum_{\gamma \in \Gamma} z^{\gamma} = \sum_{\beta \in \mathcal{P}(\mathcal{B})} z^{\beta} = \prod_{B \in z_{\mathcal{B}}} (1 + z_B) \;).$$

3) A difference arises for the two cases of interest $G = \mathcal{K}$ and $G = \Gamma$. In fact in the latter case, the small polynomial is not, in general, the partition function of small subsystems as in the first case.

Among all possible coverings of $\mathcal{B}$ there exists simple coverings which are useful to discuss zeroes of the partition function because of the Poisson formulae which we introduce in the next section.

8.3. Properties of Small Polynomials

The Poisson formulae for abelian groups

$$\sum_{\beta \in G} f(\beta) = |G| \sum_{\bar{\beta} \in G^{\perp}} \tilde{f}(\bar{\beta}) \tag{8.3}$$

where $G^{\perp} = \{ \bar{\beta} \subset \mathcal{B} \, ; \, \sigma_\beta(\bar{\beta}) = 1 \;\; \forall \beta \in G \}$

and $\tilde{f}(\bar{\beta}) = 2^{-|\mathcal{B}|} \sum_{\beta \in \mathcal{P}(\mathcal{B})} f(\beta)\, \sigma_\beta(\bar{\beta})$

applied to the polynomial $M(z_{\mathcal{B}})$ gives

$$M(z_{\mathcal{B}}) = |G|\, 2^{-|\mathcal{B}|} \prod_{B \in \mathcal{B}} (1 + z_B) \sum_{\bar{\beta} \in G^{\perp}} \tilde{z}^{\bar{\beta}} = |G| \cdot 2^{-|\mathcal{B}|} \prod_{B \in \mathcal{B}} (1 + z_B)\, \tilde{M}(\tilde{z}_{\mathcal{B}}) \tag{8.4}$$

where $\tilde{z}_B = \dfrac{1 - z_B}{1 + z_B} \qquad z_B = \dfrac{1 - \tilde{z}_B}{1 + \tilde{z}_B} \equiv \tilde{\tilde{z}}_B$

and we have [109] :

Proposition 1

Let G be a subgroup of $\mathcal{P}(\mathcal{B})$, then the zeroes of $M(z_{\mathcal{B}}) = \sum_{\beta \in G} z^{\beta}$ such that $z_B \neq -1$ for all $B \in \mathcal{B}_\Lambda$ are related to the zeroes of the polynomial

$$\tilde{M}(\tilde{z}_{\mathcal{B}_\Lambda}) = \sum_{\bar{\beta} \in G^{\perp}} \tilde{z}^{\bar{\beta}}$$

by the transformation Eq. (8.4).

In particular for lattice systems $\{\Lambda, K\}$, the Poisson formulae Eq. (8.3) relates the H.T. expansion$(G = \mathcal{K}_\Lambda)$ to the L.T. expansion$(G = \Gamma_\Lambda = \mathcal{K}^{\perp})$ and conversely (+) [5]. Moreover, it provides an elegant method to discuss the zeroes of small polynomials defined by the following covering of $\mathcal{B}$: let $\{g_i^{\perp}\}$ be a set of minimal elements of $G^{\perp}$ which generates $G^{\perp}$, i.e. $g_i^{\perp} \in G^{\perp}$ such that $g^{\perp} \subset g_i^{\perp}$ and $g^{\perp} \in G^{\perp}$ implies $g^{\perp} = \phi$ or $g^{\perp} = g_i^{\perp}$. Then with $\mathcal{B}_i = g_i^{\perp}$ we have $G_i^{\perp} = \{\phi, g_i^{\perp}\}$ and proposition 1 relates the zeroes of $M(z_{\mathcal{B}_i})$ such that $z_B \neq -1$ directly to the zeroes of

$$\tilde{M}(\tilde{z}_{\mathcal{B}_i}) = 1 + \prod_{B \in g_i^{\perp}} \tilde{z}_B \tag{8.5}$$

where $\tilde{z}_B$ and z_B are related by Eq. (8.4).

Furthermore, to discuss the zeroes of $\tilde{M}(\tilde{z}_{\mathcal{B}_i})$ we have the following proposition which follows immediately from Eq. (8.5) and from the fact that the homographic transformation Eq. (8.4) mapps the line $\arg \tilde{z} = \delta$ onto the circle of center $z = i\,\mathrm{ctg}\,\delta$ which radius $r = (1 + \mathrm{ctg}^2 \delta)^{1/2}$ and the circle $|\tilde{z}| = \rho$ onto the circle of center $z = \frac{1+\rho^2}{1-\rho^2}$ with radius $r = \frac{2\rho}{1-\rho^2}$

(+) This relation will be further discussed in Sec.2.3 of Part III.

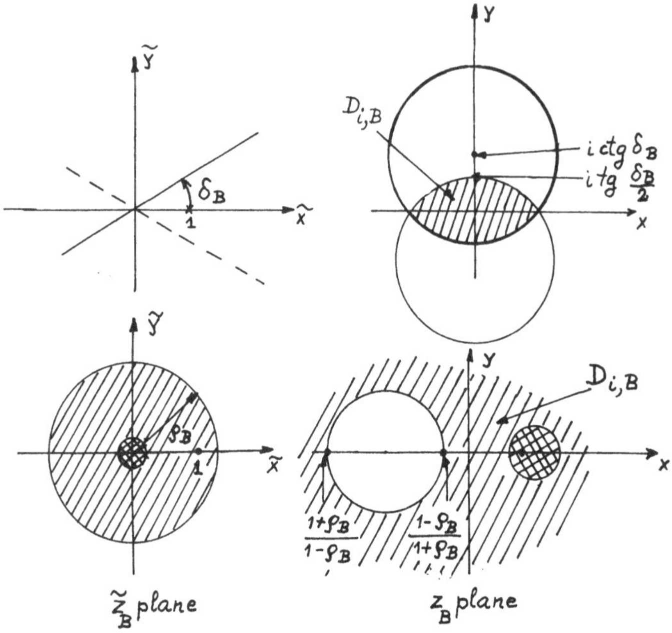

Fig. 8.1. : The homographic transformation Eq. (8.4) and domains $D_{i,B}$ such that $M(z_{\mathcal{B}_i}) \neq 0$ if $z_B \in D_{i,B}$

Proposition 2

For the covering $\mathcal{B} = \bigcup_i \mathcal{B}_i$ defined by means of minimal generators of $G^{\perp}$, then $M(z_{\mathcal{B}_i})$ is different from zero

1) either if $\delta_B^{min} \leq \arg \tilde{z}_B \leq \delta_B^{max} \quad \forall B \in \mathcal{B}_i$

where $(2k-1)\pi < \sum_{B \in \mathcal{B}_i} \delta_B^{min} < \sum_{B \in \mathcal{B}_i} \delta_B^{max} < (2k+1)\pi \quad k = 0,1,2,\cdots$

i.e. $z_B \in D_{i,B}$ for all $B \in \mathcal{B}_i$ where $D_{i,B}$ is the open set intersection of the interior of the two circles of center $z_B = \pm i\, ctg\, \delta_B$ and radius $r = (1 + ctg^2 \delta_B)^{1/2}$ with $\delta_B \geq 0$ and $\sum_{B \in \mathcal{B}_i} \delta_B = \pi$ (see Fig. 8.1.).

2) or if $|\tilde{z}_B| = \frac{1-z_B}{1+z_B} < \rho_B \quad \forall B \in \mathcal{B}_i$ where $\prod_{B\in\mathcal{B}_i} \rho_B = 1$

i.e. $z_B \in \mathcal{D}_{i,B}$ for all $B \in \mathcal{B}_i$ where $\mathcal{D}_{i,B}$ is the open set outside (resp. inside) of the circle centered at $z_B = \frac{1+\rho_B^2}{1-\rho_B^2}$ of radius $\frac{2\rho_B}{1-\rho_B^2}$ if $\rho_B > 1$ (resp. $\rho_B < 1$) with $\prod_{B\in\mathcal{B}_i} \rho_B = 1$.

Notice that in order to apply Ruelle's theorem, we must take $\delta_B > 0$ or $\rho_B > 1$ for all B undergoing contraction (otherwise $0 \in R_{i,B}$).

We emphasize that the hard point in applying the Asano-Ruelle method lies in the discussion of the small polynomials [106] : an arbitrary covering of $\mathcal{B}_\Lambda$ satisfying the Asano Condition gives in general small polynomials which are not easy to handle with, especially if one is interested in an optimal domain of analyticity.

8.4. Analyticity Properties of the Free Energy

Let $\{\mathcal{L}, K\}$ be an infinite system which is $\mathbb{Z}^\nu$ -invariant, i.e. $K(T_a B) = K(B)$ for all $a \in \mathbb{Z}^\nu$, $B \in \mathcal{B}$ (Sec. 2.6). The free energy of the infinite system is defined by

$$-\beta f = \lim_{\Lambda\to\mathcal{L}} \frac{1}{|\Lambda|} \cdot \log Z(\Lambda, K)$$

where $\Lambda \to \mathcal{L}$ in the sense of van Hove [1] . We then have (Ch. 2) :

$$-\beta f = \ln 2 + \sum_{B\in\mathcal{B}_0} \log \cosh K(B) + \lim_{\Lambda\to\mathcal{L}} \log \sum_{\beta\in\mathcal{K}_\Lambda} W(\beta)$$

$$= \sum_{B\in\mathcal{B}_0} K(B) + \lim_{\Lambda\to\mathcal{L}} \frac{1}{|\Lambda|} \log \sum_{\beta\in\Gamma_\Lambda} \omega(\beta)$$

where $\mathcal{B}_0$ is a fundamental set of bonds (Sec. 2.6).

We first consider the H.T. expansion of the finite system $\{\Lambda, \mathcal{B}_\Lambda\}$ (*) i.e. $G = \mathcal{K}$, $G^{\perp} = \Gamma$ and $z_B = \tanh K(B)$; for the covering $\mathcal{B}_\Lambda = \bigcup_{x\in\Lambda} \gamma(x)$, $\gamma(x) = \{B \in \mathcal{B}_\Lambda \,;\, B \ni x\}$, then $\tilde{M}(\tilde{z}_{\gamma(x)}) = 1 + \prod_{B\ni x} \tilde{z}_B$ (**) . In this case z_B can undergo contraction only if $|B| \geq 2$ and will undergo at most $|B|-1$ contractions. Using propositions 1 and 2, together with Ruelle's theorem, we conclude :

Theorem 2

1) $Z(\Lambda, \mathcal{B}_\Lambda) \neq 0$ if $|\arg e^{-2h}| \leq \delta_0$, $|\tanh K(B)| < \left(tg \frac{\delta_B}{2}\right)^{|B|}$ if $|B| \geq 2$

(8.6)

where $\{\delta_B\}_{B\in\mathcal{B}}$ are arbitrary real numbers such that

$$\delta_0 \geq 0 \;,\; \delta_B > 0 \quad \delta_{T_a B_0} = \delta_{B_0} \quad \forall a \quad \delta_0 + \sum_{\substack{B\ni r \\ |B|\geq 2}} \delta_B = \pi$$

2) $Z(\Lambda, \mathcal{B}_\Lambda) \neq 0$

if $|e^{-2h}| < \rho_0 < 1$; $z_B = \tanh K(B) \notin -[-\{z \,;\, |z - \frac{1+\rho_B^2}{1-\rho_B^2}|^2 < \frac{2\rho_B}{\rho_B^2 - 1}\}]^{*|B|}$

(8.7)

where $\delta_B > 1$ $\rho_{T_a B_0} = \rho_{B_0}$ and $\rho_0 \prod_{\substack{B\ni r \\ |B|\geq 2}} \rho_B = 1$

Using Vitali's theorem, we then have

Theorem 3

The free energy of $\{\mathcal{L}, K\}$ is an analytic function of the complex variables $z_B = \tanh K(B)$, $B \in \mathcal{B}_0$ in the domains defined by Eq. (8.6) or Eq. (8.7).

(*) $\mathcal{B}_\Lambda$ corresponds either to "open" or "γ" boundary conditions.

(**) We have assumed that $\{\Lambda, \mathcal{B}_\Lambda\}$ is such that $\gamma(x)$ is minimal, which is expected to be always satisfied for the type of boundary conditions considered.

Next we investigate the <u>L.T. expansion</u> of $\{\Lambda, \mathcal{B}_\Lambda\}$, i.e. $G = \Gamma$, $G^{\perp} = \mathcal{K}$ and $z_B = e^{-2K(B)}$. Taking the covering $\mathcal{B}_\Lambda = \bigcup_i \varkappa_i$ defined by minimal generators $\varkappa_i$ of $\mathcal{K}$, one has for the small polynomials

$$\tilde{M}(z_{\varkappa_i}) = 1 + \prod_{B \in \varkappa_i} \tilde{z}_B$$

and proposition 2 combined with Ruelle's theorem yields :

<u>Theorem 4</u>

$Z(\Lambda, \mathcal{B}_\Lambda) \neq 0$ in the domain

$|\arg \tanh K(B')| \leq \delta_{i,B'}$ if $z_{B'}$ does not undergo contraction

$$|z_B = e^{-2K(B)}| < \prod_{\varkappa_i \ni B} tg \frac{\delta_{i,B}}{2} \qquad \text{if } z_B \text{ undergoes contraction} \tag{8.8}$$

where $\delta_{i,B'} \geqslant 0$, $\delta_{i,B} > 0$ $\delta_{i,T_a B_o} = \delta_{i,B_o}$, $\sum_{B \in \varkappa_i} \delta_{i,B} = \pi$

To obtain analyticity of the free energy in the thermodynamic limit in the low temperature region, it is necessary to choose the generators $\varkappa_i$ such that the number of contractions of a given bond becomes independent of Λ if Λ is large enough. Clearly, we can always find generators $\varkappa_i$ having this property but one has to show that they generate $\mathcal{K}$. This problem was raised by J. Slawny and solved for some particular models [108] . Later W. Holsztyński [110] showed that the H.T. group of systems $\{\mathbb{Z}^\nu, K\}$ with translationally invariant interactions can always be generated by such elements. Again, with Vitali's theorem we conclude:

<u>Theorem 5</u>

The free energy of $\{\mathbb{Z}^\nu, K\}$ (*) is an analytic function of

---

(*) It is expected that theorem 5 also holds for systems $\{\mathcal{L}, K\}$ with translationally invariant interactions.

the complex variables $z_B = e^{-2K(B)}$, $B \in \mathcal{B}_0$, in the domain defined by Eq. (8.8).

8.5 Examples

In this section, we briefly illustrate the results expressed in Theorem 2 and 4 with two examples. For further examples we refer to [106-109].

Example 1 : ν - dimensional Ising model with translationally invariant nearest neighbour interactions and magnetic field.

For this model, the fundamental family of bonds is given by

$$\mathcal{B}_0 = \{(x_0), (x_0, y_1), \cdots\cdots, (x_0, y_\nu)\}$$

where (x_0, y_i), $i = (1, \cdots, \nu)$ are the ν independent nearest neighbour bonds pointing from x_0 with $\delta_{T_a(x_0, y_i)} = \delta$ for all $a \in \mathbb{Z}^\nu$ and $i = 1, 2, \cdots \nu$, Theorem 2 reads :

Corollary :

1) For the ν dimensional Ising model with translationally invariant nearest neighbour interactions J and external field h, $M(z_{\mathcal{B}_\Lambda}) \neq 0$ for complex h, J such that

$$|\arg e^{-2\beta h}| \leq \delta_0 \quad |\tanh \beta J| < \left(tg\left(\frac{\pi - \delta_0}{4\nu}\right)\right)^2$$

2) For $h \in \mathbb{R}$ the free energy is analytic in $x = \tanh \beta J$ in the domain $|x| < \left(tg \frac{\pi}{4\nu}\right)^2$

3) For real interactions J such that $|\tanh \beta J| < (tg\,\delta)^2$ the free energy is analytic in $z = e^{-2\beta h}$ if $|\arg z| < \pi - 4\nu\delta$.

In particular for $\nu = 2$ and $J \in \mathbb{R}$, $|\tanh \beta J| < (\sqrt{2}-1)^2$ yields $e^{-2\beta J} > 1/\sqrt{2}$ and we recover a result of Sarbach and Rys.[111] Moreover for $\nu = 2$ and $h = 0$ we recover Ruelle's result (proposition 2.4 in [106,b]); indeed, the self-duality of the model and the analyticity for $\tanh \beta J < (\sqrt{2}-1)^2$ implies analyticity for $|e^{-2\beta J}| < (\sqrt{2}-1)^2$.

Example 2 Triangular model with 3-body forces and external field. (See Ch.2) We consider the case of open boundary condition and take $\delta_B = \delta$ for all $B \in \mathcal{B}$ with $|B| = 3$; Theorem 2 yields :

Corollary

1) For the triangular model with 3-body forces J and external field h, $M(z_{\mathcal{B}}) \neq 0$ for complex h, J such that

$$|\arg e^{-2\beta h}| \leq \delta_0 \quad |\tanh \beta J| < \left(tg\left(\frac{\pi - \delta_0}{6}\right)\right)^3$$

2) For $h \in \mathbb{R}$ the free energy is analytic in $x = \tanh \beta J$ in the domain $|x| < (2-\sqrt{3})^3$.

3) For $h = 0$, the free energy is analytic in the domain $|\tanh \beta J| < (2-\sqrt{3})^3$ and $e^{-2\beta|J|} < (2-\sqrt{3})^3$.

(The last statement (3) of the corollary follows from the self duality of the model with h=0).

To derive analyticity properties in $z = e^{-2\beta h}$ consider the generators $\mathcal{X}_a$ of $\mathcal{K}$ defined by :

$$\mathcal{X}_a = \{ B_a, r_{a_1}, r_{a_2}, r_{a_3} \}, \quad |B_a| = 3 \quad r_{a_i} \in B_a \quad i = 1,2,3$$

Since B_a does not undergo contraction, Theorem 5 implies

Corollary

For the triangular model with ferromagnetic 3-body interaction the free energy is analytic in $z = e^{-2\beta h}$ for $|z| < 3^{-3}$.

Moreover, using the symmetry relation $Z(h,-K) = Z(-h,K)$ we conclude analyticity of the free energy of the anti-ferromagnetic model for $|z| > 3^{-3}$.

8.6 Conclusions

Theorem 2 and 4 yield general analyticity domains for finite systems $\{\Lambda, K\}$ at high and low temperatures. They imply by means of Vitali's theorem analyticity domains in the high temperature region for the free energy of arbitrary infinite systems $\{\mathcal{L}, K\}$ provided the free energy is well defined. General analyticity domains for $\{\mathcal{L}, K\}$ in the L.T. region are limited to systems such that $\mathcal{K}$ can be generated by finite elements [108, 110]. This is true for systems $\{\mathcal{L}, K\}$ having translationally invariant interactions and the methods presented in this chapter are not limited to spin $\frac{1}{2}$ systems. They can be generalised to spin $\frac{1}{2}$ systems with constraints and to higher spin systems (part II and III).

The next chapter is devoted to an extension of the above method to discuss analyticity of the correlation functions and unicity of states of $\{\mathcal{L}, K\}$.

CHAPTER 9 - ANALYTICITY AND UNIQUENESS OF THE INVARIANT EQUILIBRIUM STATE

9.1. Introduction

In this chapter we combine the results of the last chapter with the following proposition of J.Slawny [108] valid for $\mathbb{Z}^\nu$ invariant systems : With $X \in \mathcal{P}_f(\mathcal{L})$, let K_X be the function on $\mathcal{P}_f(\mathcal{L})$ defined by : $K_X(Y) = 1$ if $Y \in \{T_a X\}_{a\in\mathbb{Z}^\nu} = [X]$, $K_X(Y) = 0$ otherwise; if the free energy $p(K + \lambda K_X)$ is differentiable in λ at $\lambda = 0$ then for all $\mathbb{Z}^\nu$ invariant equilibrium states ω we have

$$\omega[\sigma_X] = \frac{d}{d\lambda} p(K + \lambda K_X)\Big/_{\lambda=0} \qquad (9.1)$$

The results we shall derive in this chapter are different in the two temperature regions : in the high temperature region we obtain analyticity and uniqueness of the translation invariant equilibrium state for any $\mathbb{Z}^\nu$ invariant system; on the other hand in the low temperature region we obtain analyticity and uniqueness of the invariant equilibrium state (i.e. state invariant under the full symmetry group of the hamiltonian - (Ch. 7)) but only for ferromagnetic crystal lattice $\mathcal{L} = \mathbb{Z}^\nu$. The bounds are then compared with the results obtained by other methods for the Ising ferromagnet.

9.2. Unicity of the $\mathbb{Z}^\nu$-invariant equilibrium state at high temperatures

To study the unicity of the equilibrium state of systems $\{\mathcal{L}, K\}$ at high temperatures, we apply theorem 3 of Sec. 8.4. to all systems $\{\mathcal{L}, K+\lambda K_X\}$, with $X \in \mathcal{P}_f(\mathcal{L})$. For these systems, the fundamental family of bonds (Sec. 2.6.) is $\mathcal{B}_o' = \mathcal{B}_o \cup X$. In the following, any primed quantity is taken with respect to $\mathcal{B}_0'$ rather than $\mathcal{B}_0$. Since $\mathcal{K}'^{\perp} = \Gamma'$ is always generated by $\{\gamma'_r\}_{r\in\Lambda}$ we obtain

Theorem 1

For $\{\mathcal{L}, K\}$ a $\mathbb{Z}^\nu$ invariant system with finite range interactions

1) at high temperatures, i.e. at temperatures T such that

$$\left|\tanh \frac{J(B)}{kT}\right| < \left(tg\, \frac{\delta_B}{2}\right)^{|B|}, \quad \sum_{\substack{B\ni r\\ |B|\geq 2}} \delta_B = \pi$$

there exists a unique equilibrium state ω invariant under translations. Moreover the correlation functions $\omega[\sigma_X]$ are analytic functions of z_B, $B\in\mathcal{B}_0$ in the domain defined by Eq.(8.6.) (*).

2) For systems with non zero external magnetic field such that $z = |e^{-2h}| < \rho_0 < 1$, there exists a unique equilibrium state ω invariant under translations in the domain defined by Eq. (8.7.). Moreover in this domain the correlation functions are analytic functions of z_B, $B\in\mathcal{B}_0$.

Proof :

1) Follows from Eq. (9.1.) and the fact that for all $X\in\mathcal{P}_f(\mathcal{L})$ and any $\varepsilon>0$, the free energy $p(K+\lambda K_X)$ is analytic in the domain

$$|\arg e^{-2h}| \leq \delta_0, \quad |\tanh K(B)| < \tanh\left(\frac{\delta_B}{2}\right)^{|B|} \text{ if } |B|\geq 2$$

$$|\tanh \lambda| < \left(\tanh \frac{\varepsilon}{2q}\right)^{|X|}$$

where $\delta_0 + \sum_{B\ni r, |B|\geq 2} \delta_B = \pi - \varepsilon$ and $q =$ number of translates of X containing the site r.

(*) We assumed for all $X\in\mathcal{P}_f(\mathcal{L})$ that $\{\Lambda, K+\lambda K_X\}$ is such that $\gamma(x)$ is minimal for all $x\in\Lambda$.

Therefore $\omega[\sigma_X] = \frac{d}{d\lambda} p(K+\lambda K_X)\Big|_{\lambda=0}$ has the same value for all translationally invariant equilibrium states. Letting $\varepsilon \to 0$ we conclude the proof.

2) This part is obtained in a similar manner taking

$$\rho_0 = \prod_{\substack{B \ni r \\ |B| \geq 2}} \rho_B = \left(\frac{1}{\rho}\right)^q \qquad |\tanh \lambda| < \frac{\rho-1}{\rho+1}$$

and letting $\rho \searrow 1$.

As a consequence of part 2) of the above Theorem, we also recover Ruelle's result [59] concerning the unicity of the translationally invariant equilibrium state for ferromagnetic systems with two body forces and non zero external field (see ref. [109]).

9.3. Unicity of the invariant equilibrium state at low temperature for ferromagnetic systems

As in the previous section, analyticity of the correlation functions of the system $\{\mathcal{L}, K\}$ at low temperature is shown by a discussion of $\{\mathcal{L}, K+\lambda K_X\}$. First we have to find for any $X \in \mathcal{P}_f(\mathcal{L})$ a set of minimal generators of $\Gamma'^{\perp} = \mathcal{K}'$. In the following, it is essential that for any $a \in \mathbb{Z}^\nu$ the bond $T_a X$ do not undergo contraction: i.e. $T_a X$ belongs to one and only one generator of $\mathcal{K}'$. This is achieved if there exist $B_1, \cdots, B_m$ such that $(X, B_1, \cdots B_m) = \kappa' \in \mathcal{K}'$, i.e. if $X \in \overline{\mathcal{B}}$. In conclusion we can apply Theorem 5 of Sec. 8.4. only to systems $\{\mathcal{L}, K+\lambda K_X\}$ with $X \in \overline{\mathcal{B}}$. Therefore analyticity of the correlation functions and unicity of the $\mathbb{Z}^\nu$ invariant equilibrium state can only be shown for symmetric states. Moreover the use of Theorem 5 limits our considerations to ferromagnetic crystal systems $\{\mathbb{Z}^\nu, K\}$. The analogue of Theorem 1 reads at low temperatures:

Theorem 2

Let $\{\mathbb{Z}^\nu, K\}$ be a $\mathbb{Z}^\nu$- invariant, ferromagnetic crystal systems with finite range interactions; at low temperatures such that

$$e^{-2\frac{J(B)}{kT}} < \prod_{\mathcal{H}_i \ni B} \tanh\left(\frac{\delta_{i,B}}{2}\right) \quad \text{with} \quad \sum_{B \in \mathcal{H}_i} \delta_{i,B} = \pi$$

there exists a unique invariant equilibrium state ω. Moreover the correlation functions $\omega[\sigma_X]$ are analytic functions of z_B , $B \in \mathcal{B}_0$, in the domain defined by Eg. (8. 8)

Proof :

This follows from Eg. (9.1.) and from the fact that for any $X \in \bar{\mathcal{B}}$ the group $\mathcal{K}'$ associated with $\mathcal{B}' = \mathcal{B} \cup \{T_a X\}$ is generated by $\mathcal{K}$ and $T_a \mathcal{H}'$ where $\mathcal{H}' = (X, B_1, \cdots B_m)$. Therefore

$$M(z_{\mathcal{B}}) \neq 0 \quad \text{for} \quad |z_B| < r_B \quad \forall B \in \mathcal{B}$$

and

$$\tilde{M}(\tilde{z}_{\mathcal{H}'}) \neq 0 \text{ for } \quad |\tilde{z}_X| < \bar{\rho} < 1 \quad |\tilde{z}_{B_i}| < \bar{\rho}^{-\frac{1}{m}}$$

yields

$$\tilde{M}(z_{\mathcal{B}'}) \neq 0 \quad \text{if} \quad \begin{cases} |z_B| < r_B & B \notin \bigcup_{i=1}^{m} [B_i] \\ |z_{B_i}| < \left|\frac{1-\bar{\rho}^{-\frac{1}{m}}}{1+\bar{\rho}^{-\frac{1}{m}}}\right|^{n_B} \cdot r_B \, , & B \in \bigcup_{i=1}^{m} [B_i] \\ |z_Y - 1| < \bar{\rho} & Y \in [X] \end{cases}$$

and n_B is the number of generators $\mathcal{H}_i$ containing the bond $\mathcal{B}$. Letting $\bar{\rho} \searrow 0$ we conclude that the unicity of the equilibrium state ω invariant under translation and such that $\omega[\sigma_X] = 0$ if $X \notin \bar{\mathcal{B}}$ which is precisely the unicity of the invariant equilibrium state.

We remark that if X undergo contraction, no conclusion can be drawn in the limit $\lambda \to 0$.

9.4. Comparison with other results

To compare the domains obtained by the general Asano-Ruelle method with those given for specific models by other techniques, we consider the 2-dimensional Ising model with nearest neighbour interactions and zero magnetic field.

For this model, Theorem 1 yields unicity of the $\mathbb{Z}^2$ invariant state for $T > T_1$ where $\tanh \beta_1 J < (tg \frac{\pi}{8})^2 = 0{,}17$. This result is not as good as the one given in Chap. 6 where $\tanh \beta J < 0{,}32$. (see applications). On the other hand, Theorem 2 gives unicity of the symmetric $\mathbb{Z}^2$ invariant state for $T < T_2$ where $e^{-2\beta_2 J} < (tg \frac{\pi}{8})^2 = 0{,}17$; again this result can be compared with the one given by Gallavotti-Miracle-Sole [96] i.e. $e^{-2\beta J} < 0{,}38$.

PART II

SPIN $\frac{1}{2}$ LATTICE SYSTEMS WITH CONSTRAINTS

CHAPTER 1 - DEFINITIONS AND GROUP STRUCTURE FOR SYSTEMS WITH CONSTRAINTS.

1.1. Definition and General Properties. [113]

Spin lattice systems with constraints are systems such as those discussed in Part I but submitted to the condition that not all spin configurations are allowed; the set of admissible configurations is then characterised by the "constraints". These systems appear for example in the study of lattice-gas with extended hard core, ferro-electric models, models supporting metastable states [114] ; one arrives also at such systems in the study of duality when applied to models with some interactions going to zero (e.g. interaction introduced to break a symmetry), or in the study of lattice systems by means of the renormalisation group techniques; recently systems with constrains have also been introduced in the investigation of Gauge Field on a lattice [115] .

In this Part II we shall show that it is possible to extend the techniques of Part I to discuss such systems; in fact the only new idea is to consider the family of bonds associated with the constraints - or infinite bonds - together with the family of bonds associated with the interactions - or finite bonds; the groups of interest are then introduced as before by means of the kernel and image of two group homomorphisms. However the results which we shall derive will not be as general as those of Part I and in each case it will be necessary to specify the class of systems for which they hold; in particular the H.T. expansion will be useful only for those systems such that the admissible configurations is a group;

on the other hand the unicity of the invariant equilibrium state will be obtained only for a class of systems which contains for example all the systems with hard core in an external field or the previously mentioned class of systems whose admissible configurations is a group.

A spin $\frac{1}{2}$ lattice system with constraints $\{\mathcal{L}, \mathcal{C}, K\}$ is defined as a countable set $\mathcal{L}$ of points in $\mathbb{R}^{\nu}$ - the "lattice $\mathcal{L}$" - together with a function $\mathcal{C} : \mathcal{P}(\mathcal{L}) \to \{0,1\}$ - the "constraints" - defining the admissible configurations, and a real or complex function $K: \mathcal{P}_f(\mathcal{L}) \to \mathbb{C}$ defining the "interactions", such that $|K(X)| < \infty$ for all $X \in \mathcal{P}_f(\mathcal{L})$.

The admissible configurations of the system are defined by :

$$\mathcal{P}^a(\mathcal{L}) = \{X \subset \mathcal{L}; \ \mathcal{C}(X) = 1\} \tag{1.1}$$

and for any finite $\Lambda \subset \mathcal{L}$, the hamiltonian $H_{\Lambda,Y}$ of the finite system Λ with boundary condition $Y \subset \mathcal{L}$ is given by :

$$-\beta H_{\Lambda,Y} = \sum_{\substack{B \in \mathcal{B} \\ B \cap \Lambda \neq \phi}} K(B)\, \sigma_B(Y^c)\, \sigma_B \tag{1.2}$$

$$Y^c = Y \cap (\mathcal{L}/\Lambda)$$

where σ_B is the function defined in chapter 1 of Part I, and $\mathcal{B}$ is the set of finite bonds defined by the support of K :

$$\mathcal{B} = \{B \in \mathcal{P}_f(\mathcal{L}) \ ; \ K(B) \neq 0\} \tag{1.3}$$

The internal symmetry group $\mathcal{S}$ for the interactions is defined by the condition

$$\forall s \in \mathcal{S} \qquad H_{\Lambda,Y}(X) = H_{\Lambda,sY}(sX) \qquad \forall \ \Lambda \in \mathcal{P}_f(\mathcal{L}),\ Y \subset \mathcal{L},\ X \subset \Lambda$$

and is given by :

$$\mathcal{S} = \{ S \subset \mathcal{L}; \ \sigma_B(S) = +1 \ \forall \ B \in \mathcal{B}\} = \mathcal{B}^{\perp} \tag{1.4}$$

Similarly we introduce the internal symmetry group $\mathcal{S}_\infty$ for the constraints by the condition

$$\forall \ s_\infty \in \mathcal{S}_\infty \qquad \mathcal{C}(X) = \mathcal{C}(s_\infty X) \qquad \forall \ X \subset \mathcal{L} \tag{1.5}$$

and the <u>internal symmetry group $\mathcal{S}'$</u> for the system by

$$\mathcal{S}' = \mathcal{S} \cap \mathcal{S}_\infty \tag{1.6}$$

It follows from these definitions that for any $X_0 \in \mathcal{P}^a(\mathcal{L})$

$$\mathcal{P}^a(\mathcal{L}) \supset X_0 \cdot \mathcal{S}_\infty \supset X_0 \cdot \mathcal{S}' \tag{1.7}$$

For infinite systems the constraint $\mathcal{C}$ is assumed to satisfy the following locality condition : there exists a family $\{LC\}$ of <u>Local Constraints</u> $\mathcal{C}_C$ where $C \in \{LC\} \subset \mathcal{P}_f(\mathcal{L})$ has finite diameter and

$$\mathcal{C}_C(X) = \mathcal{C}_C(X \cap C) \tag{1.8}$$

such that :

$$\mathcal{C} = \prod_{C \in \{LC\}} \mathcal{C}_C \tag{1.9}$$

In this case $\mathcal{P}^a(\mathcal{L})$ is a compact subset of $\mathcal{P}(\mathcal{L})$ (*) and

$$\mathcal{P}^a(\mathcal{L}) = \{X \subset \mathcal{L} \ ; \ \mathcal{C}_C(X) = 1 \quad \text{for all} \quad C \in \{LC\}\} \tag{1.10}$$

Without any loss of generality [113] we can always assume that for any local constraint $\mathcal{C}_C$ and any $X' \subset C$, there exists $X'' \subset \mathcal{L}/C$ such that $\mathcal{C}(X'X'') = \mathcal{C}_C(X')$.

By analogy with Eq. (1.2), the Fourier decomposition $\mathcal{C}_C = \sum_{Y \subset C} \tilde{\mathcal{C}}_C(Y)\sigma_Y$ defines the <u>set $\mathcal{B}_\infty$ of infinite bonds</u> and the <u>set $\mathcal{B}'$ of bonds</u> as

$$\mathcal{B}_\infty = \{B_\infty \in \mathcal{P}_f(\mathcal{L}) \ ; \ \tilde{\mathcal{C}}_C(B_\infty) \neq 0 \quad \text{for some} \quad C \in \{LC\}\} \tag{1.11}$$

$$\mathcal{B}' = \mathcal{B} \cup \mathcal{B}_\infty$$

Again "bonds" and "internal symmetry group" are related by the following equation which corresponds to Eq. (1.4)

$$\mathcal{S}_\infty = \mathcal{B}_\infty^\perp \qquad\qquad \mathcal{S}' = \mathcal{B}'^\perp \tag{1.12}$$

and moreover we have the following property [113] :

(*) Note that $\mathcal{P}(\mathcal{L})$ is compact for the topology induced by $\{-1, +1\}^{|\mathcal{L}|}$ and $\mathcal{C}_C$ is continuous.

Property 1

1. $\mathscr{C}_C(X) = \mathscr{C}_C(XS)$ for all $X \subset \mathscr{L}$ and all $C \in \{LC\}$ iff $S \in \mathscr{S}_\infty$

2. The following conditions are equivalent :
   i) $\mathscr{P}^a(\mathscr{L})$ is a subgroup of $\mathscr{P}(\mathscr{L})$
   ii) $\mathscr{P}^a(\mathscr{L}) = \mathscr{S}_\infty$
   iii) $\mathscr{P}^a(\mathscr{L}) \subset \mathscr{S}_\infty$
   iv) The local constraints can be chosen of the form $\mathscr{C}_{B_\infty} = \frac{1}{2}(1 + \sigma_{B_\infty})$

Remarks

We shall always assume that, for infinite systems, there exists an action of $\mathbb{Z}^\nu$ on $\mathscr{L}$, i.e. for all $a \in \mathbb{Z}^\nu$ and $x \in \mathscr{L}$, $\tau_a x \in \mathscr{L}$. Moreover in most cases the local constraints will be defined by means of a fundamental family of local constraints $\{LC\}_0$, i.e. for any local constraint $\mathscr{C}_C$ there exists $C_0 \in \{LC\}_0$ and $a \in \mathbb{Z}^\nu$ such that $\mathscr{C}_C(X) = \mathscr{C}_{C_0}(\tau_a^{-1} X)$.

In the next section we shall show how some well known models can be rephrased within our general formalism.

1.2. Examples of Systems with Constraints

Considering two different models with constraints, e.g. a ferro-electric model and a hard core model, there is at first sight no common structure. The aim of this section is to show that a large class of models with constraints have indeed a common basic structure, which is the one introduced in the preceding section. For different models with constraints, we shall now explicitely compute the local constraints, the infinite bonds and the Hamiltonian of the equivalent model in terms of spin, called the Ising equivalent model.

1.2.1. Systems Satisfying the Subgroup Property

A system $\{\mathcal{L}, \mathcal{C}, k\}$ satisfies the <u>subgroup property</u> if $P^a(\mathcal{L})$ is a subgroup of $P(\mathcal{L})$; from property 1 (Sec. 1.1), this is equivalent to

$$\mathcal{C}_{B_\infty} = \tfrac{1}{2}(1 + \sigma_{B_\infty}) \qquad \text{for all } B_\infty \in \mathcal{B}_\infty \qquad (1.13)$$

and therefore these systems can be considered as limit of systems without constraints but with interactions K' such that $K'(B_\infty) \to +\infty$ for all $B_\infty \in \mathcal{B}_\infty$.

Example 1

Let us consider the model $\{\mathbb{Z}^2, \mathcal{C}, k\}$ defined on the square lattice $\mathbb{Z}^2$ with four body constraints distributed on the shaded squares as defined on Fig. 1 a) and finite magnetic field. For this model a fundamental family $\{\mathcal{L}c\}_0$ is respresented by a shaded square.

a)

$$\mathcal{B} = \{(x)\}_{x \in \mathbb{Z}^2}$$

$$\mathcal{B}_\infty = \{ B_\infty = \boxed{\diagup} \}$$

$$\mathcal{C}_{B_\infty} = \tfrac{1}{2}(1 + \sigma_{B_-})$$

b)

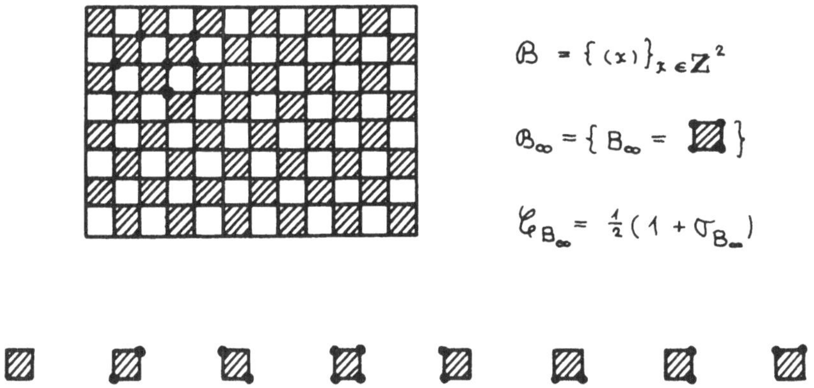

Fig. 1 : a) Distribution of local constraints (shaded squares) on $\mathbb{Z}^2$ together with an inadmissible configuration $X \notin P^a(\mathbb{Z}^2)$ (a dot on site x means $\sigma_x(X) = -1$).

b) List of configurations $X \cap B_\infty$ satisfying $\mathcal{C}_{B_\infty}(X) = 1$.

By definition (Eq. 1.10), any admissible configuration X must satisfy $\frac{1}{2}\left(1+\sigma_{B_\infty}(X)\right)=1$, *i.e.* $\sigma_{B_\infty}(X)=1$ and therefore one must have on each shaded square one of the eight local configurations of Fig. 1 b).

Property 1 of section 1.1 gives $\mathcal{S}_\infty = \mathcal{P}^a(\mathbb{Z}^2)$, the presence of an external field implies $\mathcal{S} = \{\phi\}$, hence $\mathcal{S}' = \{\phi\}$. Further properties of this model [26,113] will be discussed in Chapter 4.

Example 2

Let us discuss the model $\{\mathbb{Z}^2, \mathcal{C}, K\}$ with nearest neighbour (n.n.) interactions defined on the same lattice but with constraints such that only the first four configurations listed in Fig. 1 b) are locally admissible configurations on the shaded squares. Taking $C =$  $= (1, 2, 3, 4)$, we have :

$\tilde{\mathcal{C}}_C(\phi) = \frac{1}{4}$; $\tilde{\mathcal{C}}_C(i) = \tilde{\mathcal{C}}_C(ij) = \tilde{\mathcal{C}}_C(ijk) = 0$ i,j,k $\in$ C and i,j = n.n.

$\tilde{\mathcal{C}}_C(ik) = \tilde{\mathcal{C}}_C(C) = \frac{1}{4}$, i,k $\in$ C i,k = next nearest neighbour (n.n.n.)

hence

$$\mathcal{C}_C = \frac{1}{4}\left(1+\sigma_{13}+\sigma_{24}+\sigma_{1234}\right) = \frac{1}{2}\left(1+\sigma_{13}\right)\cdot\frac{1}{2}\left(1+\sigma_{24}\right)$$

and the model satisfies the subgroup property; moreover, the simplest choice of the set of infinite bonds is :

$$\mathcal{B}_\infty = \{B_\infty = (xy) \ ; (xy) = \text{n.n.n. on shaded squares}\}$$

Again $\mathcal{S}_\infty = \mathcal{P}^a(\mathcal{L})$ whereas $\mathcal{S} = \{\phi, \mathbb{Z}^2\}$, therefore $\mathcal{S}' = \mathcal{S}$.

1.2.2. General Hard Core Lattice Systems

Another important type of constraints (which do not yield a subgroup of admissible configurations) are the "Hard Core Constraint", constraints satisfying the condition that

$\mathcal{C}(X) = 1$ implies $\mathcal{C}(Y) = 1$ for all $Y \subset X$.

Let us note that we can always choose the family $\{L\ C\}$ such that for any $C \in \{LC\}$, $\mathscr{C}_C(C) = 0$ and $\mathscr{C}_C(T) = 1$ for all $T \subset C$, $T \neq C$. In this case the local constraints can be written as

$$\mathscr{C}_C = 1 - \prod_{x \in C} \tfrac{1}{2}(1 - \sigma_x) \qquad (1.14)$$

and therefore, the set of infinite bonds becomes

$$\mathscr{B}_\infty = \{ B_\infty \subset C \ ; \ C \in \{LC\} \} \qquad (1.15)$$

thus implying $\mathscr{S}_\infty = \{\phi\}$ and $\mathscr{S}' = \{\phi\}$.

Example 3

The simplest hard core systems are those having "two point exclusions" (*) on an arbitrary lattice $\mathscr{L}$. Then any $C \in \{LC\}$ is of the form $C = (x,y)$, representing an "exclusion" between the sites x, y and Eq. (1.14) gives

$$\mathscr{C}_{(x,y)} = \tfrac{1}{4}\left(3 + \sigma_x + \sigma_y - \sigma_{xy}\right)$$

This type of hard core systems can also be viewed as a limiting case of an antiferromagnetic two spin interacting Ising model with a field. $\mathscr{C}_C = \lim_{K \to +\infty} e^{K(\mathscr{C}_C - 1)}$ indicates that the field and the antiferromagnetic two body coupling constant have to go to infinity with ratio minus one in order to obtain this hard core constraint.

In particular we shall consider in the following the special model defined by the lattice $\mathbb{Z}^2$ and the two point "exclusions" between nearest neighbour (n.n.) and some next nearest neighbour (n.n.n.) as defined on Fig. 2 a). As we shall see this model with constraints can also be interpreted as a vertex model on the square lattice shown in Fig. 2 b). Moreover we shall show in Sec. 5.2.3 that this model has no phase transition i.e. its free energy density is analytic.

(*) In the spin language "two points (x,y) exclusion" means that $\sigma_x = -1$ implies $\sigma_y = +1$ and $\sigma_y = -1$ implies $\sigma_x = +1$.

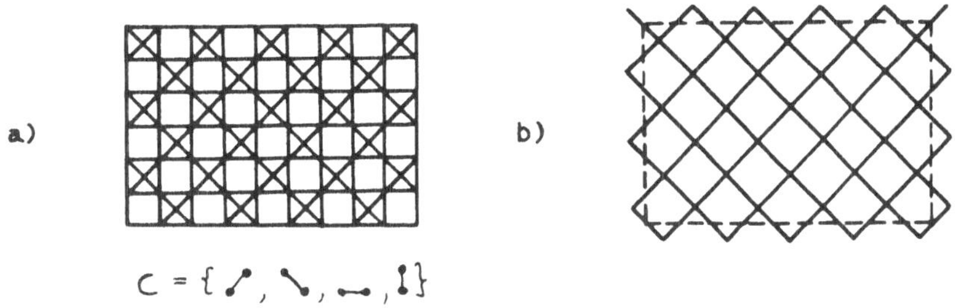

Fig. 2 : a) square lattice $\mathcal{L}$ with two point exclusions $\mathcal{C}_c$

b) lattice $\mathcal{L}_1$ of the equivalent vertex model.

1.2.3. K - Vertex Models k ≤ 16

Eight and sixteen vertex models and their relationship with Ising models with up to four spin interactions have been most completely discussed in the review article of E. Lieb and F.Y. Wu [116] . The general sixteen and eight vertex models are equivalent to Ising models without constraints. The Ising equivalent of the 16-vertex model has 1,2,3 and 4 spin interactions whereas the 8-vertex model is equivalent to an Ising model with 2 and 4 spin interactions only. It was pointed out in [116] that any 6-vertex model is equivalent to an Ising model with infinite interactions. For this reason, this correspondance has not been considered very useful. However in our approach we separate interactions and constraints and this equivalence remains of interest.

Here we focus our attention on the Ising equivalent model of 5-, 6-, and 7 vertex models defined on a square lattice (and we shall look at these k - vertex models as Ising model with constraints and finite interactions). Generalisations of the method to triangular lattices and lattices of higher dimensions are immediate.

As in [116] , a vertex model is defined on a square lattice

$\mathcal{L} = \mathbb{Z}^2$ and a configuration is given by arrow arrangements on the edges of the lattice. The two possible arrow configurations correspond to the two spin variable values ± 1 defined on a lattice point of the lattice $\mathcal{L}_1$ whose points are located on the centers of the edges of $\mathcal{L}$. This situation, illustrated on Fig. 3 a), gives the usual one to one correspondence between arrow configurations on $\mathcal{L}$ and spin configurations on $\mathcal{L}_1$.

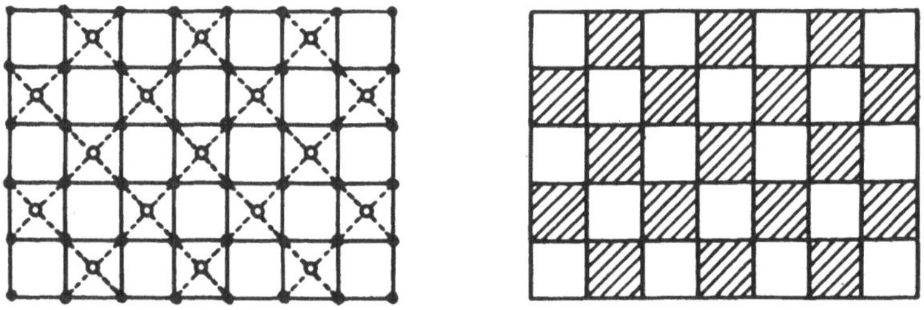

Fig. 3 : a) The arrow lattice $\mathcal{L} = \{\circ\}$ and the Ising lattice $\mathcal{L}_1 = \{\bullet\}$

b) distribution of constraints (shaded squares) on $\mathcal{L}_1$.

On the 4 edges joining a given vertex of $\mathcal{L}$, there are 16 different arrow configurations to which we can associate 16 spin configurations $X_j \subset \mathcal{L}_1$, $j = 1, 2, \ldots, 16$ (Fig. 4); the most general vertex model has 16 energy parameters e_j with Boltzmann factor $w_j = e^{-\beta e_j}$.

General k - vertex models are obtained by taking 16-k of these Boltzmann factors equal to zero, i.e. 16-k vertex configurations are not allowed. In terms of the Ising equivalent, this can be done by introducing the local constraints $\mathcal{C}_C$ defined on the shaded squares $(|C| = 4)$ of the lattice $\mathcal{L}_1$ as shown in Fig. 3 b). The special hard core model defined in example 3 is thus a 5 - vertex model with $w_1, w_9, w_{10}, w_{11}, w_{12} \neq 0$.

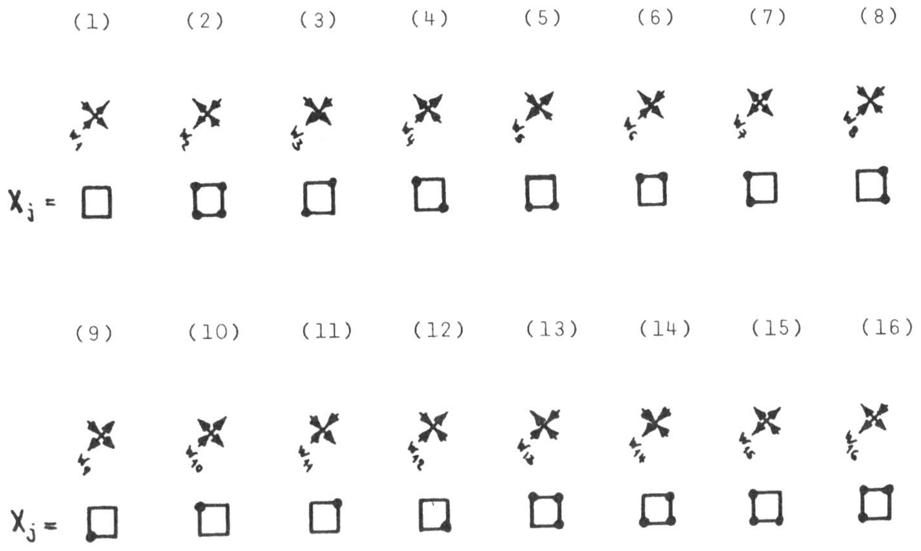

Fig. 4 : Vertex and corresponding Ising configurations X_j. An arrow projection pointing to the left corresponds to $\sigma_x = -1$. A dot on $\mathcal{L}_1$ represents a spin value $\sigma_x = -1$ ($\sigma_x = +1$ otherwise).

Example : General 6 - vertex model

The general 6 vertex model is defined by $w_{i_1}, \dots, w_{i_6} \neq 0$ $w_{i_7} = \dots = w_{i_{16}} = 0$ for any choice of 6 different indices $i_1, \dots, i_6 \subset \{1, \dots, 8\}$. To fix the ideas, we consider the standard 6 - vertex model, where $\omega_7 = w_8 = 0$ (for any other choice of $i_1, \dots, i_6$, we can repeat step by step the discussion below). With $C = (1\,2\,3\,4) = {}_{1}^{4}\square_{2}^{3}$ the local constraint $\mathcal{C}_C$ is defined by :

$$\mathcal{C}_C(\phi) = \mathcal{C}_C(C) = \mathcal{C}_C(13) = \mathcal{C}_C(24) = \mathcal{C}_C(12) = \mathcal{C}_C(34) = 1$$

and $\mathcal{C}_C(T) = 0$ for all other $T \subset C$.

Computing the Fourier coefficients of $\mathscr{C}_C$, we find

$$\tilde{\mathscr{C}}_C(\phi) = \tilde{\mathscr{C}}_C(C) = 3/8 \qquad \tilde{\mathscr{C}}_C(T) = 0 \qquad T \subset C \quad |T| = \text{odd}$$

$$\tilde{\mathscr{C}}_C(12) = \tilde{\mathscr{C}}_C(34) = \tilde{\mathscr{C}}_C(13) = \tilde{\mathscr{C}}_C(24) = -\tilde{\mathscr{C}}_C(14) = -\tilde{\mathscr{C}}_C(23) = \frac{1}{8}$$

hence

$$\mathscr{C}_C = \frac{1}{8}\left(3 + \sigma_{12} + \sigma_{34} + \sigma_{13} + \sigma_{24} - \sigma_{14} - \sigma_{23} + 3\,\sigma_{1234}\right) = \frac{1}{8}\Big[(1+\sigma_{12})(1+\sigma_{34}) + (1+\sigma_{13})(1+\sigma_{24}) + (1-\sigma_{14})(1-\sigma_{23})\Big]$$

and the set of infinite bonds becomes

$$\mathscr{B}_\infty = \{(xy)\,;\ x,y = n.n.\} \cup \{(xy)\,;\ x,y = n.n.n. \text{ on } C\} \cup \{C\}_{C \in \{\mathcal{L}C\}}$$

Notice that multiplying the admissible configurations of this standard 6-vertex model by X_1 [resp. X_2] where X_1 is such that $X_1 \cap C = (12)$ $\forall\, C$ [resp. $X_2 \cap C = (13)$], we obtain 6-vertex models with $w_3 = w_4 = 0$ [resp. $w_6 = w_7 = 0$] and local constraints defined by

$$\mathscr{C}_C^{X_1}(Y) = \mathscr{C}_C(X_1 Y) \qquad \mathscr{C}_C^{X_2}(Y) = \mathscr{C}_C(X_2 Y)$$

respectively. Therefore the set of infinite bonds of these two 6-vertex models coincides with the one of the standard 6-vertex model.

The local hamiltonian H_C of the spin system associated with the 6-vertex model is defined by the relation (*)

$$H_C(X_j) = e_j \qquad j = 1, \dots, 16$$

To obtain the finite bonds of this system we notice that $\sigma_C(X) = 1$ for all admissible configurations and thus for any $T \subset C$ and $X \in \mathscr{P}^a(\mathscr{L}_1)$, $\sigma_T(X) = \sigma_{C/T}(X)$. It then follows that we can write

$$-H_C = J_0 + \sum_{x \in C} J_x \sigma_x + \frac{1}{2} \sum_{(xy) \subset C} J_{xy}\, \sigma_{xy}$$

where

$$J_B = -\frac{1}{8} \sum_{j=1}^{16} \sigma_B(X_j)\, e_j$$

Explicitely we have :

$$J_0 = -\frac{1}{8}\left(e_1 + e_2 + e_3 + e_4 + e_5 + e_6 + e_7 + e_8\right)$$

$$J_1 = -\frac{1}{8}\left(e_1 - e_2 - e_3 + e_4 - e_5 + e_6 - e_7 + e_8\right)$$

$$J_2 = -\frac{1}{8}\left(e_1 - e_2 + e_3 - e_4 - e_5 + e_6 + e_7 - e_8\right)$$

$$J_3 = -\frac{1}{8}\left(e_1 - e_2 - e_3 + e_4 + e_5 - e_6 + e_7 - e_8\right)$$

(*) By convention we choose $e_j = 0$ if $w_j = 0$.

$$J_4 = -\frac{1}{8}(e_1 - e_2 + e_3 - e_4 + e_5 - e_6 - e_7 + e_8)$$

$$J_{12} = -\frac{1}{8}(e_1 + e_2 - e_3 - e_4 + e_5 + e_6 - e_7 - e_8) = J_{34} \qquad (1.16)$$

$$J_{23} = -\frac{1}{8}(e_1 + e_2 - e_3 - e_4 - e_5 - e_6 + e_7 + e_8) = J_{14}$$

$$J_{13} = -\frac{1}{8}(e_1 + e_2 + e_3 + e_4 - e_5 - e_6 - e_7 - e_8) = J_{24}$$

As an illustration, let us compute the Hamiltonian of the standard 6-vertex model with the following energy assignments :

$$e_1 = \varepsilon_2 - h - v \qquad e_2 = \varepsilon_2 + h + v$$

$$e_3 = \varepsilon_1 + \varepsilon_2 - h + v \qquad e_4 = \varepsilon_1 + \varepsilon_2 + h - v$$

$$e_5 = e_6 = \varepsilon_1 \qquad e_7 = e_8 = 0$$

where h (v) represents a horizontal (vertical) electric field. Computing the J_B's, we obtain for the local Hamiltonian

$$-H_C = -\frac{\varepsilon_1+\varepsilon_2}{4} + \frac{v}{2}(\sigma_1+\sigma_3) + \frac{h}{2}(\sigma_2+\sigma_4) + \frac{\varepsilon_1}{4}(\sigma_{23}+\sigma_{14}) - \frac{\varepsilon_2}{4}(\sigma_{13}+\sigma_{24})$$

and we immediately see that the presence of a horizontal and vertical field implies a staggered field on the corresponding Ising lattice $\mathcal{L}_1$. Hence $\mathcal{L}_1$ is divided into two sublattices

$$\mathcal{L}_1 = \mathcal{L}_h \cup \mathcal{L}_v$$

each point of a sublattice having only points of the other sublattice as nearest neighbours (n.n.). The Hamiltonian of a given volume $\Lambda \in \mathcal{P}_f(\mathcal{L}_1)$ is then $H_\Lambda = \sum_{\substack{C \in \{LC\} \\ C \subset \Lambda}} H_C$

$$-H_\Lambda = -\frac{\varepsilon_1+\varepsilon_2}{2} N_{C,\Lambda} + v \sum_{x \in \mathcal{L}_v \cap \Lambda} \sigma_x + h \sum_{x \in \mathcal{L}_h \cap \Lambda} \sigma_x + \frac{\varepsilon_1}{4} \sum_{\substack{x,y \text{ n.n.} \\ \text{vertical}}} \sigma_{xy} - \frac{\varepsilon_2}{4} \sum_{\substack{x,y,\text{ n.n.n} \\ \text{on } C;\ C \subset \Lambda}} \sigma_{xy}$$

where $N_{C,\Lambda}$ is the number of shaded squares within Λ . The set of finite bonds is then

$$\mathcal{B} = \{x\}_{x \in \mathcal{L}_1} \cup \{(xy);\ x,y \text{ n.n. vertical on } \mathcal{L}_1\} \cup \{(xy);\ x,y = \text{n.n.n. on } C\}_{C \in \{LC\}}$$

Particular cases are the KDP and F models, for these models we have

KDP : $\varepsilon_1 > 0$, $\varepsilon_2 = h = v = 0$ $\quad \mathcal{B}^{KDP} = \{(xy);\ x,y\ \text{n.n. vertical}\}$

F : $\varepsilon_2 > 0$, $\varepsilon_1 = h = v = 0$ $\quad \mathcal{B}^{F} = \{(xy);\ x,y\ \text{n.n.n. on } C\}_{C \in \{\iota c\}}$

1.3. Thermodynamics, Gibbs States and some Properties of Finite and Infinite Systems

With Λ a sequence of finite volume which tends to $\mathcal{L}$ in the sense of van Hove [1] the thermodynamics of $\{\Lambda, \mathcal{C}_\Lambda, K\}$ and $\{\mathcal{L}, \mathcal{C}, K\}$ is deduced from the "reduced partition function" and the corresponding "reduced free energy".

$$Z_\Lambda^{red.} = \mathrm{Tr}\,\{ e^{-\beta(H_\Lambda - H_\Lambda(\Phi))} \} = \sum_{X \subset \Lambda} \mathcal{C}_\Lambda(X) \prod_{B \in \mathcal{B}} e^{K(B)[\sigma_B(X) - 1]}$$

$$-\beta f_\Lambda^{red.} = \frac{1}{|\Lambda|} \ln Z_\Lambda^{red.} = - F_\Lambda^{red} \tag{1.17}$$

$$f^{red} = \lim_{\Lambda \to \mathcal{L}} f_\Lambda^{red}$$

For finite systems the Gibbs states $\omega_{(\Lambda, \mathcal{C}_\Lambda, K)}$ are defined by the correlation functions

$$\omega_{(\Lambda, \mathcal{C}_\Lambda, K)}[\sigma_X] = \langle \sigma_X \rangle_{(\Lambda, \mathcal{C}_\Lambda, K)} = Z_\Lambda^{red\,-1} \sum_{T \subset \Lambda} \mathcal{C}_\Lambda(T)\, \sigma_X(T) \prod_{B \in \mathcal{B}} e^{K(B)[\sigma_B(T) - 1]} \tag{1.18}$$

and we have the following property :

Property 2

1 - the hamiltonian H_Λ and the constraint $\mathcal{C}_\Lambda$ are both invariant under the group of automorphisms $\{\tau_s\}_{s \in \mathcal{S}'}$ where for any function A on $\mathcal{P}(\Lambda)$, $(\tau_Y A)(X) = A(YX)$.

2 - the Gibbs states $\omega_{(\Lambda, \mathcal{C}_\Lambda, K)}$ are invariant under the group of transformations $\{\tau'_s\}_{s \in \mathcal{S}'}$ induced by $\{\tau_s\}$, where $(\tau'_s \omega)[A] = \omega[\tau_s^{-1} A]$.

Moreover for any state ω of the finite or infinite systems we have the following results :

Property 3 [113]

1) $\omega[\mathscr{C}_c \sigma_T] = \omega[\sigma_T]$ for all $c \in \{\iota c\}$ and all $T \in \mathscr{P}_f(\mathscr{L})$
2) ω is entirely defined by the correlation functions $\omega[\sigma_T]$ with $T \in \mathscr{P}_f^a(\mathscr{L})$.
3) ω is invariant under the subgroup $\mathscr{S}_\omega$ of $\mathscr{P}(\mathscr{L})$ if and only if $\omega[\sigma_X] = 0 \quad \forall X \notin \mathscr{S}_\omega^\perp$.

Proof of (2)

Notice that (1) also holds for any local constraint $\mathscr{C}_R$ such that $\mathscr{C}_R \mathscr{C} = \mathscr{C}$. Then defining for any $X \in \mathscr{P}_f(\mathscr{L})$, $X \notin \mathscr{P}^a(\mathscr{L})$ the local constraint $\mathscr{C}_X$ by $\mathscr{C}_X(Y) = 1$ for all $Y \subset X$, $Y \neq X$ and $\mathscr{C}_X(X) = 0$ we obtain

$$\omega_{(\mathscr{L},\mathscr{C},K)}[\sigma_X] = -(-1)^{|X|} \sum_{\substack{Z \subset X \\ Z \neq X}} (-1)^{|Z|} \omega_{(\mathscr{L},\mathscr{C},K)}[\sigma_Z]$$

hence $\omega[\sigma_X]$ is expressed as a linear combination of $\omega[\sigma_Z]$ with $|Z| < |X|$; iterating the argument implies that for any $X \notin \mathscr{P}_f^a(\mathscr{L})$, $\omega[\sigma_X]$ can be expressed as a finite linear combination of $\omega[\sigma_{X_i}]$ with $X_i \subset X$ and $X_i \in \mathscr{P}^a(\mathscr{L})$.

The proof of (1) and (3) are straightforward.

Corollary

The Gibbs states of any finite systems are such that

$$\omega_{(\Lambda,\mathscr{C}_\Lambda,K)}[\sigma_X] = 0 \qquad \text{if } X \notin \overline{\mathscr{B}'}$$

1.4 Group Structure for Systems with Constraints

Following the ideas of Part I we introduce two groups to discuss the system $\{\mathcal{L}, \mathcal{C}, K\}$. The first group is the group $\mathcal{P}(\mathcal{L})$ introduced for systems without constraints, whereas the second group, $\mathcal{P}(\mathcal{B}')$, is in general different, due to the presence of infinite bonds. For some classes of systems with constraints to be specified later, it is however possible to work with $\mathcal{P}(\mathcal{B})$ rather than with $\mathcal{P}(\mathcal{B}')$. In fact, is is for these systems that the most general results will be obtained [11].

The basic groups being $\mathcal{P}(\mathcal{L})$ and $\mathcal{P}(\mathcal{B}')$, we introduce the two homormophism π' and γ' defined by :

$$\pi' : \mathcal{P}(\mathcal{B}') \longrightarrow \mathcal{P}(\mathcal{L}) \qquad \gamma' : \mathcal{P}(\mathcal{L}) \longrightarrow \mathcal{P}(\mathcal{B}')$$
$$\beta \longmapsto \pi'(\beta) = \prod_{B\in\beta} B \qquad X \longmapsto \gamma'(X) = \{B' ; \sigma_{B'}(X) = -1\}$$

If we are working with $\mathcal{P}(\mathcal{L})$ and $\mathcal{P}(\mathcal{B})$, these two homomorphism are the usual mappings π and γ previously defined in Part I. The structure of $\mathcal{P}(\mathcal{B}')$ and $\mathcal{P}(\mathcal{L})$ being the same as the one of $\mathcal{P}(\mathcal{B})$ and $\mathcal{P}(\mathcal{L})$, the homomorphisms π' and γ' have the same properties as the homomorphisms π and γ which we recall here for the sake of clarity :

1° : $\langle \pi'(\beta) ; X \rangle_{\mathcal{L}} = \langle \beta ; \gamma'(X) \rangle_{\mathcal{B}'}$ for all $X \in \mathcal{P}_f(\mathcal{L}), \beta \in \mathcal{P}(\mathcal{B}')$ and for all $X \in \mathcal{P}(\mathcal{L})$, $\beta \in \mathcal{P}_f(\mathcal{B}')$.

2° : The kernel $\mathcal{K}'$ of π' and the image Γ' of γ' are subgroups of $\mathcal{P}(\mathcal{B}')$ related by

$$\mathcal{K}' = (\Gamma'^{f})^{\perp} \qquad \Gamma' = (\mathcal{K}'_f)^{\perp}$$

where $\Gamma'^{f} = \gamma'[\mathcal{P}_f(\mathcal{L})]$ and $\mathcal{K}'_f = \mathcal{K}' \cap \mathcal{P}_f(\mathcal{B}')$

3° : The kernel $\mathcal{S}'$ of γ' and the image $\overline{\mathcal{B}'}$ of π' restricted to $\mathcal{P}_f(\mathcal{B}')$ are subgroups of $\mathcal{P}(\mathcal{L})$, $\mathcal{P}_f(\mathcal{L})$ related by :

$$\mathcal{S}' = \mathcal{B}'^{\perp} \qquad \overline{\mathcal{B}'} = \mathcal{S}'^{\perp}$$

4° : Finally, the different groups are related by

$$(\mathcal{P}(\mathcal{B}')/\mathcal{K}')^{\wedge} \cong \Gamma'^{f} \cong (\mathcal{I}m\,\pi')^{\wedge} \qquad (\mathcal{P}_f(\mathcal{B}')/\mathcal{K}'_f)^{\wedge} \cong \Gamma' \cong \overline{\mathcal{B}'}^{\wedge}$$

$$(\mathcal{P}(\mathcal{L})/\mathcal{S}')^{\wedge} \cong \overline{\mathcal{B}'} \cong \Gamma'^{\wedge} \qquad (\mathcal{P}_f(\mathcal{L})/\mathcal{S}'_f)^{\wedge} \cong \mathcal{I}m\,\pi' \cong \mathcal{S}'^{\perp}_f \cong \Gamma'$$

where $\mathcal{S}'_f = \mathcal{S}' \cap \mathcal{P}_f(\mathcal{L})$

In the following we shall also use the following subgroups of $\mathcal{P}(\mathcal{B})$ defined by means of γ: $\mathcal{P}(\mathcal{L}) \to \mathcal{P}(\mathcal{B})$.

$$\Gamma^{a} = \{\gamma(X);\ X \in \overline{\mathcal{P}^{a}(\mathcal{L})}\} \qquad \Gamma^{a,f} = \{\gamma(X);\ X \in \overline{\mathcal{P}^{a}_f(\mathcal{L})}\}$$

$$\mathcal{K}^{a}_f = \Gamma^{a\,\perp} \qquad \mathcal{K}^{a} = (\Gamma^{a,f})^{\perp}$$

where the orthogonal is taken in $\mathcal{P}(\mathcal{B})$. In the special case where $\Gamma^{a,f}$ is a subgroup of some $\mathcal{P}(\hat{\mathcal{B}})$, $\hat{\mathcal{B}} \subset \mathcal{B}$, the orthogonal will be taken in $\mathcal{P}(\hat{\mathcal{B}})$.

CHAPTER 2 - EXPANSIONS FOR THE PARTITION FUNCTION

2.1 The Low Temperature (L.T.) Expansion

From the definition of the mapping γ' introduced in Sec. 1.4 together with Eq. (1.12) we obtain the L.T. expansion of the reduced partition function Eq. (1.17) :

$$Z^{red}(\Lambda, \mathscr{C}_\Lambda, K) = |\mathscr{S}'| \sum_{\beta' \in \gamma'[\mathscr{P}^a(\Lambda)]} \prod_{B \in \beta'} e^{-2K(B)} \qquad (2.1)$$

where $\gamma'[\mathscr{P}^a(\Lambda)] = \{\gamma'(X) \, ; \, X \in \mathscr{P}^a(\Lambda)\} \subset \mathscr{P}(\mathscr{B}')$

Since $e^{-2K(B)} = 1$ for any $B \in \mathscr{B}'/\mathscr{B}$, only $\beta' \cap \mathscr{B}$ contributes to the Boltzmann factor of the above L.T. expansion. However the reduction to $\mathscr{P}(\mathscr{B})$ is only useful for systems satisfying $\tilde{\mathscr{B}} \supset \mathscr{B}_\infty$. Furthermore we emphasize that $\gamma'[\mathscr{P}^a(\Lambda)]$ is a subgroup of $\mathscr{P}(\mathscr{B}')$ if and only if $\mathscr{P}^a(\Lambda)$ is a group. To write Eq. (2.1) in terms of the group Γ', we define constraints $\mathscr{C}_\Lambda(\beta')$ on $\mathscr{P}(\mathscr{B}')$ by :

$$\mathscr{C}_\Lambda(\beta) = \mathscr{C}_\Lambda(\beta \cap \mathscr{B}_\infty) = \begin{cases} 1 & \text{if } \beta \cap \mathscr{B}_\infty \in \gamma_\infty[\mathscr{P}^a(\Lambda)] \\ 0 & \text{otherwise} \end{cases} \qquad (2.2)$$

where $\gamma_\infty : \mathscr{P}(\mathscr{L}) \to \mathscr{P}(\mathscr{B}_\infty)$ is defined in the same manner as γ' or γ. These constraints have the property

$$\beta \in \Gamma' \text{ and } \mathscr{C}_\Lambda(\beta) = 1 \Leftrightarrow \beta \in \gamma'[\mathscr{P}^a(\Lambda)] \qquad (2.3)$$

Indeed $\beta \in \Gamma'$ implies $\beta = \gamma'(X)$ with $X \in \mathscr{P}(\Lambda)$, whereas $\mathscr{C}_\Lambda(\beta) = 1$ yields $\beta \cap \mathscr{B}_\infty = \gamma_\infty(Y)$ with $Y \in \mathscr{P}^a(\Lambda)$, therefore $\gamma_\infty(XY) = \{\phi\}$ which is equivalent to $XY \in \mathscr{S}_\infty$. Together with Eq. (1.12) we conclude $X \in \mathscr{P}^a(\Lambda)$ and therefore $\beta \in \gamma'[\mathscr{P}^a(\Lambda)]$. Eq. (2.1) then becomes

$$Z^{red}(\Lambda, \mathscr{C}_\Lambda, K) = |\mathscr{S}'| \sum_{\beta \in \Gamma'} \mathscr{C}_\Lambda(\beta) \prod_{B \in \beta} e^{-2K(B)} \qquad (2.4)$$

and similarly, with $\beta \in \mathcal{P}(\mathcal{B}')$, the correlation functions are given by

$$\omega_{(\Lambda, \mathcal{E}_\Lambda, K)}[\sigma_{\gamma'(\beta)}] = \frac{\sum_{\beta' \in \gamma'[\mathcal{P}^a(\Lambda)]} \sigma_{\beta'}(\beta) \prod_{B \in \beta'} e^{-2K(B)}}{\sum_{\beta' \in \gamma'[\mathcal{P}^a(\Lambda)]} \prod_{B \in \beta'} e^{-2K(B)}} = \frac{\sum_{\beta' \in \Gamma'} \mathcal{E}_\Lambda(\beta') \sigma_{\beta'}(\beta) \prod_{B \in \beta'} e^{-2K(B)}}{\sum_{\beta' \in \Gamma'} \mathcal{E}_\Lambda(\beta') \prod_{B \in \beta'} e^{-2K(B)}} \tag{2.5}$$

For two classes of systems to be specified in property 1, the above L.T. expansions can be restricted to subgroups of $\mathcal{P}(\hat{\mathcal{B}}) \subset \mathcal{P}(\mathcal{B}')$, where $\hat{\mathcal{B}} \subset \mathcal{B}'$:

Property 1 :

If a) $\mathcal{P}^a(\Lambda)$ is a subgroup of $\mathcal{P}(\Lambda)$ (*)
or if b) $\bar{\mathcal{B}} \supset \mathcal{B}_\infty$
then the sum over $\beta \in \gamma'[\mathcal{P}^a(\Lambda)] \subset \mathcal{P}(\mathcal{B}')$ can be replaced by the sum over $\beta \in \gamma[\mathcal{P}^a(\Lambda)] \subset \mathcal{P}(\mathcal{B})$ in the L.T. expansion Eq. (2.1)

i.e. $$Z^{red}_{(\Lambda, \mathcal{E}_\Lambda, K)} = |\mathcal{S}| \sum_{\beta \in \gamma[\mathcal{P}^a(\Lambda)]} \prod_{B \in \beta} e^{-2K(B)} \tag{2.6}$$

moreover if $\mathcal{P}^a(\Lambda)$ is a group, we have

$$\Gamma^a = \gamma[\mathcal{P}^a(\Lambda)] \equiv \Gamma' \cap \mathcal{P}(\hat{\mathcal{B}}) \tag{2.7}$$

where $$\hat{\mathcal{B}} = \mathcal{B}' / \bar{\mathcal{B}}_\infty \tag{2.8}$$

proof :

a) This follows from the fact that $\gamma'(y) \in \mathcal{P}(\mathcal{B}'/\bar{\mathcal{B}}_\infty)$ iff $\sigma_B(y) = +1$ for all $B_\infty \in \mathcal{B}_\infty$; therefore with $\mathcal{S}_\infty = \mathcal{P}^a(\Lambda)$ $\gamma'(y) \in \mathcal{P}(\mathcal{B}'/\bar{\mathcal{B}}_\infty)$ iff $y \in \mathcal{P}^a(\Lambda)$.

(*) in this case $\gamma'[\mathcal{P}^a(\Lambda)]$ is a subgroup of $\mathcal{P}(\mathcal{B}/\bar{\mathcal{B}}_\infty)$.

b) $\bar{\mathcal{B}} \supset \mathcal{B}_\infty$ implies $\mathcal{S} \subset \mathcal{S}_\infty$ and $\mathcal{S} = \mathcal{S}'$, therefore

$$\sum_{s \in \mathcal{S}} \mathcal{C}_\Lambda(Xs) = |\mathcal{S}'|\, \mathcal{C}_\Lambda(X)$$

Notice that in the case $\bar{\mathcal{B}} \supset \mathcal{B}_\infty$, for any $X \in \mathcal{P}(\Lambda)$, the components of $\gamma'(X) \cap \mathcal{B}_\infty$ are completely determined by those of $\gamma'(X) \cap \mathcal{B}$ and the reduction to $\mathcal{P}(\mathcal{B})$ contains the full information on the system.

If $\{\mathcal{L}, \mathcal{C}, K\}$ satisfies the subgroup property, we have a structure which is entirely similar to the case of systems without constraints. Defining $z_B = e^{-2K(B)}$ for all $B \in \mathcal{B}$, we have for any $\beta \in \mathcal{P}(\mathcal{B})$

$$Z^{red}_{(\Lambda, \mathcal{C}_\Lambda, K)} \sim \sum_{\beta' \in \Gamma^a} \prod_{B \in \beta'} z_B \tag{2.9}$$

$$\omega_{(\Lambda, \mathcal{C}_\Lambda, K)}[\sigma_{\pi\beta}] = \frac{\sum_{\beta' \in \Gamma^a} \sigma_\beta(\beta') \prod_{B \in \beta'} z_B}{\sum_{\beta' \in \Gamma^a} \prod_{B \in \beta'} z_B}$$

2.2. The High Temperature (H.T.) Expansions

For systems without constraints, the H.T. expansion obtained by means of Poisson formulae from the L.T. expansion [5] coincides with the one obtained in performing a Fourier expansion of the Boltzmann factors (see Sec. 8.3 Part I). For systems with constraints, these two H.T. expansions are in general different.

To derive the first H.T. expansion, we use the L.T. expansion Eq. (2.4); since $\left(\mathcal{P}(\mathcal{B}')/\Gamma'\right)^\wedge \cong \Gamma'^\perp \cong \mathcal{K}'$, the Poisson formulae gives

$$Z^{red}_{(\Lambda, \mathcal{C}_\Lambda, K)} = \frac{|\mathcal{S}'|\,|\Gamma'|}{|\mathcal{P}(\mathcal{B}_\infty)|} \prod_{B \in \mathcal{B}'/\mathcal{B}_\infty} \left(\frac{1 + e^{-2K(B)}}{2}\right) \sum_{\varkappa' \in \mathcal{K}'} \tilde{\mathcal{C}}_\Lambda(\varkappa') \prod_{B \in \varkappa' \cap \mathcal{B}'/\mathcal{B}_\infty} \tilde{z}_B \tag{2.10}$$

where

$$\tilde{z}_B = \tanh K(B) \tag{2.11}$$

$$\tilde{\mathcal{C}}_\Lambda(\beta) = \sum_{\beta_\infty \in \gamma_\infty[\mathcal{P}^a(\Lambda)]} \sigma_\beta(\beta_\infty) \prod_{B \in \beta_\infty} e^{-2K(B)}$$

In general, $\tilde{\mathscr{C}}_\Lambda(\beta')$ is not a constraint on $\mathcal{P}(\mathcal{B}')$. If $\mathcal{B} \cap \mathcal{B}_\infty \neq \phi$, $\tilde{\mathscr{C}}_\Lambda(\beta')$ also depends on the interactions. In this case, we have no control on $\tilde{\mathscr{C}}_\Lambda(\beta')$ at high temperature $(K(B) \to 0)$ and Eq. (2.10) is not a useful H.T. expansion. If $\mathcal{B}_\infty \supset \mathcal{B}$, we have $\prod_{B \in \varkappa' \cap \mathcal{B}'/\mathcal{B}_\infty} \tilde{z}_B = 1$ and we come back to the L.T. expansions (see example).

A second H.T. expansion can be obtained in terms of $\tilde{z}_B = \tanh K(B)$ and of the correlation functions $\omega_{(\Lambda, \mathscr{C}_\Lambda, 0)}[\sigma_T]$ of the system without interactions $\{\Lambda, \mathscr{C}_\Lambda, 0\}$. With the identity $e^{K(B)\sigma_B(x)} = \cosh K(B)[1+\tilde{z}_B\sigma_B(x)]$ we get from Eq. (1.17) :

$$Z^{red}(\Lambda, \mathscr{C}_\Lambda, K) = |\mathcal{P}^a(\Lambda)| \prod_{B \in \mathcal{B}_\Lambda} \left(\frac{1+e^{-2K(B)}}{2}\right) \sum_{\beta \in \mathcal{P}(\mathcal{B}_\Lambda)} \omega_{(\Lambda, \mathscr{C}_\Lambda, 0)}[\sigma_{\pi\beta}] \prod_{B \in \beta} \tilde{z}_B \qquad (2.12)$$

where

$$\omega_{(\Lambda, \mathscr{C}_\Lambda, 0)}[\sigma_{\pi\beta}] = |\mathcal{P}^a(\Lambda)|^{-1} \sum_{X \in \mathcal{P}^a(\Lambda)} \sigma_{\pi\beta}(X)$$

Notice that this expansion is also a good H.T. expansion in the case $\mathcal{B} \cap \mathcal{B}_\infty \neq \phi$. Moreover, there is no problem to determine the degeneracy factors in Eq. (2.12) if one is interested in the power series in the variable $\tilde{z}_B$. The whole combinatorial problem is in the determination of the expansion coefficients $\omega_{(\Lambda, \mathscr{C}_\Lambda, 0)}[\sigma_{\pi\beta}]$, which can also be interpreted as the correlation functions at infinite temperature $[K(B) = 0 \ \forall B \in \mathcal{B} \Leftrightarrow T = \infty]$. These correlation functions (with $\pi\beta \neq \phi$) generally do not vanish as for systems without constraints.

For systems satisfying the subgroup property, we have [113] :

<u>Property 2</u> :

If $\mathcal{P}^a(\Lambda)$ is a subgroup of $\mathcal{P}(\Lambda)$, then both H.T. expansions coincide and we have

$$Z^{red}(\Lambda, \mathscr{C}_\Lambda, K) = |\mathcal{P}^a(\Lambda)| \prod_{B \in \hat{\mathcal{B}}} \left(\frac{1+e^{-2K(B)}}{2}\right) \sum_{\varkappa \in \mathcal{K}^a} \prod_{B \in \varkappa} \tilde{z}_B \qquad (2.13)$$

where $\mathcal{K}^a = \{\varkappa = \varkappa' \cap \hat{\mathcal{B}} \ ; \ \varkappa' \in \mathcal{K}'\} \equiv (\Gamma^a)^\perp \qquad (2.14)$

Example : 1-dimensional hard core lattice system

Let us consider the 1 dimensional system defined on $\Lambda = \{x_1, \dots, x_{|\Lambda|}\}$ with periodic boundary conditions, i.e. $x_{|\Lambda|+1} = x_1$ and a Hamiltonian given by

$$-\beta H_\Lambda = \sum_{x \in \Lambda} h_x \sigma_x$$

together with nearest neighbour exclusion constraints

$$\mathscr{C}_{(k,k+1)} = \frac{1}{4}\left(3 + \sigma_{x_k} + \sigma_{x_{k+1}} - \sigma_{x_k}\sigma_{x_{k+1}}\right) \qquad k = 1, \dots |\Lambda|$$

For this model $\mathcal{B} = \{(x_i)\}_{i=1..|\Lambda|}$ and $\mathcal{B}_\infty = \{(x_i), (x_i, x_{i+1})\}_{i=1..|\Lambda|}$ hence

$$\mathcal{B} \subset \mathcal{B}_\infty \qquad and \qquad \mathcal{B}' \equiv \mathcal{B}_\infty \tag{2.15}$$

implying $\pi_\infty \equiv \pi'$, $\gamma_\infty \equiv \gamma'$. The homomorphisms π', π and γ', γ are different : With $(x_i) \in \mathcal{P}^a(\Lambda)$ we have

$$\gamma((x_i)) = \{(x_i)\} \qquad \gamma'((x_i)) = \{(x_i), (x_{i-1}, x_i), (x_i, x_{i+1})\}$$

and with $\beta = \{(x_i), (x_{i+1}), (x_{i-1}, x_i)\} \in \mathcal{P}(\mathcal{B}')$:

$$\pi(\beta \cap \mathcal{B}) = (x_i, x_{i+1}) \qquad \text{whereas} \qquad \pi'(\beta) = (x_{i-1}, x_{i+1})$$

Moreover, the kernels of the above homomorphism are $\mathcal{S} \equiv \mathcal{S}' \equiv \mathcal{S}_\infty = \{\phi\}$ respectively $\mathcal{K} = \{\phi\}$ whereas $\mathcal{K}'$ is generated by $\varkappa'_i = \{(x_i), (x_{i+1}), (x_i, x_{i+1})\}$; Eq. (2.15) implies in particular $\bar{\mathcal{B}} = \bar{\mathcal{B}'} = \bar{\mathcal{B}}_\infty = \mathcal{P}(\Lambda)$.

The L.T. expansion for this model, Eq. (2.1) yields $Z^{red}(\Lambda, \mathscr{C}_\Lambda, K) = \sum_{X \in \mathcal{P}^a(\Lambda)} \prod_{x \in X} e^{-2h_x}$. The H.T. expansion however, need a bit more discussion :

a) $\mathcal{B}' \equiv \mathcal{B}_\infty$, together with $|\mathcal{P}(\mathcal{B}_\infty)| = 2^{2|\Lambda|}$, $|\mathcal{S}'||\Gamma'| = 2^{|\Lambda|}$ implies that Eq. (2.10) takes the form

$$Z^{red}(\Lambda, \mathscr{C}_\Lambda, K) = 2^{-|\Lambda|} \sum_{\varkappa' \in \mathcal{K}'} \tilde{\mathscr{C}}_\Lambda(\varkappa') \tag{2.16}$$

since $\tilde{\mathscr{C}}_\Lambda(\varkappa') = Z^{red}(\Lambda, \mathscr{C}_\Lambda, K)$ for all $\varkappa' \in \mathcal{K}'$. (2.16) is no H.T. expansion.

b) The H.T. expansion in the form of Eq. (2.12) requires an explicite knowledge of the correlation functions of the hard core nearest neighbour system without interactions. Using the transfermatrix, one finds

$$\omega_{(\Lambda, \mathscr{C}_\Lambda, 0)}[\sigma_x] = \begin{cases} \frac{1}{\sqrt{5}} \tanh\left(|\Lambda| \ln\left(\frac{1+\sqrt{5}}{2}\right)\right) & \text{if } |\Lambda| = \text{even} \\ \frac{1}{\sqrt{5}} \operatorname{cotanh}\left(|\Lambda| \ln\left(\frac{1+\sqrt{5}}{2}\right)\right) & \text{if } |\Lambda| = \text{odd} \end{cases}$$

e.g. if $|\Lambda|$ = even, Eq. (2.12) becomes

$$Z^{red}(\Lambda, \mathscr{C}_\Lambda, K) = |\mathscr{P}^a(\Lambda)| \prod_{i=1}^{|\Lambda|} \left(\frac{1+e^{-2h_i}}{2}\right) \left\{ 1 + \frac{1}{\sqrt{5}} \tanh\left(|\Lambda| \ln\left(\frac{1+\sqrt{5}}{2}\right)\right) \sum_{x\in\Lambda} \tanh h_x + \right.$$
$$\left. + \dots \right\}$$

CHAPTER 3 - PARTIAL TRACE METHOD AND EQUILIBRIUM EQUATIONS

As in Ch. 6 of Part I, we shall use the method of partial traces to transform the partition function and the correlation functions of a given model into the corresponding functions of another model which is either a model with or without constraints. This will always be possible if the constraints and the set of finite bonds $\mathcal{B}$ satisfies some properties to be specified below. Besides the transformation into another model, this method enables us to derive for any system with constraints a set of linear equations for the observables $\langle \sigma_X \rangle$, $X \in \mathcal{P}_f^a(\mathcal{L})$, valid for finite and infinite systems [113]. These equations can then be taken as definition of "Equilibrium States" and therefore the properties of the solutions will yield physical properties of equilibrium states.

3.1. Partial Trace Transformation

The partial trace method [78] was first applied to a class of systems with constraints by L.K. Runnels [57] . He showed that some hard core lattice gas models are related by partial trace to Ising models without constraints but with many body interactions. Here we shall consider more general cases and show that partial trace gives in general models with constraints.

3.1.1. General Framework

For any system $\{\Lambda, \mathcal{C}_\Lambda, K\}$ let $\Lambda_0 \subset \Lambda$ be a sublattice such that

i) $|C \cap \Lambda_0| \leq 1$ for all $C \in \{\ell C\}$ (3.1)

ii) $|B \cap \Lambda_0| \leq 1$ for all $B \in \mathcal{B}$

moreover, for any $x_0 \in \Lambda_0$, we define the subset $\Delta_{x_0} \subset \Lambda/\Lambda_0$ by

$$\Delta_{x_0} = \bigcup_{\substack{B \in \mathcal{B}' \\ B \ni x_0}} B \cdot x_0 \tag{3.2}$$

The following result is the extension to systems with constraints of Theorem 1 in Sec. 6.1 of Part I :

Theorem 1

Let $\{\Lambda, \mathcal{C}_\Lambda, K\}$ be a general lattice system with constraints such that there exists Λ_0 satisfying Eq. (3.1), then

$$Z(\Lambda, \mathcal{C}_\Lambda, K) = Z(\Lambda/\Lambda_0, \mathcal{C}'_{\Lambda/\Lambda_0}, K')$$

where $\mathcal{C}'_{\Lambda/\Lambda_0} = \prod_{\substack{C \in \{\ell C\} \\ C \cap \Lambda_0 = \phi}} \mathcal{C}_C \prod_{x_0 \in \Lambda_0} \mathcal{C}_{x_0}$ $\qquad \mathcal{C}_{x_0}$ given by Eq. (3.4)

and $K'(X') = K(X') + \sum_{x_0 \in \Lambda_0} k(x_0; X')$, $\forall X' \subset \Lambda/\Lambda_0$; $k(x_0;)$ given by Eq. (3.6)

Proof

With the assumptions of the theorem it follows that :

$$Z(\Lambda, \mathcal{C}_\Lambda, K) = \sum_{X' \subset \Lambda/\Lambda_0} \prod_{\substack{C \in \{\ell C\} \\ C \cap \Lambda_0 = \phi}} \mathcal{C}_C(X') \prod_{\substack{B \in \mathcal{B} \\ B \cap \Lambda_0 = \phi}} e^{K(B)\sigma_B(X')} \prod_{x_0 \in \Lambda_0} F_{x_0}(X')$$

where

$$F_{x_0}(X') = \prod_{\substack{C \in \{\ell C\} \\ C \ni x_0}} \mathcal{C}_C(X')\, e^{\sum_{B \ni x_0} K(B)\sigma_B(X')} + \prod_{\substack{C \in \{\ell C\} \\ C \ni x_0}} \mathcal{C}_C(X' x_0)\, e^{-\sum_{B \ni x_0} K(B)\sigma_B(X')} \tag{3.3}$$

since $F_{x_0}(X')$ can vanish, we have to introduce the new constraints $\mathcal{C}_{x_0}$ on Λ/Λ_0 by

$$\mathcal{C}_{x_0}(X') = \prod_{\substack{C \in \{\ell C\} \\ C \ni x_0}} \mathcal{C}_C(X') + \prod_{\substack{C \in \{\ell C\} \\ C \ni x_0}} \mathcal{C}_C(X' x_0) - \prod_{\substack{C \in \{\ell C\} \\ C \ni x_0}} \mathcal{C}_C(X')\, \mathcal{C}_C(X' x_0) \tag{3.4}$$

moreover using the lemma of Sec. 7.1 (Part I) we have :

$$F_{x_0} = \mathcal{C}_{x_0} \exp\Big[\sum_{R \subset \Delta_{x_0}} k(x_0; R)\, \sigma_R\Big] \tag{3.5}$$

where

$$k(x_0; R) = 2^{-|\Delta_{x_0}|} \sum_{\substack{R' \subset \Delta_{x_0} \\ \mathcal{C}_{x_0}(R') = 1}} [\ln F_{x_0}(R')]\, \sigma_R(R') \cdot \delta_{R \cap \Delta_{x_0}, R} \tag{3.6}$$

Let us remark that an application of this technique to the general 6-vertex model yields up to 6 body interactions and complicated expressions for the new constraints $\mathscr{C}'_{\Lambda/\Lambda_0}$ and no new information is gained for these models. In the next sections we restrict therefore the application to general hard core lattice systems and to systems satisfying the subgroup property.

3.1.2. Partial Trace Method for General Hard Core Lattice Systems

General Hard core lattice systems were defined in Sec. 1.2.2.. The constraints given by Eq. (1.14) implies that Eq. (3.3) takes the form

$$F_{x_0}(X') = \exp\Big[\sum_{B\ni x_0} K(B)\sigma_B(X')\Big] + \prod_{\substack{C\in\{LC\}\\ C\ni x_0}} \mathscr{C}_C(X'x_0)\exp\Big[-\sum_{B\ni x_0} K(B)\sigma_B(X')\Big] \tag{3.7}$$

hence F_{x_0} cannot vanish; therefore no supplementary constraints have to be introduced. For the supplementary interactions we find

$$k(x_0;R) = \sum_{B\ni x_0} K(B)\,\delta_{R,\,B\cdot x_0} + 2^{-|\Delta_{x_0}|}\sum_{\substack{T\subset\Delta_{x_0}\\ \mathscr{C}_\Lambda(Tx_0)=1}} \sigma_R(T)\,\ln\Big[1+\exp-2\sum_{B\ni x_0}K(B)\sigma_B(T)\Big]\cdot\delta_{R\cap\Delta_{x_0},\,R} \tag{3.8}$$

In conclusion, we have :

Corollary 1 :

Any general hard core lattice system $\{\Lambda, \mathscr{C}_\Lambda, K\}$ with a sublattice satisfying Eq. (3.1) is equivalent to a hard core lattice system with the family of local constraints $\{LC\}' = \{C\in\{LC\}\ ;\ C\cap\mathscr{L}_0 = \phi\}$ (if $\{LC\}' = \{\phi\}$, we just have a system without constraints) and interactions $K' = K + \sum_{x_0\in\mathscr{L}_0} k(x_0;\ )$ with $k(x_0;\ )$ given by Eq. (3.8).

Example 1 : Hard Core Model with n.n. Exclusions on the Lattice $\mathbb{Z}^\nu$

Choosing Λ_0 such that each point of Λ_0 has only nearest neighbours (n.n.) on Λ/Λ_0, then $\Delta_{x_0} = \{x_1, \dots, x_{2\nu}\}$

is the set of 2ν points surrounding a given point $x_0 \in \Lambda_0$. For the set of finite bonds

$$\mathcal{B} = \{(x,y);\ x,y \text{ n.n.}\}_{x,y \in \mathbb{Z}^\nu} \cup \{(x)\}_{x \in \mathbb{Z}^\nu}$$

the conditions of Corollary 1 are satisfied. Since $C \cap \Lambda_0 \neq \phi$ for all $C \in \{LC\}$, we end up with a system without constraints. For translationally invariant, isotropic coupling constants $K((x,y)) = K$ for all $(x,y) \in \mathcal{B}$ and $K((x)) = h$ for all $(x) \in \mathcal{B}$, Eq. (3. 8) gives then for any $R \subset \Delta_{x_0}$

$$k(x_0;R) = h\,\delta_{R,\phi} + K\,\delta_{|R|,1} + 2^{-2\nu} \ln(1+e^{-2h-4\nu K})$$

and it follows that the new interactions are <u>ferromagnetic</u>, $k(x_0;R) \geqslant 0$ if $|R| \geqslant 2$. Moreover, the Hamiltonian of the equivalent Ising system becomes :

$$\begin{aligned} -\beta H_{\Lambda/\Lambda_0}(X') = {} & h\,|\Lambda_0| + 2^{-2\nu}|\Lambda_0| \ln(1+e^{-2h-4\nu K}) + \\ & + \sum_{x \in \Lambda/\Lambda_0} \left(h + 2\nu K + \nu\, 2^{-(2\nu-1)} \ln(1+e^{-2h-4\nu K})\right) \sigma_x(X') + \\ & + \sum_{x_0 \in \Lambda_0} \sum_{\substack{R \subset \Delta_{x_0} \\ |R| \geqslant 2}} 2^{-2\nu} \ln(1+e^{-2h-4\nu K})\ \sigma_R(X') \end{aligned}$$

<u>Example 2</u> : <u>n.n. Hard Core Exclusions on the Hexagonal Lattice</u>

The sublattices of the hexagonal lattice are defined in Fig. 1 a). Δ_{x_0} is then a 3 point set as shown in Fig. 1 b). For the set of finite bonds

$$\mathcal{B} = \{\Delta_{x_0}\}_{x_0 \in \Lambda_0} \cup \{(x,y)\}_{\substack{x,y \in \Lambda \\ \text{n.n.}}} \cup \{(x)\}_{x \in \Lambda}$$

where the diluted 3 body interactions are shown in Fig. 1 a), the conditions of Corollary 1 are satisfied; Again, the equivalent Ising system has no constraints. Choosing translationally invariant, isotropic coupling constants $K(\Delta_{x_0}) = K_3$, $K((x,y)) = K$ for all $(x,y) \in \mathcal{B}$, $K((x)) = h$ for all $(x) \in \mathcal{B}$, we find for any $R \subset \Delta_{x_0}$

$$k(x_0;R) = h\,\delta_{R,\phi} + K\,\delta_{|R|,1} + \tfrac{1}{8} \ln(1+e^{-2h-6K})$$

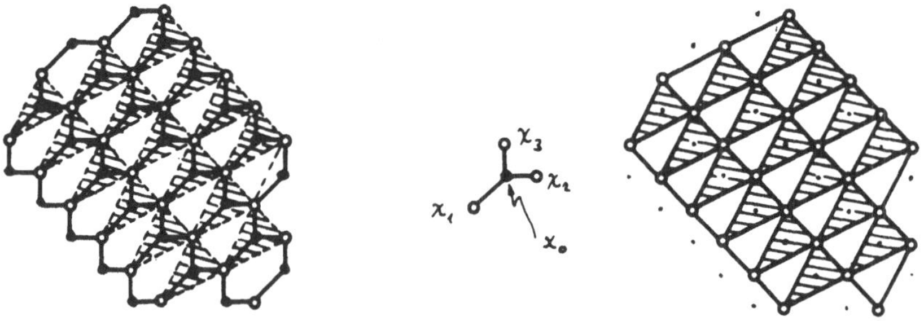

(a) (b) (c)

Fig. 1 a) The hexagonal lattice Λ , the sublattices $\Lambda_0 = \{\circ\}$ $\Lambda / \Lambda_0 = \{\bullet\}$ and the 3 body finite bonds (shaded);

b) $\Delta_{x_0} = \{x_1, x_2, x_3\}$ for $x_0 \in \Lambda_0$

c) The remaining lattice Λ / Λ_0 with the supplementary 2 body bonds and 3 body bonds (shaded).

The equivalent Ising system has then the Hamiltonian defined on the triangular lattice Λ / Λ_0 (Fig. 1 c)).

$$
\begin{aligned}
-\beta H_{\Lambda/\Lambda_0}(X') = {} & h\,|\Lambda_0| + \tfrac{1}{8}\,|\Lambda_0| \ln\left(1 + e^{-2h-6K}\right) + \sum_{x \in \Lambda/\Lambda_0} \left(h + 3K + \tfrac{3}{8} \ln\left(1 + e^{-2h-6K}\right)\right) \sigma_x(X') \\
& + \sum_{\substack{x,y \in \Lambda/\Lambda_0 \\ n.n.}} \tfrac{1}{8} \ln\left(1 + e^{-2h-6K}\right) \sigma_{xy}(X') + \sum_{x_0 \in \Lambda_0} \left(K_3 + \tfrac{1}{8} \ln\left(1 + e^{-2h-6K}\right)\right) \sigma_{\Delta_{x_0}}(X')
\end{aligned}
\tag{3.9}
$$

We observe that in the particular case $K_3 = K = 0$ the n.n. two body and 3 body interactions are ferromagnetic; therefore the statement in [57] p. 44 is wrong. (Runnels claimed the 3 body coupling constant to have the opposite sign of the two body one). Finally, we remark that particular cases of models without constraints and Hamiltonian such as given by Eq. (3.9) have already been investigated in part I.

3.1.3. Partial Trace Method Applied to Systems Satisfying the Subgroup Property

Let $\{\Lambda, \mathscr{C}_\Lambda, K\}$ be a system satisfying the subgroup property and such that there exists $\Lambda_0 \subset \Lambda$ satisfying Eq. (3.1). As we have seen in Sec. 1.2.1. the constraints are of the form

$\mathcal{C}_{B_\infty} = \frac{1}{2}(1+\sigma_{B_\infty})$ with $B_\infty \in \mathcal{B}_\infty$. Defining for any $x_0 \in \Lambda_0$

$$\mathcal{B}_{x_0} = \{ B \in \mathcal{B}_\infty \; ; \; B \ni x_0 \} \equiv \gamma_\infty (x_0) \tag{3.10}$$

Eq. (3.3) implies

$$F_{x_0}(X') = \prod_{B \in \mathcal{B}_{x_0}} \tfrac{1}{2}(1+\sigma_B(X'))\, e^{\sum_{B \ni x_0} K(B)\sigma_B(X')} + \prod_{B \in \mathcal{B}_{x_0}} \tfrac{1}{2}(1-\sigma_B(X'))\, e^{-\sum_{B \ni x_0} K(B)\sigma_B(X')} =$$

$$= 2^{-(|\mathcal{B}_{x_0}|-1)} \Big\{ \cosh\Big(\sum_{B \ni x_0} K(B)\sigma_B(X')\Big) \cdot \sum_{\substack{\beta \in \mathcal{P}(\mathcal{B}_{x_0}) \\ |\beta| = \text{even}}} \sigma_{\pi\beta}(X') + \sinh\Big(\sum_{B \ni x_0} K(B)\sigma_B(X')\Big) \cdot \sum_{\substack{\beta \in \mathcal{P}(\mathcal{B}_{x_0}) \\ |\beta| = \text{odd}}} \sigma_{\pi\beta}(X') \Big\} =$$

$$= e^{\sum_{B \ni x_0} K(B)\sigma_{B\tilde{\mathcal{B}}}(X')}\; 2^{-(|\mathcal{B}_{x_0}|-1)} \sum_{\beta \in \mathcal{P}(\mathcal{B}_{x_0}),\, |\beta| = \text{even}} \sigma_{\pi\beta}(X')$$

where $\tilde{\mathcal{B}}$ is any element in $\mathcal{B}_{x_0}$.

With $\tilde{\mathcal{B}}_{x_0} = \{ \tilde{B}_{x_0} \}$ any set of generators of the group $\mathcal{G}_{x_0} = \{ \pi\beta \; ; \; \beta \in \mathcal{P}(\mathcal{B}_{x_0}) \, , \, |\beta| = \text{even} \}$ we finally have

$$F_{x_0}(X') = e^{\sum_{B \ni x_0} K(B)\sigma_{B\tilde{\mathcal{B}}}(X')} \prod_{\tilde{B}_{x_0} \in \tilde{\mathcal{B}}_{x_0}} \tfrac{1}{2}(1 + \sigma_{\tilde{B}_{x_0}}(X')) \tag{3.11}$$

and we have

Corollary 2

Any lattice system $\{\Lambda, \mathcal{C}_\Lambda, K\}$ with constraints satisfying the subgroup property and having a sublattice $\Lambda_0 \subset \Lambda$ satisfying Eq. (3.1) is equivalent to a system with constraints satisfying the subgroup property with infinite bonds $\{\tilde{\mathcal{B}}_{x_0}\}_{x_0 \in \Lambda_0}$ and interactions $K' = K + \sum_{x_0 \in \Lambda_0} k(x_0; \;)$ where $k(x_0; X') = \sum_{B \ni x_0} K(B)\sigma_{\tilde{B}B}(X')$ and $\tilde{B}$ is any arbitrary element of $\mathcal{B}_{x_0}$.

Example 3 :

Consider the system $\{\mathcal{L}, \mathcal{C}, K\}$ with 3 body constraints on the shaded triangles of Fig.2 a) and an external field h .

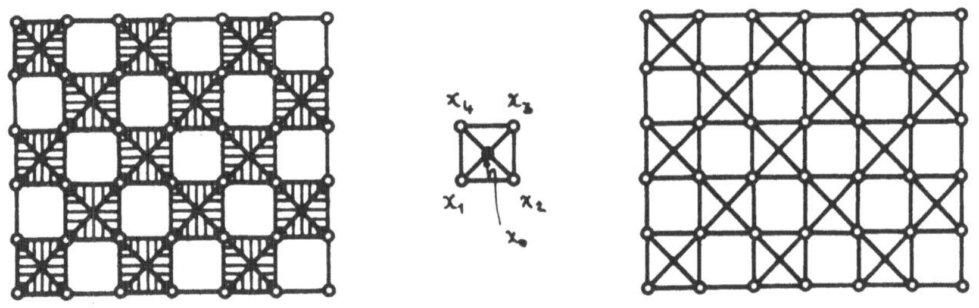

(a) (b) (c)

Fig. 2 : a) the sublattices $\Lambda_0 = \{\bullet\}$, $\Lambda/\Lambda_0 = \{\circ\}$ and the infinite bonds (shaded triangles)
b) $x_0 \in \Lambda_0$ and $\Delta_{x_0} = \{x_1, x_2, x_3, x_4\}$
c) the remaining lattice Λ/Λ_0 and the infinite bonds (diagonal bonds) of the equivalent model.

We then have :

$$\mathcal{B}_{x_0} = \{ (x_0 x_1 x_2), (x_0 x_2 x_3), (x_0 x_3 x_4), (x_0 x_1 x_4) \}$$

and

$$\mathcal{G}_{x_0} = \{ \phi , (x_1 x_3) , (x_2 x_4), (x_1 x_2 x_3 x_4) \}$$

hence $\mathcal{G}_{x_0}$ can be generated by $\tilde{\mathcal{B}}_{x_0} = \{(x_1 x_3),(x_2 x_4)\}$ and the remaining infinite bonds are the diagonal bonds shown in Fig. 2 c). Choosing $\tilde{B} = (x_0 x_1 x_2)$, we have with $B = (x_0) \in \mathcal{B}$ that $B\tilde{B} = (x_1 x_2)$ is a new finite bond. Since $\sigma_{x_1 x_2}(X) = \sigma_{x_3 x_4}(X)$ for any $X \in \mathcal{P}^a(\Lambda/\Lambda_0)$ Eq. (3.11) becomes in this case :

$$F_{x_0}(X') = e^{\frac{h}{2}(\sigma_{x_1 x_2}(X') + \sigma_{x_3 x_4}(X'))} \frac{1}{2}(1 + \sigma_{x_1 x_3}(X')) \frac{1}{2}(1 + \sigma_{x_2 x_4}(X'))$$

and the equivalent model satisfies the subgroup property and has horizontal n.n. interactions.

3.2. Equations for the Correlation Functions and Equilibrium Equations

To complete this chapter we give in this section a summary of results obtained for systems $\{\mathcal{L}, \mathcal{C}, K\}$. For more detailed informations and the proofs we refer to [113] .

With the same motivation as in Ch. 6 of Part I, we can use the partial trace method to derive a set of linear equations for the finite volume correlation functions for $\{\Lambda, \mathcal{C}_\Lambda, K\}$ and generalize them to discuss properties of states of systems $\{\mathcal{L}, \mathcal{C}, K\}$; these equations are equivalent to the D.L.R. equilibrium equations for systems with constraints satisfying the subgroup property, and for general hard core systems.

3.2.1. Integral Equations for the Correlation Functions

Following [113] , we introduce for any lattice system $\{\mathcal{L}, \mathcal{C}, K\}$ the equilibrium equations for the correlation functions as

$$\begin{aligned} &\omega[\mathcal{C}\sigma_X] = \omega[\sigma_X] \\ &\omega[(\tau_Z \mathcal{C})\sigma_X] = \omega\Big[(\tau_Z \mathcal{C})\,\sigma_X \tanh\Big(\sum_{B\in\gamma(Z)} K(B)\sigma_B\Big)\Big] \end{aligned} \tag{3.12}$$

for all $X, Z \in \mathcal{P}_f(\mathcal{L})$ such that $\sigma_X(Z) = -1$

where $\tau_Z \mathcal{C} = \prod_{C\in\{\mathcal{L}C\}} \tau_Z \mathcal{C}_C$ and $(\tau_Z \mathcal{C}_C)(X) = \mathcal{C}_C(ZX)$

It is easily verified [113] that the Gibbs States of finite systems are solutions of these equations. In this section we shall discuss the connection between these equations and the so called "DLR Equilibrium Equations" :

$$\begin{aligned} &\mathcal{C}(XY)\,\mu_\Lambda[X;dY] = \mu_\Lambda[X;dY] \\ &\mathcal{C}(XY)\,\mu_\Lambda[XZ;dY] = \mathcal{C}(XYZ)\,e^{-2\sum_{B\in\gamma(Z)} K(B)\sigma_B(XY)}\,\mu_\Lambda[X;dY] \end{aligned} \tag{3.13}$$

for all $\Lambda \in \mathcal{P}_f(\mathcal{L})$, $X\cup Z \subset \Lambda$, $Y \subset \mathcal{L}/\Lambda$.

A first connection between these two sets of equations is given by the following result :

Lemma 1

Any solution of the DLR equations Eq. (3.13) yields a solution of the equations for the correlation functions Eq. (3.12).

To establish the converse statement, we first define minimal subsets Z of $\mathcal{L}$ to be subsets of $\mathcal{L}$ such that for any $X \subset \mathcal{L}/Z$ there exist at most 2 subsets Z' of Z such that $\mathcal{C}(Z'X)=1$. For systems without constraints as with hard core constraints, the minimal subsets are one point subsets of $\mathcal{L}$. For systems with constraints satisfying the subgroup property, the minimal subsets Z will be admissible configurations such that $Z' \subset Z$ and Z' admissible implies $Z' = Z$ or $Z' = \phi$.

Lemma 2

Any state ω solution of the equilibrium equation Eq. (3.12) for all Z minimal has probability measure $\mu_\Lambda[X; dY]$ satisfying the DLR equations Eq. (3.13) for all Z minimal.

This establishes the equivalence of both sets of equilibrium equations provided that Z is minimal. However, this equivalence should also be true without any restriction on Z. We proved it for two classes of systems :

Theorem 2

For constraints such that the admissible configurations form a subgroup of $\mathcal{P}(\mathcal{L})$ and for constraints satisfying the hard core conditions, any state which is solution of Eq. (3.12) defines an equilibrium state. Moreover, any state which is solution of Eq. (3.12) with Z minimal, will be a solution of the full set of equations (i.e. for arbitrary Z).

Corollary

For any lattice system $\{\mathcal{L}, \mathcal{C}, K\}$ satisfying the subgroup property, the equilibrium states can be defined as the solutions of :

$$\omega[\sigma_X \sigma_{B_\infty}] = \omega[\sigma_X]$$

$$\omega[\sigma_X] = \omega[\sigma_X \tanh(\sum_{B \in \gamma'(Z)} K(B)\sigma_B)] \qquad (3.14)$$

for any $B_\infty \in \mathcal{B}_\infty, X \in \mathcal{P}_f(\mathcal{L}), Z \in \mathcal{P}_f^a(\mathcal{L})$ such that $\sigma_Z(X) = -1$.

satisfying the following positivity conditions :

for any $\Lambda \in \mathcal{P}_f(\mathcal{L})$, $\sum_{X \subset \Lambda} \omega[\sigma_X]\sigma_X(Y) \geq 0$ for all $Y \subset \mathcal{L}$.

Moreover, in the above equations, it is sufficient to consider a family of generators for $\mathcal{P}_f^a(\mathcal{L})$.

Finally, let us remark that if $\{\mathcal{L}, \mathcal{C}, K\}$ has the subgroup property, the equilibrium equations for the correlation functions have the same structure as the corresponding equation for systems without constraints discussed in Ch. 6 of Part I. Therefore in the same way it can be used to prove the unicity of the equilibrium state at high temperature, to obtain domains of analyticity, to derive upper bounds for the critical temperature, and to discuss Ergodic decomposition.

CHAPTER 4 - DUALITY TRANSFORMATION RESTRICTED TO FINITE BONDS

By analogy with the duality transformations for systems without constraints discussed in Part I, the concept of "Duality transformations for systems with constraints" [113] is introduced as a transformation from the original system $\{\mathcal{L}, \mathcal{C}, K\}$ into another system $\{\mathcal{L}^*, \mathcal{C}^*, K^*\}$. This transformation is defined by a surjection from the family of bonds of $\{\mathcal{L}, \mathcal{C}, K\}$ onto the bonds of $\{\mathcal{L}^*, \mathcal{C}^*, K^*\}$ which is such that it induces a bijection between the group $\mathcal{K}_\rho^a$, or $\Gamma^{a,\rho}$, onto the group $\Gamma^{a,\rho\,*}$ (HT-LT or LT-LT duality transformations) or onto the group $\mathcal{K}_\rho^{a\,*}$ (HT-HT or LT-HT duality transformations). We shall restrict ourselves to systems $\{\mathcal{L}, \mathcal{C}, K\}$ satisfying the subgroup property, and as application of these duality transformations we shall discuss the problem of the definition of phase transition.

4.1. Duality Transformations for Systems Satisfying the Subgroup Property

As remarked in Sec. 1.2.1, we can always assume that $\mathcal{B} \cap \mathcal{B}_\infty = \phi$. For such systems duality transformations can be defined by means of a surjection from the finite bonds $\mathcal{B}$ of the system $\{\mathcal{L}, \mathcal{C}, K\}$ onto the bonds $\mathcal{B}^*$ of a system without constraints. We shall show moreover that those systems $\{\mathcal{L}, \mathcal{C}, K\}$ are "equivalent" to systems without constraints.

By analogy with Part I, the duality transformations are introduced by the following definitions :

The system without constraints $\{\mathcal{L}^*, K^*\}$ is called a HT-LT dual for $\{\mathcal{L}, \mathcal{C}, K\}$ if there exists a surjection $d: B \mapsto B^*$ of the finite bonds $\mathcal{B}$ onto $\mathcal{B}^*$ which induces a bijection of $\mathcal{K}_\rho^a$ onto $\Gamma^{\rho *}$ such that :

$$\beta = \bigcup_{B \in \beta} d^{-1}[d\beta] \qquad \forall \beta \in \mathcal{K}_\rho^a \tag{4.1}$$

$$e^{-2K^*(B^*)} = \prod_{B\in d^{-1}B^*} \tanh K(B) \quad \text{for each} \quad B \in \mathcal{B}$$

i.e.

$$z_{B^*} = \prod_{B\in d^{-1}B^*} \tilde{z}_B$$

Similarly LT-HT and LT-LT duals are defined by replacing the pairs $(\mathcal{K}_f^a, \Gamma^{f*})$, respectively by $(\Gamma^{a,f}, \mathcal{K}_f{}^*)$, and by $(\Gamma^{a,f}, \Gamma^{f*})$, with corresponding change in the exp and tanh functions.

Finally, the system without constraints $\{\mathcal{L}^*, K^*\}$ is called a HT-HT dual for $\{\mathcal{L}, \mathcal{C}, K\}$ if there exists a surjection $d : B \mapsto B^*$ of the finite bonds $\mathcal{B}$ onto $\mathcal{B}^*$ which induces a bijection of $\mathcal{K}_f^a$ onto $\mathcal{K}^{f*}$ such that $\tilde{z}_B = \prod_{B\in d^{-1}B^*} \tilde{z}_{B^*}$ and $\Gamma^{a,f}$ is mapped into Γ^{f*}.

Introducing the observables $\mu_B = e^{-2K(B)\sigma_B}$ we obtain from the above definitions the following results.

Property 1

If $\{\Lambda^*, K^*\}$ is a HT-LT dual for $\{\Lambda, \mathcal{C}_\Lambda, K\}$, respectively LT-LT, HT-HT, then the partition functions and the correlation functions satisfy the following Duality Relations :

$$Z^{red}(\Lambda, \mathcal{C}_\Lambda, K) = |\mathcal{S}'|\, |\mathcal{S}^*|^{-1}\, |\Gamma^a| \prod_{B\in\mathcal{B}} (1+\tilde{z}_B)^{-1}\, Z^{red}(\Lambda^*, K^*) \qquad \text{HT-LT}$$

and

$$\omega_{(\Lambda, \mathcal{C}_\Lambda, K)}\Big[\prod_{B\in\beta} \mu_B\Big] = \omega^*_{(\Lambda^*, K^*)}\Big[\prod_{B\in\beta} \sigma_{dB}\Big] \quad \text{for all } \beta \in \mathcal{P}_f(\mathcal{B})$$

$$Z^{red}(\Lambda, \mathcal{C}_\Lambda, K) = |\mathcal{S}'|\, |\mathcal{S}^*|^{-1}\, Z^{red}(\Lambda^*, K^*) \qquad \text{LT-LT}$$

and

$$\omega_{(\Lambda, \mathcal{C}_\Lambda, K)}\Big[\prod_{B\in\beta} \sigma_B\Big] = \omega^*_{(\Lambda^*, K^*)}\Big[\prod_{B\in\beta} \sigma_{dB}\Big]$$

$$Z^{red}(\Lambda, \mathcal{C}_\Lambda, K) = |\mathcal{S}'|\, |\mathcal{S}^*|^{-1}\, |\Gamma^a|\, |\Gamma^*|^{-1} \frac{\prod_{B^*\in\mathcal{B}^*}(1+\tilde{z}_{B^*})}{\prod_{B\in\mathcal{B}}(1+\tilde{z}_B)} \cdot Z^{red}(\Lambda^*, K^*) \qquad \text{HT-HT}$$

and

$$\omega_{(\Lambda, \mathcal{C}_\Lambda, K)}\Big[\prod_{B\in\beta} \mu_B\Big] = \omega^*_{(\Lambda^*, K^*)}\Big[\prod_{\substack{B^*\in\beta^* \\ |d^{-1}B^*| = odd}} \mu_{B^*}\Big]$$

In the case of HT-HT and LT-LT duality, the relations between the invariant equilibrium states of $\{\mathcal{L}^*, K^*\}$ and $\{\mathcal{L}, \mathcal{C}, K\}$ are expressed by the following property :

Property 2

1. With $\{\mathcal{L}^*, K^*\}$ any HT-HT dual for the finite or infinite system $\{\mathcal{L}, \mathcal{C}, K\}$ such that d is a bijection, then every equilibrium state $\omega^*_{(\mathcal{L}^*, K^*)}$ invariant under $\mathcal{S}^*$ defines an equilibrium state $\omega_{(\mathcal{L}, \mathcal{C}, K)}$ which is invariant under $\mathcal{S}'$ by the relation :

$$\omega_{(\mathcal{L}, \mathcal{C}, K)}[\prod_{B \in \beta} \sigma_B] = \omega^*_{(\mathcal{L}^*, K^*)}[\prod_{B^* \in \beta^*} \sigma_{B^*}] \qquad \forall \beta \in \mathcal{P}_f(\mathcal{B})$$

$$\omega_{(\mathcal{L}, \mathcal{C}, K)}[\prod_{B \in \beta} \sigma_B \cdot \sigma_{B_\infty}] = \omega_{(\mathcal{L}, \mathcal{C}, K)}[\prod_{B \in \beta} \sigma_B] \qquad \forall B_\infty \in \mathcal{B}_\infty$$

$$\omega_{(\mathcal{L}, \mathcal{C}, K)}[\sigma_X] = 0 \qquad \forall X \notin \overline{\mathcal{B}'}$$

2. With $\{\mathcal{L}^*, K^*\}$ any LT-LT dual for the finite or infinite system $\{\mathcal{L}, \mathcal{C}, K\}$ such that d is a bijection, then there exists a bijection between $\mathcal{S}^*$-invariant equilibrium states $\omega^*_{(\mathcal{L}^*, K^*)}$ and $\mathcal{S}'$-invariant equilibrium states $\omega_{(\mathcal{L}, \mathcal{C}, K)}$.

For the proof, we refer to [113] .

It was shown in Part I that for LT-LT duality, the mapping d is always a bijection. This is not true for systems with constraints :

Property 3

With $\{\mathcal{L}^*, K^*\}$ any LT-LT dual for $\{\mathcal{L}, \mathcal{C}, K\}$, then d is a bijection if and only if for all $B \in \mathcal{B}$ and all $\overline{B}_\infty \in \overline{\mathcal{B}}_\infty$ the condition $B\overline{B}_\infty \in \mathcal{B}$ implies $\overline{B}_\infty = \phi$.

Proof

Suppose that there exists $B_1, B_2 \in \mathcal{B}$ such that $B_1 \cdot B_2 \in \overline{\mathcal{B}}_\infty$, then for any $\beta \in \Gamma^{a,f}$, $\beta \ni B_1$ if and only if

$\beta \supset B_2$. Therefore, for any LT-LT duality transformation $B_1, B_2 \in \mathcal{B}_\infty$ implies $dB_1 = dB_2$.

In particular, it follows from Property 3, that for systems without constraints, LT-LT duals can only be defined by bijections and that Property 2 holds for any LT-LT dual.

To conclude our dicussion on duality transformation, we shall now show that it is always possible to construct HT-LT and LT-LT duals for systems $\{\mathcal{L}, \mathcal{C}, K\}$ satisfying the subgroup property. The general method follows the one discussed in Part I for HT-LT duality for systems without constraints.

Let G be any subgroup of $\mathcal{P}_f(\mathcal{B})$ and $\{g_i\}$ a family of generators for G such that any element g in G is a finite product of generators ; with each g_i, we associate a point $r_i^* \in \mathcal{L}^*$ and we define the mapping $d : B \mapsto B^*$ by the relation :

$$B^* = \{ r_i^* \ ; \ g_i \ni B \}$$

It follows immediately that :

i) $B \in g_i$ implies $d^{-1}[dB] \subset g_i$

ii) $\bigcup_{B \in g_i} dB = \{B^* ; B^* \ni r_i^*\} = \gamma^*(r_i^*)$

therefore the mapping d induces a surjection of the group G onto the group Γ^{f*} satisfying the condition of Eq. (4.1); moreover d is a bijection from $\mathcal{B}$ onto $\mathcal{B}^*$ if and only if G separates $\mathcal{B}$. We conclude :

Theorem

Any lattice system $\{\mathcal{L}, \mathcal{C}, K\}$ satisfying the subgroup property and such that the group $\Gamma^{a,f}$ separates the finite bonds $\mathcal{B}$ is "equivalent" to a system without constraints $\{\mathcal{L}^*, K^*\}$ in the sense that there exists a bijection between symmetric equilibrium states of $\{\mathcal{L}, \mathcal{C}, K\}$ and symmetric equilibrium states of $\{\mathcal{L}^*, K^*\}$. Moreover, if the residual entropy density is zero and if the free energy density of both systems exist, then they will be proportional.

We emphasize that for systems $\{\mathcal{L}, \mathcal{C}, K\}$ this theorem cannot be extended to the set of all equilibrium states (see Sec. 4.2).

<u>Example : Model with 4 body constraints and 2 body forces on $\mathbb{Z}^2$</u>

Consider the usual n.n. Ising model together with a family of constraints $\mathcal{C}_C = \frac{1}{2}(1+\sigma_C)$ where C are 4- point subsets of $\mathbb{Z}^2$ as represented on Fig. 1.

<u>Fig. 1</u> : $\mathbb{Z}^2$ with 4-body constraints and 2 body forces.

The admissible configurations $\mathcal{P}^a(\mathcal{L})$ are a subgroup of $\mathcal{P}(\mathcal{L})$ and we have :

$$\mathcal{B} = \{<i,j>\} \quad ; \quad \mathcal{B}_\infty = \{ ▨ \} \quad ; \quad \mathcal{B}' = \mathcal{B} \cup \mathcal{B}_\infty \quad ; \quad \overline{\mathcal{B}'} = \mathcal{P}_f^{even}(\mathbb{Z}^2)$$

$$\mathcal{S} = \mathcal{B}^\perp = \{\phi, \mathbb{Z}^2\} \; ; \; \mathcal{S}_\infty = \mathcal{B}_\infty^\perp = \mathcal{P}^a(\mathcal{L}) \quad ; \quad \mathcal{S}' = \mathcal{B}'^\perp = \{\phi, \mathbb{Z}^2\}$$

The LT group $\Gamma^{a,f} \subset \mathcal{P}(\mathcal{B})$, and the HT group $\mathcal{K}_f^a \subset \mathcal{P}(\mathcal{B})$ are generated by the elements shown respectively on Fig. 2 a) and b).

<u>Fig. 2</u> : a) Generators of $\Gamma^{a,f}$ b) Generators of $\mathcal{K}_f^a$

We notice at once that the condition $B\bar{B}_\infty \in \mathcal{B}$ does not imply $\bar{B}_\infty = \phi$; therefore LT-LT Duality Transformations will be defined by a mapping d which is not a bijection; in fact parallel bonds situated on the same C will be mapped onto the same B^* .

On the other hand, HT-LT Duality Transformations may be defined by means of a bijection d . In particular applying the general method to the above generators of $\mathcal{H}_f^a$ we obtain as HT-LT dual, the system with 2-body and 3-body forces represented on Fig. 3 ; moreover, this model can furthermore be reduced to a model with 2-body forces only by means of a partial trace.

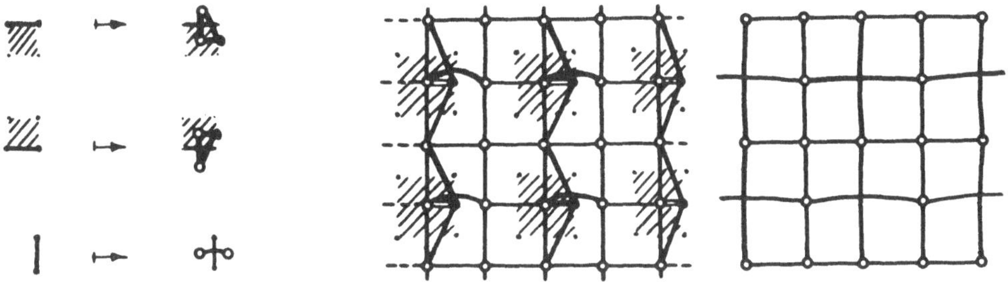

(a) (b) (c)

Fig. 3 : a) The HT-LT duality mapping $d : \mathcal{B} \to \mathcal{B}^*$
b) HT-LT dual for the model with $\mathcal{C}_C = \frac{1}{2}(1+\sigma_C)$
c) Model obtained from b) by partial trace on the sites.

The same result can also be obtained directly by means of the LT-HT Duality transformation defined on Fig. 4, which gives :

$$Z^{red}(\Lambda, K) = Z^{red}(\Lambda^*, K_1^*, K_2^*)$$

with $\tanh K_1^* = e^{-2K}$ and $\tanh K_2^* = e^{-4K}$

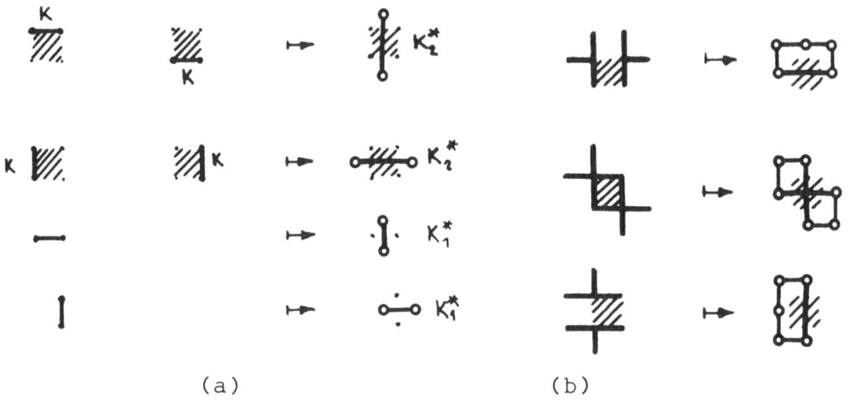

Fig. 4 : a) The LT-HT Duality Transformation d
b) Image of generators of $\Gamma^{a,f}$ under the mapping d.

4.2. On the Definition of Phase Transitions [26]

As we have already discussed in Ch. 3 and 4 of Part I, several definitions can be introduced to characterise a phase transition. In particular by analogy with thermodynamics one can define a phase transition by the singularity of some Gibbs Potential as function of its natural variables; on the other hand to prove the existence of a phase transition (Ch. 4 of Part I) one usually defines a phase transition by means of an instability of the state with respect to boundary perturbation which is equivalent to the non uniqueness of equilibrium state. For a long time it had been suspected that these two definitions coincide; in this section we shall use the result of Sec. 4.1 to show that there exists phase transitions associated with a singularity of the free energy for systems such that the equilibrium state is unique at all temperatures.

Let us apply the results of the previous section to the model defined in example 1 of Sec. 1.2.1.. Provided that

$h = \beta H > 0$ this model is a HT-LT dual for the n.n. Ising model without field [2] , and therefore the reduced free energies are related by

$$f_{H}^{red}(T) = f_{Ising,\, J_1 = H}^{red}(T)$$

We thus conclude that there exists a phase transition at $T_c = T_c^{(Ising)}$ associated with a singularity of $f_H^{red}(T)$. However we shall now show that there exists a unique equilibrium state for the infite system invariant under some subgroups $\mathcal{T}$ of $\mathbb{Z}^2$ such that $|\mathbb{Z}^2/\mathcal{T}| < \infty$. To establish this result, consider the LT-LT duality transformation defined by the bijection d shown in Fig. 5 a).

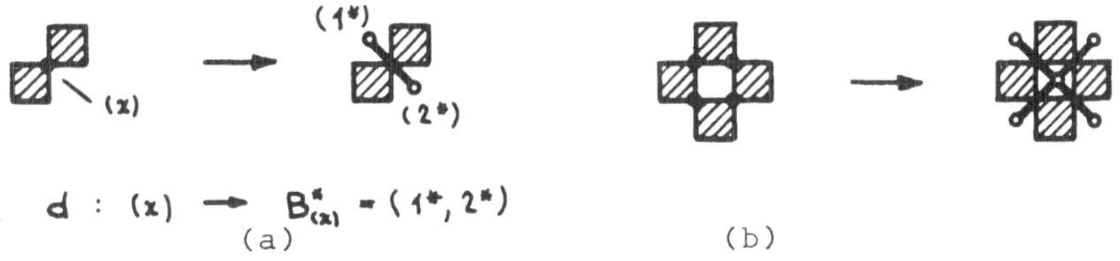

Fig. 5 : a) the duality map $d : \mathcal{B} \to \mathcal{B}^*$
b) a generator of $\Gamma^{a,f}$ and its image under d.

As shown in Fig. 5 b) the map d induce a bijection between the generators of $\Gamma^{a,f}$ and the generators of Γ^{*f} of the n.n. Ising model without field. Therefore the n.n. Ising model without field is a LT-LT dual restricted to finite bonds for our model; moreover from property 2 of Sec. 4.1. it follows that there exists a bijection between $\mathcal{S}'$- invariant equilibrium states ω for our model and $\mathcal{S}^*$-symmetric states $\omega^{*(Ising)}$ of the Ising model. From the unicity of the $\mathbb{Z}^2$-invariant, symmetric, equilibrium state of the Ising model [97] and the fact that $\mathcal{S}' = \{\phi\}$ we conclude that there exists a unique equilibrium state of our model invariant under some subgroup $\mathcal{T}$ of the translation group such that $|\mathbb{Z}^2/\mathcal{T}| < \infty$.

In conclusion although there exists a phase transition associated with a singularity of the free energy, the equilibrium state is unique at all temperatures. Moreover this phase transition which is not associated with a symmetry breakdown is related to a phase transition with spontaneous symmetry breakdown of an equivalent system.

Let us also remark that equilibrium state can also be defined as convex combination of thermodynamic limit of Gibbs States with specified boundary conditions. It is then possible to show that the correlation functions of our model are mapped onto the even point correlation functions of the n.n. Ising model without field and well defined boundary conditions. With this definition of equilibrium states (which is equivalent to the previous one in view of the results of Ch. 3), the unicity of the equilibrium state follows again from the results of reference [97].

To conclude the discussion of this model we shall make the following comments concerning the state and the nature of this phase transition.

1 - Using the LT-LT duality transformation one can show that the equilibrium state is invariant under the translation group of the constraints but is not $\mathbb{Z}^2$- invariant.

2 - The phase transition is characterised by a "Long range order parameter" $\eta(T)$ such that $\eta = 0$ for $T \geq T_c$ and $\eta > 0$ for $T < T_c$, in fact $\eta(T) = \lim_{n \to \infty} \omega[\sigma_1 \dots \sigma_n] = [m^{Ising}_{J_2 = H}(T)]^2$ where $(1, \dots, n)$ are the sites situated on a diagonal.

3 - The phase transition can be furthermore characterised by a "local order parameter $\mu(T)$" such that $\mu(T) > 0$ for $T > T_c$ and $\mu(T) = 0$ for $T \leq T_c$; in fact $\mu(T) = \lim_{K_4 = K(B_4) \to \infty} \langle \mu_{B_4} \rangle_{(h, K_4)} = m^{Ising}_{J_2 = H}(T^*)$ where $e^{-2\frac{H}{kT^*}} = \tanh \frac{H}{kT}$; it should be stressed that $\mu(T)$ is not an ordinary order parameter since (*)

$$\mu(T) = \lim_{\lambda \to 0} \frac{1}{\sqrt{2\lambda}} \{ \omega_{f, \lambda = 0}[\sigma_{B_4}] - \omega_{f, \lambda}[\sigma_{B_4}] \}$$

while for standard order parameter such as the spontaneous magnetisation, one has :

$$-m(T) = \lim_{h \to 0} \{ \omega_{f, h=0}[\sigma_x] - \omega_{f, h}[\sigma_x] \}$$

(*) Where $\omega_{f, \lambda}$ is the equilibrium state associated with free loundary condition and $\tanh 2K_4 = 1 - \lambda$

4 - From the general relation $\langle \mu_{B_4} \rangle_{(h,K_4)} = m^{Ising}(K_2 = h^*, K_1 = K_4^*)$ (Ch. 4) it follows that for $h < 0$ we have the following relation :

$$\mu(T) = m^{Ising}(K_2 = |h|^* + i\frac{\pi}{2})$$

As we have seen in Part I (Sec. 2.6) this last quantity can be calculated explicity and we obtain the following diagram

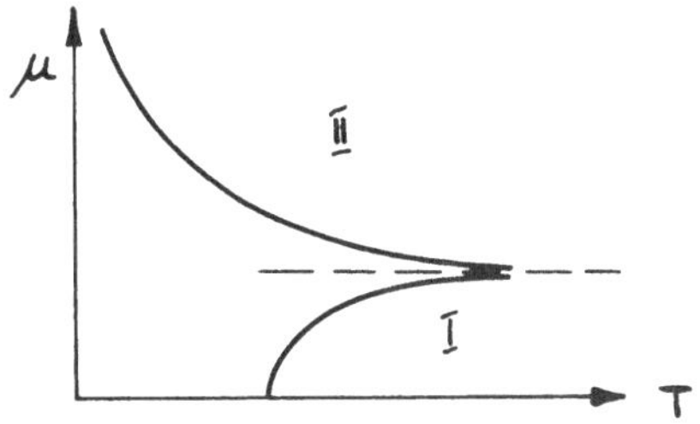

Fig. 6 : Local order parameter μ for $h > 0$ (I) and $h < 0$ (II).

5 - It is interesting to note that this model is equivalent to an 8 vertex model with $\omega_1 = \omega_2^{-1}$, $\omega_i = 0$ $i = 3, 4, \ldots, 8$ and the behaviour of $\mu(T), h>0$, is reminiscent of the behaviour of the vertical polarisation of the F-model with respect to the vertical field.

Another model which shows similar properties is the example 3 of Sec. 2.6 (Part I); for the special case $h \neq 0$ $K_1 = 0$, $K_2 = K_3 = K \to +\infty$ a HT-LT duality transformation yields a hexagonal Ising model without field and we derive the same conclusions as before.

In conclusion some model with K-body interaction and external field can give rise to a phase transition only in the limit where some interaction become infinite.

CHAPTER 5 - ASANO CONTRACTIONS AND UNICITY OF STATE

Properties of zeroes of the partition function for several particular systems with constraints e.g. hard core, ferroelectric and other models have already been worked out by several authors[124,100,101,105,117-121]. To our knowledge, all these investigations were limited to a discussion of analyticity domains for the free energy only and the problem of the unicity of the state and analyticity of the correlation functions was disregarded. In this chapter we shall extend the Asano-Ruelle method to systems with constraints and derive general theorems to obtain analyticity properties and discuss the unicity of equilibrium states [122].

As we shall see the generalisation is achieved by extending the contraction process not only to the set of finite bonds but to the total set of bonds $\mathcal{B}' = \mathcal{B} \cup \mathcal{B}_\infty$. Unfortunately, the presence of constraints does not allow a full generalisation in the sense that some results obtained for systems without constraints are not available for systems with constraints, thus showing the limits of the Asano-Ruelle method when applied to systems with constraints.

The most complete information is derived for systems satisfying the subgroup property. We derive necessary and sufficient conditions to construct coverings of the set of bonds which satisfies the Asano condition. Without any supplementary conditions on the system, analyticity of the free energy and unicity of the state at high temperature can be obtained by investigations on the H.T. expansion. For ferromagnetic systems, the L.T. expansion yields the same results at low temperature, provided the interactions are translationally invariant and the constraints are either translationally invariant or satisfies $\overline{\mathcal{B}} \supset \mathcal{B}_\infty$.

For systems not satisfying the subgroup property, the situation becomes worse. A first conclusion is that no general results can be obtained by the Asano-Ruelle method in the high temperature region. The reason is that the polynomial associated with the H.T. expansion cannot be built up by means of Asano contractions. For the L.T. expansion, the contraction process is well defined and general results can be derived. Once more, the most general results are limited to systems satisfying $\bar{\mathcal{B}} \supset \mathcal{B}_{\infty}$, for which we can disregard the infinite bonds in the contraction process. If this last condition is satisfied, analyticity of the free energy for ferromagnetic systems with suitable translationally invariance properties is derived at low temperature. In some examples, we show how to deal with systems not satisfying $\bar{\mathcal{B}} \supset \mathcal{B}_{\infty}$. To derive the unicity of the state at low temperature, however, requires a supplementary condition. This condition is explicitly given in the case of general hard core lattice systems. In particular, we show that this condition is always satisfied for hard core systems with a field.

5.1. Systems Satisfying the Subgroup Property

In this section, we investigate systems $\{\mathcal{L}, \mathcal{C}, K\}$ with constraints satisfying the subgroup property. We recall that the system is said $\mathbb{Z}^{\nu}$-invariant if there exists an action of $\mathbb{Z}$ on $\mathcal{L}$ such that $K(T_a B) = K(B)$ and $\mathcal{C}_C(T_a X) = \mathcal{C}_{T_a C}(X)$.

As shown in Ch. 2, the reduced partition function $Z^{red}(\Lambda, \mathcal{C}_\Lambda, K)$ can be written either as a sum over the group Γ^a or over the group $\mathcal{H}^a$, subgroups of $\mathcal{P}(\hat{\mathcal{B}})$.

Defining

$$z_B = e^{-2K(B)} \qquad z^{\beta} = \prod_{B\in\beta} z_B$$

$$\tilde{z}_B = \tanh K(B) \qquad \tilde{z}^{\beta} = \prod_{B\in\beta} \tilde{z}_B \tag{5.1}$$

the reduced partition function, Eqs. (2.1) and (2.13) becomes

$$Z^{red}(\Lambda, \mathcal{C}_\Lambda, K) = |\mathcal{B}'| \sum_{\beta \in \Gamma^a} z^\beta = |\mathcal{B}'| \, |\Gamma^a| \prod_{B \in \beta} (1+\tilde{z}_B)^{-1} \sum_{\beta \in \mathcal{K}^a} \tilde{z}^\beta \quad (5.2)$$

As in Part I (Ch. 8) to study analyticity properties we are interested in domains of the complex plane not containing zeroes of polynomials $\mathcal{M}(z_{\mathcal{B}})$ in the complex variables $z_{\hat{\mathcal{B}}} = \{z_B\}_{B \in \hat{\mathcal{B}}}$ of the form

$$\mathcal{M}(z_{\hat{\mathcal{B}}}) = \sum_{\beta \in G} z^\beta \quad (5.3)$$

where G is <u>any</u> subgroup of $\mathcal{P}(\hat{\mathcal{B}})$.

Since Asano contraction (Ch. 8, Part I), is defined for arbitrary sets $\mathcal{B}$ and since $\mathcal{K}^a$ and Γ^a are subgroups of $\mathcal{P}(\hat{\mathcal{B}})$, the results of sections 8.2, 8.3 and 8.4 of Part I remain valid for systems with constraints satisfying the subgroup property if we replace $\mathcal{B}$ by $\hat{\mathcal{B}} = \mathcal{B}'/\bar{\mathcal{B}}_\infty$.

Let us first apply these results to the <u>H.T. expansion</u> of $Z^{red}(\Lambda, \mathcal{C}_\Lambda, K)$. In this case, we have

$$\begin{aligned} G &= \mathcal{K}^a & \tilde{z}_B &= \tanh K(B) \\ G^\perp &= \Gamma^a & \hat{\tilde{z}}_B &= z_B = e^{-2K(B)} \end{aligned} \quad (5.4)$$

With $\{X_i\}$ any family of generators for $\mathcal{P}^a(\Lambda)$, the family $\{\gamma'(X_i)\}$ then generates Γ^a and a covering of $\hat{\mathcal{B}}$ satisfying the Asano condition is given by

$$\hat{\mathcal{B}} = \bigcup_{i=1}^{n} \{\gamma'(X_i)\} \quad (5.5)$$

Moreover with a covering of $\hat{\mathcal{B}}$ defined by minimal generators $\gamma'(X_i)$ of Γ^a, it follows :

$$G_i^\perp = G^\perp \cap \mathcal{P}(\mathcal{B}_i) = \Gamma^a \cap \mathcal{P}(\mathcal{B}_i) = \{\phi, \gamma'(X_i)\} \quad (5.6)$$

and the corresponding small polynomials are

$$\mathcal{M}_i(z_{\mathcal{B}_i}) = 1 + \prod_{B \in \gamma'(X_i)} z_B \quad (5.7)$$

Using now Section 8.3 and Ch. 9 of Part I it follows :

Theorem 1

Let $\gamma'(X_i)_{i=1..n}$ be a family of minimal genenerators of Γ^a and $\mathcal{B}_i = \gamma'(X_i)$ then :

i) The reduced partition function of the finite system $\{\Lambda, \mathcal{C}_\Lambda, K\}$ with constraints satisfying the subgroup property is different from zero in the complex domain :

$$|\tanh K(B)| \leq \left(tg \frac{\delta_B}{2}\right)^{N_B} \quad N_B = \text{number of } \mathcal{B}_i \text{ containing } B$$
$$|\arg e^{-2K(B)}| \leq \delta_B \qquad \text{if } N_B = 1 \tag{5.8}$$

where δ_B, $B \in \hat{\mathcal{B}}$, are arbitrary real numbers such that $\delta_B > 0$ if $N_B > 1$, $\delta_B \geq 0$ if $N_B = 1$ and $\sum_{B \in \gamma'(X_i)} \delta_B = \pi$.

ii) If the infinite system is $\mathbb{Z}^\nu$-invariant then the reduced free energy is an analytic function of the complex variables $z_B = \tanh K(B)$, $B \in \mathcal{B}_0$, where $\mathcal{B}_0$ is a fundamental family of bonds, in the domain defined by Eq. (5.8). Moreover, there exists a unique $\mathbb{Z}^\nu$- invariant equilibrium state and the correlation functions are analytic in the high temperature domain defined by Eq. (5.8).

Second, for the L.T. expansion of $Z^{red}(\Lambda, \mathcal{C}_\Lambda, K)$ one has

$$G = \Gamma^a \qquad z_B = e^{-2K(B)}$$
$$G^\perp = \mathcal{K}^a \qquad \hat{z}_B = \tilde{z}_B = \tanh K(B) \tag{5.9}$$

Since the group $\mathcal{K}^a$ is obtained from $\mathcal{K}'$ by removing the bonds in $\bar{\mathcal{B}}_\infty$, the results obtained in Part I immediately generalise to systems with constraints satisfying the subgroup property and we have

Theorem 2

For any system $\{\mathbb{Z}^\nu, \mathcal{C}, K\}$ satisfying the subgroup property with $\mathbb{Z}^\nu$ invariant, finite range ferromagnetic interactions and constraints $\mathcal{C}$ defined by translations $b \in \mathcal{T} \subset \mathbb{Z}^{\nu}$ of a fundamental family $\mathcal{B}_{\infty,0}$ of infinite bonds:

i) The partition function of the finite system $\{\Lambda, \mathcal{C}_{\Lambda}, K\}$ is different from zero in the low temperature domain

$$\begin{aligned} &|z_B = e^{-2K(B)}| \leq \prod_{\varkappa_i \ni B} tg \frac{\delta_{i,B}}{2} && \text{if } z_B \text{ undergoes contraction} \\ &|\arg \tanh K(\tilde{B})| \leq \delta_{i,\tilde{B}} && \text{otherwise} \end{aligned} \tag{5.10}$$

where $\{\varkappa_i\}$ is a family of minimal generators for $\mathcal{K}^{a}$ and $\delta_{i,B}$, $B \in \hat{\mathcal{B}}$, are arbitrary real numbers such that $\delta_{i,\tilde{B}} \geq 0$, $\delta_{i,B} > 0$ and for all i $\sum_{B \in \varkappa_i} \delta_{i,B} = \pi$.

ii) If $\overline{\hat{\mathcal{B}}} \supset \mathcal{B}_{\infty}$ or if $\mathcal{T} = \mathbb{Z}^{\nu}$, the reduced free energy of the infinite system is an analytic function of the complex variables $z_B = e^{-2K(B)}$, $B \in \hat{\mathcal{B}}_0$, where $\hat{\mathcal{B}}_0$ is a fundamental family of bonds, in the domain defined by Eq. (5.10)

iii) Under the same condition as ii), if $e^{-2K(B)} \leq \prod_{\varkappa_i \ni B} tg \frac{\delta_{i,B}}{2}$ with $\sum_{B \in \varkappa_i} \delta_{i,B} = \pi$, there exists a unique state invariant under $\mathcal{T}$ and $\mathcal{S}' = \mathcal{B}'^{\perp}$; moreover the correlation functions are analytic in the domain defined by Eq. (5.10).

For the proof of this theorem we refer to [122, 123].

Remark :

This theorem gives a non vanishing analyticity domain at low temperature for infinite systems $\{\mathbb{Z}^{\nu}, \mathcal{C}, K\}$ with ferromagnetic interactions $K(B) > 0$. The domain Eq. (5.10) is not the best one and can generally be enlarged by a more careful study of the small polynomials.

Example :

Consider the usual ferromagnetic Ising model with $\mathcal{B} = \{B = (x,y) ; x,y = n.n.\}$ defined on $\mathbb{Z}^{2}$ together with 4 body constraints $C = (x_1, x_2, x_3, x_4)$ which are defined by the shaded unit squares on Fig. 1 . These constraints are not Z_-^2 invariant, but they have the property $\bar{\mathcal{B}} \supset \mathcal{B}_\infty$.

$$\mathcal{B} = \{ (x,y) ; x, y = n.n. \}$$
$$\mathcal{B}_\infty = \{ C = \underset{x_4\ x_2}{\overset{x_3}{\boxtimes}}{}^{x_3} \}$$
$$\mathcal{B}' = \mathcal{B} \cup \mathcal{B}_\infty$$
$$\hat{\mathcal{B}} = \mathcal{B}$$

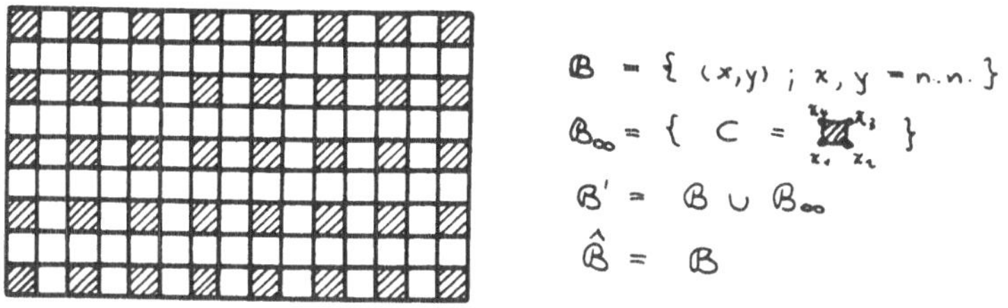

Fig. 1 : Distribution of local constraints $\mathscr{C}_c$ (C = shaded square) on the square lattice Z^2 .

The constraints are then defined by $\mathscr{C}_c = \frac{1}{2} (1 + \sigma_c)$; the locally admissible configurations are listed in Fig. 2 .

(a)

(b)

(c)

(d)

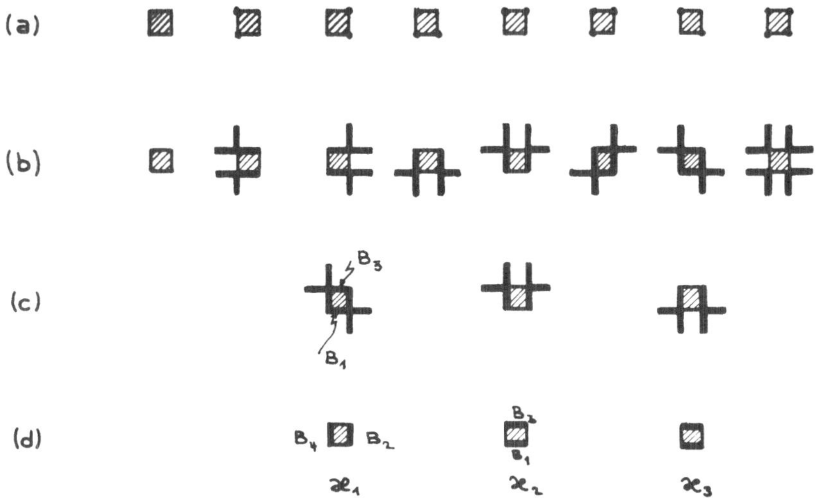

Fig. 2 : (a) Admissible configuration on the shaded unit squares (generators of $\mathcal{P}^a (Z^2)$)

(b) corresponding elements of $\Gamma^a \subset \mathcal{P}(\hat{\mathcal{B}})$

(c) minimal generators of Γ^a

(d) minimal generators of $\mathcal{K}^a$

Let us first consider the H.T. expansion i.e. $\mathcal{G} = \mathcal{K}^a$. Since $\bar{\mathcal{B}} \supset \mathcal{B}_\infty$ the infinite bonds can be disregarded. For the choice of minimal generators for $\mathcal{G}^\perp = \Gamma^a$ indicated on Fig. 2 c), we have $N_{B_1} = N_{B_3} = 1$, $N_B = 3$ otherwise. With $\delta_{B_1} = \delta_{B_3} = 0$, $\delta_B = \frac{\pi}{6}$ otherwise, theorem 1 implies analyticity of the free energy in the domain

$$\begin{aligned} &|\tanh K(B)| \leq \left(tg \tfrac{\pi}{12}\right)^3 = 0.019\ldots \qquad B \neq B_1, B_3 \\ &|\arg e^{-2K(B_1)}| = |\arg e^{-2K(B_3)}| = 0 \end{aligned} \tag{5.11}$$

Notice that another choice of minimal generators does not improve this domain . Remembering the analyticity domain for the Ising model without constraints, $|\tanh K(B)| < (\sqrt{2}-1)^2 = 0.17\ldots$, we observe that the presence of constraints considerably reduces the high temperature domain in which analyticity can be proven.

For the L.T. expansion, we can choose two minimal generators of $\mathcal{G}^\perp = \mathcal{K}^a$ on the shaded unit squares and one on the unshaded unit squares of $\mathbb{Z}^2$ (see Fig. 2 d)). Theorem 2 yields with

$$\begin{aligned} \delta_{i,B} &= \frac{\pi}{2} \qquad \text{if} \quad B \in \bigcup_a T_a \varkappa_1 \ \bigcup_a T_a \varkappa_2 \\ \delta_{i,B} &= \frac{\pi}{4} \qquad \text{if} \quad B \in \bigcup_a T_a \varkappa_3 \end{aligned}$$

$$\mathcal{M}(z_{\mathcal{B}}) \neq 0 \quad \text{if} \quad \begin{cases} |z_B| < \sqrt{2}-1 & \text{for } B \subset C,\ C \in \{LC\} \\ |z_B| < (\sqrt{2}-1)^2 = 0.17\ldots & \text{otherwise} \end{cases} \tag{5.12}$$

For a non uniform choice of $\delta_{i,B}$ if $B \in \tau_a \varkappa_3$, we obtain a uniform domain

$$\mathcal{M}(z_{\mathcal{B}}) \neq 0 \ \text{if} \quad |z_B| < \frac{\sqrt{2\sqrt{2}-1}-1}{\sqrt{2}} = 0.24\ldots \quad \text{for all } B \in \mathcal{B} \tag{5.13}$$

moreover the reduced free energy of $\{\mathbb{Z}^2, \mathcal{C}, K\}$ is analytic within the same domain.

We remark that the constraints enlarge the low temperature domain in which analyticity can be proven (for the same model, but without contraints one has $\mathcal{M}(z_{\mathcal{B}}) \neq 0$ if $|z_B| < (\sqrt{2}-1)^2$) .

In conclusion, the presence of constraints improves the low temperature domain but reduces the high temperature domain if one compares the domain obtained for the model with constraints to those of the same model but without constraints. This result can be explained by the fact that the presence of constraints satisfying the subgroup property reduces the number of terms in the L.T. expansion whereas the number of terms in the H.T. expansion increases (provided $\tilde{\mathcal{B}} \supset \mathcal{B}_\infty$). Hence less contractions are needed to build up the L.T. expansion but more contractions are needed to build the the H.T. expansion by means of Asano contractions.

5.2. Systems Without the Subgroup Property

In Sec. 2.2. we derived two H.T. expansions for the reduced partition function. The H.T. expansion given by Eq. (2.10) has coefficients $\mathcal{C}_\Lambda(\beta')$ which have no factorization properties if the system does not satisfy the subgroup property. Therefore the polynomial associated with this expansion cannot be obtained by Asano contractions. Unfortunately the same conclusion holds for the second H.T. expansion expressed in Eq. (2.11). Hence we conclude that for systems $\{\mathcal{L}, \mathcal{C}, K\}$, not satisfying the subgroup property, there is no general possibility to discuss analyticity properties in the high temperature region by the Asano-Ruelle method. This negative statement is certainly due to the fact that even at very high temperature, the action of the constraints of these systems is not weakened by the temperature.

We therefore focus our attention on the L.T. expansion of $Z^{red}(\Lambda, \mathcal{C}_\Lambda, K)$ and give conditions on the constraints and the interactions to obtain analogous results in the low temperature region as for systems without constraints and systems satisfying the subgroup property.

First, for <u>any</u> system $\{\mathcal{L}, \mathcal{E}, K\}$, we shall derive sufficient conditions to find coverings satisfying the Asano condition. Then we discuss the class of systems $\bar{\mathcal{B}} \supset \mathcal{B}_\infty$; for these systems, the presence of infinite bonds can be ignored and analyticity domains for infinite systems with specified translational invariance properties are derived. Afterwards we investigate briefly models not satisfying $\bar{\mathcal{B}} \supset \mathcal{B}_\infty$, models for which infinite bonds may undergo contractions. Next, we study general hard core systems and show which supplementary conditions imply the unicity of the state of hard core systems.

5.2.1. Covering Sets Satisfying the Asano Condition

<u>Property 1</u>

Let $\mathcal{B} = \bigcup_i \mathcal{B}_i$ be a finite covering of the arbitrary finite set $\mathcal{B}$; $\mathcal{M}$ be any subset of $\mathcal{P}(\mathcal{B})$ and $\mathcal{M}_i = \{ \beta \cap \mathcal{B}_i ; \beta \in \mathcal{M} \}$. Then if

(a) the subgroup of $\mathcal{P}(\mathcal{B})$ generated by $\{ \mathcal{M}_i^\perp \equiv \mathcal{M}^\perp \cap \mathcal{P}(\mathcal{B}_i) \}$ coincides with $\mathcal{M}^\perp$;

and

(b) for any β in $\bar{\mathcal{M}}$, the conditions $\beta \cap \mathcal{B}_i \in \mathcal{M}_i$ for all i imply $\beta \in \mathcal{M}$

then the polynomial $\mathcal{M}(z_\mathcal{B}) = \sum_{\beta \in \mathcal{M}} z^\beta$ is the Asano Contraction of the family of polynomials :

$$\mathcal{M}_i (z_{\mathcal{B}_i}) = \sum_{\beta_i \in \mathcal{M}_i} z^{\beta_i} \tag{5.14}$$

<u>Proof</u> :

Using the same argument [108] as given in Part I, it is straightforward to show that :
condition (a) implies that for all β such that $\beta \cap \mathcal{B}_i \in \bar{\mathcal{M}}_i$ for all i , we have $\beta \in \bar{\mathcal{M}}$. Hence conditions (a) and (b) then imply $\beta \in \mathcal{M}$, which is the Asano condition.

We should remark that for systems not satisfying the subgroup property, condition (a) is no longer sufficient to satisfy the Asano condition. On the other hand, conditions (a) and (b) are only sufficient but not necessary. We also emphasize that property 1 is general and valid for any arbitrary set $\mathcal{B}$ and in consequence also for the total set of bonds $\mathcal{B}' = \mathcal{B} \cup \mathcal{B}_\infty$. (See Sec. 5.2.3.)

It should be remarked that the small polynomial $M_i(z_{\mathcal{B}_i})$ is the reduced partition function $Z^{red}(\Lambda_i, \mathcal{C}_{\Lambda_i}, \mathcal{B}_i)$ of the subsystem $\{\Lambda_i, \mathcal{C}_{\Lambda_i}, \mathcal{B}_i\}$ where $\mathcal{C}_{\Lambda_i} = \prod_{C \subset \Lambda_i} \mathcal{C}_C$.

Example : Analyticity Property for Systems of Nonself-Intersecting Lattice Polygons [124]

We consider the ν-dimensional lattice $(\mathbb{Z}^\nu, \mathcal{B})$ where $\mathcal{B}$ is the family of bonds defined by nearest neighbours pairs of sites. For any finite $\Lambda \subset \mathbb{Z}^\nu$ we associate the generating function

$$Z_\Lambda(z_{\mathcal{B}}) = \sum_{\beta \in \mathcal{M}} z^\beta \qquad z^\beta = \prod_{B \in \beta} e^{-2K(B)}$$

where $\mathcal{M}$ is the subset of closed graphs such that each site $x \in \Lambda$ has coordination 0 or 2, i.e. $\mathcal{M}$ is the set of non self-intersecting multipolygons. $\mathcal{M}$ is not a subgroup of $\mathcal{P}(\mathcal{B})$ but $\bar{\mathcal{M}} = \mathcal{K}_\Lambda$ and $\mathcal{M}^\perp = \Gamma_\Lambda$. Therefore the covering of $\mathcal{B}$ defined by

$$\mathcal{B} = \bigcup_{x \in \Lambda} \gamma(x) \qquad \mathcal{M}_x = \{\beta \subset \gamma(x) ; \ |\beta| = 0, 2\}$$

satisfies the conditions (a) and (b) of property 1, and thus $Z_\Lambda(z_{\mathcal{B}})$ is the Asano contraction of

$$Z_x = \sum_{\beta \in \mathcal{M}_x} z^\beta = 1 + \sum_{(B_i, B_j) \subset \gamma(x)} z_{B_i} z_{B_j}$$

From Grace Theorem [106, b] $Z_x \neq 0$ if $z_B \notin C_x$ for all $B \ni x$ where C_x is a circular domain in the complex plane which contains all the roots of $Z_x(\{z_B = z\}) = 1 + \binom{q}{2} z^2 = 0$ i.e. $z_\pm = \pm i \binom{q}{2}^{-1/2}$ where q is the coordination number of the lattice.

Since any bond undergoes one contraction it follows from Ruelle's theorem that $Z_\Lambda \neq 0$ if $|z_B = e^{-2K(B)}| < \binom{q}{2}^{-1}$ $\forall\, B \in \mathcal{B}$.

In conclusion the "free energy" of the non interacting polygon model is an analytic function in $z = e^{-2K}$ in the domain $|z| < \binom{q}{2}^{-1}$

For the square lattice we obtain analyticity for $|z| < \frac{1}{6}$ which is a domain slightly smaller than the one obtained for the Ising model for which we had obtained $|z| < (\sqrt{2}-1)^2 = 0.17\ldots$.

Let us remark that for the square lattice this model is equivalent to an Ising model with constraints such that $\bar{\mathcal{B}} \supset \mathcal{B}_\infty$; as we shall see in the next section for such systems it is always possible to obtain analyticity domain for $|z_B| < r_B$, $r_B > 0$.

5.2.2. Systems with $\bar{\mathcal{B}} \supset \mathcal{B}_\infty$ or $\phi \in \mathcal{P}^a(\mathcal{L})$

For systems having finite bonds $\mathcal{B}$ satisfying the condition $\bar{\mathcal{B}} \supset \mathcal{B}_\infty$, where $\bar{\mathcal{B}}$ is the subgroup of $\mathcal{P}(\mathcal{L})$ generated by finite products of elements in $\mathcal{B}$, it is possible to consider coverings of $\mathcal{B}$ to build up the partition function by means of Asano contractions. Notice that this class contains in particular all lattice systems with external field. By property 1b of Sec. 2.1. the L.T. expansion is given by :

$$Z^{red}(\Lambda, \mathcal{C}_\Lambda, K) = |\mathcal{I}| \sum_{\beta \in \gamma[\mathcal{P}^a(\Lambda)]} \prod_{B \in \beta} e^{-2K(B)} \qquad (5.15)$$

With $\mathcal{M} = \gamma[\mathcal{P}^a(\Lambda)]$, the following corollary of property 1 yields coverings satisfying the Asano condition.

Corollary

Let $\bar{\mathcal{B}} \supset \mathcal{B}_\infty$ and $\mathcal{M} = \gamma[\mathcal{P}^a(\Lambda)]$, then for any covering $\mathcal{B} = \bigcup_i \mathcal{B}_i$ such that :

a) $\mathcal{M}^\perp$ coincides with the subgroup of $\mathcal{P}(\mathcal{B})$ generated by $\{\mathcal{M}_i^\perp\}$;

and

b) for any local constraint $\mathcal{C}_c$, there exists some $\mathcal{B}_i$ such that $\bar{\mathcal{B}_i} \supset \mathcal{B}_c^\infty$ ($\mathcal{B}_c^\infty \equiv$ support of $\tilde{\mathcal{C}}_c$);

the polynomial $\mathcal{M}(z_{\mathcal{B}})$ is the Asano contraction of the polynomials $\{M_i(z_{\mathcal{B}_i})\}$.

Proof :

β in $\bar{\mathcal{M}}$ implies $\beta = \gamma(X)$ with $X \in \overline{\mathcal{P}^a(\Lambda)}$; while $\beta \cap \mathcal{B}_i$ in $\mathcal{M}_i$ implies $\gamma(X) \cap \mathcal{B}_i = \gamma(X_i) \cap \mathcal{B}_i$ for some X_i in $\mathcal{P}^a(\Lambda)$. Therefore $\sigma_B(X) = \sigma_B(X_i)$ for all B in $\mathcal{B}_i$, which yields, using condition b'), $\mathcal{C}_c(X) = \mathcal{C}_c(X_i) = 1$ for all local constraints $\mathcal{C}_c$ i.e. $X \in \mathcal{P}^a(\Lambda)$ and thus $\beta \in \mathcal{M}$.

For systems $\{\mathcal{L}, \mathfrak{k}, K\}$ not satisfying the subgroup property, but with $\bar{\mathcal{B}} \supset \mathcal{B}_\infty$, the general prescription to find coverings $\mathcal{B} = \bigcup_i \mathcal{B}_i$ satisfying the Asano condition is then :

a) find $\mathcal{M}^\perp = \gamma[\mathcal{P}^a(\Lambda)]^\perp$

b') find a family of subgroup $\{G_i^\perp\}$ generating $\mathcal{M}^\perp$ then

$$\mathcal{B}_i = \bigcup_{\beta \in G_i^\perp} \beta \tag{5.16}$$

c) for any local constraint $\mathcal{C}_c$ such that none of the previous $\mathcal{B}_i$ satisfy $\bar{\mathcal{B}}_i \supset \mathcal{B}_c^\infty$, introduce :

$$\mathcal{B}_c = \bigcup_{B_\infty \in \mathcal{B}_c^\infty} \beta_{B_\infty} \tag{5.17}$$

where β_{B_∞} is a subset of $\mathcal{P}(\mathcal{B})$ satisfying $\pi(\beta_{B_\infty}) = B_\infty$ (Since $\bar{\mathcal{B}} \supset \mathcal{B}_\infty$, for any $B_\infty \in \mathcal{B}_\infty$, there exists such a β_∞ ; moreover, we then have $\bar{\mathcal{B}}_c \supset \mathcal{B}_c^\infty$).

d) then

$$\mathcal{B} = \bigcup_i \mathcal{B}_i \bigcup_c{}^* \mathcal{B}_c \bigcup_j \mathcal{B}_j \tag{5.18}$$

where $\bigcup_j \mathcal{B}_j$ is any covering of $\mathcal{B} / \bigcup_i \mathcal{B}_i \bigcup_a{}^* \mathcal{B}_c$

Now we are ready to state analyticity properties for these systems :

Theorem 3

For any system $\{\mathbb{Z}^\nu, \mathcal{C}, K\}$ with $\mathbb{Z}^\nu$- invariant, finite range interactions, and constraints $\mathcal{C}$ defined by translations $b \in \mathcal{T} \subset \mathbb{Z}^\nu$ of a fundamental family $\{L.C\}_0$ of local constraints satisfying the conditions $\phi \in \mathcal{P}^a(\mathcal{L})$ and $\bar{\mathcal{B}} \supset \mathcal{B}_\infty$,

i) there exists positive numbers r_B , $B \in \mathcal{B}$, such that the reduced partition function of the finite system $\{\Lambda, \mathcal{C}_\Lambda, K\}$ is different from zero in the complex domain :

$$|z_B| = |e^{-2K(B)}| \leq r_B \tag{5.19}$$

ii) the reduced free energy density is an analytic function of the complex variables z_B, $B \in \mathcal{B}_0$, where $\mathcal{B}_0$ is a fundamental family of finite bonds, in the domain defined by Eq. (5.19).

For the proof we refer to [122] . We emphasize that the unicity of the invariant equilibrium state and analyticity of the correlation functions within the domain given by Eq. (5.19) can be derived only if more restrictions are imposed on the system. For hard core lattice systems, these conditions are given in Sec. 5.2.3 .

Let us note that the difference between the condition of this theorem and theorem 2 lies in the fact that in theorem 3 the infinite bonds are assumed to be product of finite bonds while in theorem 2 they are assumed to be product of strictly finite bonds (i.e. $B \notin \mathcal{B}_\infty$).

Example 1 :

Let us modify the admissible configurations on the shaded squares of the example of Sec. 5.1., the finite bonds being unchanged. The admissible configurations shown in Fig. 3 do not form a group.

Fig. 3 : Locally admissible configurations and corresponding elements in Γ^a .

Its infinite bonds are then

$$\mathcal{B}_\infty = \{ (x_i, x_j), (x_1, x_2, x_3, x_4);\ x_i , x_j \text{ n.n. or n.n.n. on shaded squares} \} \qquad (5.20)$$

hence $\bar{\mathcal{B}} \supset \mathcal{B}_\infty$. Since $\overline{\gamma[\rho^a(\Lambda)]}$ is the same group as in the example of Sec. 5.1., one might also choose the same covering of $\mathcal{B}$. However, this covering does not satisfy condition b') of the corollary of Property 1. But taking $\mathcal{B}_i = \{B_1, B_2, B_3, B_4\}$ for all unit squares this condition is satisfied and

$$\mathcal{B} = \bigcup_i \mathcal{B}_i$$

On the unshaded squares, we have $\mathcal{M}_i = \{ \phi , \{B_\ell, B_{\ell'}\}_{\ell<\ell'} , \{B_1, B_2, B_3, B_4\}\}$ whereas on the shaded squares $\mathcal{M}_i = \{\phi, \{B_2, B_4\}, \{B_1, B_2, B_3, B_4\}\}$.

On the unshaded squares $M_i(z_{\mathcal{B}_i}) = 1 + \sum_{i,j} z_i z_j + z_1 z_2 z_3 z_4$, while on the shaded squares

$$M_i(z_{\mathcal{B}_i}) = 1 + z_2 z_4 + z_1 z_2 z_3 z_4$$

and we immediately have :

$$M_i(z_{\mathcal{B}_i}) \neq 0 \quad \text{if} \quad |z_2|, |z_4| < \frac{1}{\sqrt{2}} \ , \ |z_1|, |z_3| < 1$$

Writing $M_i(z_{\mathcal{B}_i}) = 1 + z_2 z_4(1 + z_1 z_3)$ it follows that

$$M_i(z_{\mathcal{B}_i}) \neq 0 \quad \text{if} \quad |z_2|, |z_4| > 1 \ , \ |z_1|, |z_3| > \sqrt{2} \ .$$

Since any bond undergoes one contraction, we conclude with Ruelle's Theorem [106] that $M(z_{\mathcal{B}}) \neq 0$ within the Ising domain $|z_B = e^{-2K(B)}| < (\sqrt{2}-1)^2$ and by symmetry $|z_B| > (\sqrt{2}+1)^2$.

A non uniform choice of $\delta_{i,\mathcal{B}}$ for $\mathcal{M}_i(z_{\mathcal{B}_i})$ yields the improved domain $\mathcal{M}(z_{\mathcal{B}}) \neq 0$ if $|z_{\mathcal{B}}| < 0.23$. Again we conclude that this domain is bigger than that of the same model but without constraints for which $|z_{\mathcal{B}}| < 0.17$.

Example 2 : Standard 6 vertex model

Here we want to discuss the zeroes in the variables $z_B = e^{-2\beta h} (e^{-2\beta v})$ of the standard 6 vertex model for real values of ε_1 and ε_2 . The Ising equivalent of this model was discussed in Sec. 1.2.3. Its infinite bonds are all even subsets of C (C = shaded unit square). Its finite interactions are the staggered field, the vertical 2 body n.n. interactions and the 2 body n.n.n. interactions on the shaded squares. The bonds of $\mathcal{B}'$ on the shaded and unshaded squares are graphically represented by :

To find a covering of the set of finite bonds $\mathcal{B}$, we have to satisfy conditions a) and b') of the corollary. The generators of $\mathcal{K}^a$ are defined on the shaded and unshaded squares

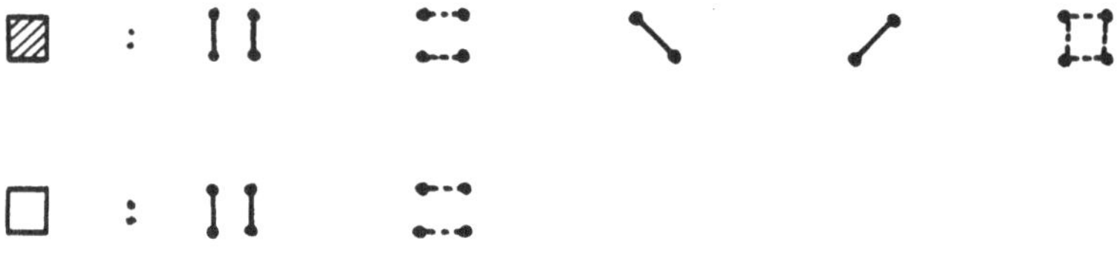

(dotted lines represents infinite bonds)

Taking as G_i the group generated by all the generators of $\mathcal{K}^a$ defined on a shaded unit square, we have :

and condition b') of the corollary is satisfied. The index i runs over all shaded unit squares *). With $z_j = e^{-2K(j)}$ $j = 1, \dots, 4$, $K(j) \in \{\beta h, \beta v\}$ and $a = e^{-\beta \varepsilon_1}$, $b = e^{\beta \varepsilon_2}$ the small polynomial becomes :

$$\mathcal{M}_i(z_{\mathcal{B}_i}) = 1 + a^2 b^2 z_1 z_2 + a^2 b^2 z_3 z_4 + a^2 z_1 z_3 + a^2 z_2 z_4 + z_1 z_2 z_3 z_4 \qquad (5.21)$$

Let $|z_j| < \rho_+ = \left(\sqrt{a^4 (b^2+1)^2 + 1} - a^2 (b^2+1)\right)^{1/2} \quad j = 1, \dots, 4$ (5.22)

then $\mathcal{R}e\, \mathcal{M}_i(z_{\mathcal{B}_i}) > 1 - 2a^2 b^2 \rho_+^2 - 2a^2 \rho_+^2 - \rho_+^4 = 0$

and $\mathcal{M}_i(z_{\mathcal{B}_i}) \neq 0 \quad if \quad |z_j| < \rho_+ \quad j = 1, \dots 4$

any field belongs to two shaded unit squares and undergoes one contraction, therefore, using the spin flip :

$$\mathcal{M}(z_{\mathcal{B}}) \neq 0 \quad if \quad |e^{-2\beta|h|}|, \; |e^{-2\beta|v|}| < \rho_+^2 \qquad (5.23)$$

In particular for the KDP and F models in a horizontal and vertical electric field.

$$\mathcal{M}^{KDP}(z_{\mathcal{B}}) \neq 0 \quad if \quad e^{-2\beta|h|}, \; e^{-2\beta|v|} < \sqrt{4a^4+1} - 2a^2$$

$$\mathcal{M}^{F}(z_{\mathcal{B}}) \neq 0 \quad if \quad e^{-2\beta|h|}, \; e^{-2\beta|v|} < \sqrt{(b^2+1)^2+1} - (b^2+1) \qquad (5.24)$$

The analyticity domain for the KDP model, Eq. (5.24) has to be compared with the result of Suzuki and Fisher [105]. These authors showed that the zeroes of the KDP model lie on the unit circle provided that $0 \leq a \leq \frac{1}{2}$ i.e. for temperatures less than the critical temperature of the KDP model without field. Our result is weaker in the region $0 \leq a \leq \frac{1}{2}$ but it is also valid for $\frac{1}{2} < a \leq 1$ and for the IKDP model; moreover it is in agreement with the numerical work of Katsura et al. [117,b]. For a discussion of these models without external field, we refer to the sec. 5.2.4.

---

*) Note that any generator of $\mathcal{K}^a$ belonging to an unshaded square also belongs to a shaded square. Hence to obtain a covering satisfying the Asano condition it is sufficient to consider the generators on shaded squares only.

5.2.3. General Hard Core Lattice Systems

Zeroes of particular hard core lattice systems with two site exclusions and external field (activity) have been investigated by L.K. Runnels and J.B. Hubbard [119]. They computed domains free of zeroes of the partition function in $z = e^{-2\beta h}$, where h represents the external field. Hence they discussed systems satisfying $\bar{\mathcal{B}} \supset \mathcal{B}_\infty$. Their coverings were given without justification but they satisfy the Asano condition.

Here, we shall go beyond the results of L.K. Runnels and J.B. Hubbard. Using property 3 of Sec. 1.3 we shall derive the unicity of the invariant state at low temperature for general hard core lattice systems with $\bar{\mathcal{B}} \supset \mathcal{B}_\infty$ and satisfying one more condition to be specified in the next theorem; let us remark that for hard core system the condition $\bar{\mathcal{B}} \supset \mathcal{B}_\infty$ is equivalent to the requirement $\bar{\mathcal{B}} = \mathcal{P}_f(\mathcal{L})$.

Theorem 4

For any ferromagnetic hard core lattice system which satisfies the conditions of Theorem 3 and such that for all $X \in \mathcal{P}_f^a(\mathcal{L})$, there exists $\beta \subset \mathcal{B}$ satisfying $X = \pi(\beta)$ and $\bigcup_{B \in \beta} B \in \mathcal{P}_f^a(\mathcal{L})$, there exists a unique $\mathcal{T}$ - invariant equilibrium state at low temperature and the correlation functions are analytic functions of $\{z_B = e^{-2K(B)}\}_{B \in \mathcal{B}_0}$ in the domain $|z_B| \leq r_B$ (Eq. 5.19).

Proof :

The method we shall use follows the same line as the one used in Chapter 9 of Part I [108,109]. For any fixed $X \in \mathcal{P}_f^a(\mathcal{L})$, we define the extended family of finite bonds $\mathcal{B}_X = \mathcal{B} \cup \{\tau_a X\}_{a \in \mathbb{Z}^\nu}$ and we denote with γ_X the mapping $\mathcal{P}(\mathcal{L}) \to \mathcal{P}(\mathcal{B}_X)$ defined in the same way as γ' or γ_∞.

(see Sec. 1.4). To find coverings of $\mathcal{B}_X$ satisfying the Asano condition, we consider for any $\Lambda \in \mathcal{P}_f'(\mathcal{L})$, $\Lambda \supset X$, the group $(\gamma_X[\mathcal{P}^a(\Lambda)])^{\perp}$ where the orthogonal is taken in $\mathcal{P}(\mathcal{B}_X)$. This group is generated by $(\gamma[\mathcal{P}^a(\Lambda)])^{\perp}$ and $\{T_a \varkappa\}_{a \in \mathbb{Z}^\nu}$, where $\varkappa = (X, B_1, \ldots, B_n)$ with $\prod_{j=1}^{n} B_j = X$. Assuming that there exists $(B_1, \ldots, B_n) \in \mathcal{P}(\mathcal{B})$ such that $B_j \subset \Lambda$, $j = 1, \ldots, n$, and $\prod_{j=1}^{n} B_j = X$, it follows from the hard core condition that the small system $\{\Lambda_i = \bigcup_j B_j, \mathcal{C}_{\Lambda_i}, K\}$ is a system without constraints; we can then repeat the arguments given in Chapter 9 of Part I, [108,109] to conclude at the analyticity of $\omega_{(\mathcal{L}, \mathcal{C}, K)}[\sigma_X]$ within the domain Eq. (5.19) for any $X \in \mathcal{P}_f^a(\mathcal{L})$. Together with Property 3 of Sec. 1.3 the proof is accomplished

Corollary

i) For any ferromagnetic hard core system with external field, $\mathbb{Z}^\nu$ - invariant interactions and $\mathcal{T}$ - invariant hard cores, there exists a unique $\mathcal{T}$ - invariant equilibrium state at low temperature.

ii) For systems with $\mathcal{T}$ - invariant 2 body exclusions and external field h only, there exists a unique invariant equilibrium state ω if $|e^{-2\beta h}| < 2^{-q}$ (q = constraint coordination number); moreover, the correlation functions are analytic functions of $z = e^{-2\beta h}$ in $|z| < 2^{-q}$.

For the proof we refer to [122] .

We emphasize that the above results cannot be derived from similar results obtained for lattice systems without constraints by letting some interactions going to infinity. Indeed the unicity of the state at low temperature has been derived in Chapter 9 of Part I [108,109] for purely ferromagnetic systems. As we have discussed in Ch. I, systems with constraints not satisfying the subgroup condition are in general not obtained as the limiting case $K(B) \to +\infty$ for some $B \in \mathcal{B}$ of ferromagnetic systems without constraints.

The above analyticity domains can be improved either by changing the covering set or by a more careful study of the small polynomials $M_{ij}(z_{\beta_{ij}}) = 1 + z_{x_i} + z_{x_j}$ where $\beta_{ij} = (x_i, x_j) =$ 2 body constraint. Indeed for this polynomial we obtain with the notation introduced in the Appendix :

$$M_{ij}(z_{\beta_{ij}}) \neq 0 \quad \text{if} \quad z_i, z_j \notin \{z ; \operatorname{Re} z \leq -\tfrac{1}{2}\} \equiv -\mathcal{D}_{1/2}$$

Therefore Ruelle's Theorem implies :

$$M(z_{\mathcal{B}_\Lambda}) \neq 0 \quad \text{if} \quad z_x \notin -\prod_{\beta_{ij} \ni x} \mathcal{D}_{1/2} \tag{5.25}$$

Theorem A on power sets proved in the Appendix then implies that $M(z_{\mathcal{B}_\Lambda}) \neq 0$ in a complex domain containing the real interval :

$$\left[0 ; \left(2 \cos \frac{\pi}{m_x}\right)^{-m_x}\right] \tag{5.26}$$

where m_x is the number of constraints containing x .

It should be stressed that the extended domain free of zeroes of $M(z_{\mathcal{B}_\Lambda})$ given by Eq. (5.25) does not improve the domain in which the unicity of state was shown. This is simply due to the fact that $M_a(z_{\tau_a x}) \neq 0$ if its variables are within circular regions and the greatest circular domain not contained in $-\prod_{\beta_{ij} \ni x} \mathcal{D}_{1/2}$ coincides with the domain given in ii) of the corollary.

<u>Example</u> 1 :

To illustrate the above discussion, we first consider the hard core system defined on the planar triangular lattice field. The corollary yields analyticity of the reduced free energy density and unicity of the state for $|z_x| < 2^{-6}$. The extended domain given in Eq. (5.25) implies analyticity of the reduced free energy density in a neighbourhood of the real interval $e^{-2\beta h} \in [0 ; (2 \cos \frac{\pi}{6})^{-6}] = [0 ; 0.037]$.

<u>Example</u> 2 : n.n. and n.n.n. exclusions

An interesting hard core model is the one with n.n. and n.n.n. exclusions on a square lattice and finite interactions given by an external field $(\mathcal{B} = \mathcal{L})$. The corollary gives analyticity of the free energy density and unicity of state for $|e^{-2\beta h}| < 2^{-8}$ and Eq. (5.26) extends the analyticity of the free energy density to a neighbourhood of the interval $[0; 0.0073]$.

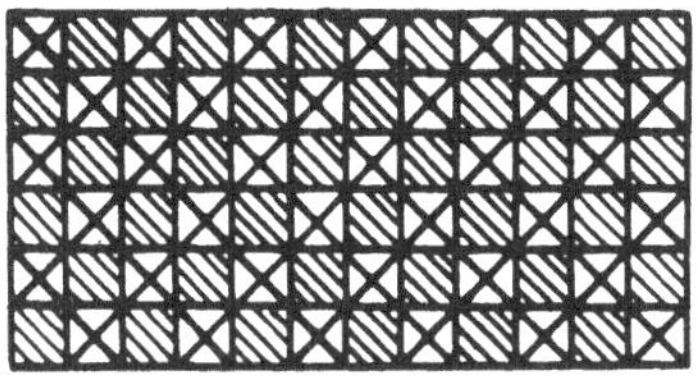

$\mathcal{B}_{i,1} = \{(x_1),(x_2),(x_3),(x_4);\ x_j \in ⊠\}$

$\mathcal{B}_{i,2} = \{(x_1),(x_2);\ x_1, x_2 \in \diagdown \text{ or } \diagup\}$

Fig. 4 : Covering of $\mathcal{B}$ for n.n. and n.n.n. hard core model on the square lattice.

The covering shown in Fig. 4 satisfies the Asano condition and consists of $\mathcal{B}_{i,1}$ defined on the shaded squares and $\mathcal{B}_{i,2}$ defined as n.n. sets on the unshaded squares. The corresponding small polynomials are :

$$M_{i,1}(z_{\mathcal{B}_{i,1}}) = 1 + \sum_{j=1}^{4} z_j \tag{5.27}$$

$$M_{i,2}(z_{\mathcal{B}_{i,2}}) = 1 + z_1 + z_2 \tag{5.28}$$

Theorems 3 and 4 then imply analyticity of the free energy density and unicity of the state for $|z_x| < \frac{1}{64}$. With the notation introduced in Eq. (5.25) and in the Appendix, we have $M(z_{\mathcal{B}}) \neq 0$ if $z_x \notin -\mathcal{D}^{*2}_{1/4} \cdot \mathcal{D}^{*2}_{1/2}$, thus extending the analyticity of the reduced free energy density to a neighbourhood of the real interval $[0; \frac{1}{16}]$.

Example 3 :

O.J. Heilmann [101] has shown that the zeroes in the activity variable of some lattice gas models with finite repulsive n.n. and n.n.n. interactions (or equivalently of an Ising model with antiferromagnetic interactions) are located on the negative real axis. We then expect that the same model with hard cores instead of the finite repulsive interactions

will have the same property; this result cannot be obtained
from Heilmann's analysis since we have to consider the simultaneous limit $h \to \infty$ and $J \to -\infty$ (Example 3, Sec. 1.2.2).
However, Heilmann's covering (Fig. 5) in the case of the
square lattice satisfies the Asano condition and can directly
be used for the corresponding hard core model (which is precisely the model introduced in Example 3, Sec. 1.2.2) without
considering any limiting process.

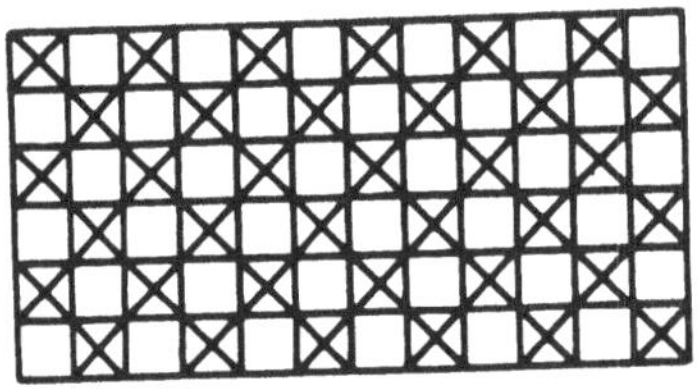

$$\mathcal{B} = \{(x)\} \cup \{(x,y)\} ; x,y = n.n. \text{ or } n.n.n.\}$$

$$\mathcal{B}_i = \{(x_1), (x_2), (x_3),(x_4) ; \quad \mathord{\underset{x_4 \quad x_2}{\overset{x_3 \quad x_1}{\boxtimes}}} \}$$

Fig. 5 : Square lattice with interaction bonds $\mathcal{B}$
of the finite repulsive lattice gas model
and a covering set $\mathcal{B}_i$ of the corresponding hard core model.

The small polynomials for the above defined covering sets
are :
$$M_i (z_{\mathcal{B}_i}) = 1 + z_1 + z_2 + z_3 + z_4$$

Following D. Ruelle's proof [106,b] of J. Heilmann's result,
and using Grace's Theorem [106,b], we have for any fixed
$\theta \in (-\frac{\pi}{2},+\frac{\pi}{2})$:

$$M_i (z_{\mathcal{B}_i}) \neq 0 \quad \text{if} \quad z_j \notin -\mathcal{D}_\theta \quad j = 1,2,3,4$$

where $\mathcal{D}_\theta = \{z ; \, Re \, z e^{i\theta} \geq 1/4 \}$

With the notations introduced in the Appendix, the boundary
function of $\mathcal{D}_\theta$ is then $g_{0,\theta} (\varphi) = \dfrac{\cos \theta}{4 \cos (\varphi - \theta)}$, $\varphi \in [-\frac{\pi}{2} + \theta, +\frac{\pi}{2} + \theta]$

Each variable z_j undergoes one contraction, Theorem A of
the Appendix then implies that the boundary of $\mathcal{D}_\theta^{*2}$ is
given by $g_\theta (\varphi) = \dfrac{\cos^2 \theta}{16} \dfrac{1}{\cos^2(\frac{\varphi}{2} - \theta)}$ for $\varphi \in [2\theta - \pi ; \, 2\theta + \pi]$
which implies $g_\theta (\pm \pi) = \infty$, $g_\theta (2\theta) = \dfrac{\cos^2 \theta}{16}$. In
consequence $t e^{i(\pi + 2\theta)} \notin \mathcal{D}_\theta^{*2}$ for any $t \in (-\frac{1}{16} \cos^2\theta, +\infty)$ (resp.

$t\,e^{2i\theta} \notin -\mathcal{D}_\theta^{*2}$). Hence $\mathcal{M}(z_{\mathcal{B}}) \neq 0$ if $\arg z_B = 2\theta$ for all $B \in \mathcal{B}$. This is valid for any $\theta \in (-\frac{\pi}{2}, +\frac{\pi}{2})$, therefore if $z_B = z$ for all $B \in \mathcal{B}$:

$$\mathcal{M}(z, \ldots, z) \neq 0 \qquad \text{if} \quad z \notin \mathbb{R}_-$$

which is the desired result. Using other techniques, this result has also been obtained by H. Kunz (H. Kunz, private communication). Thus, this hard core model has no phase transition.

5.2.4. Systems with $\bar{\mathcal{B}} \not\supset \mathcal{B}_\infty$

Let us now briefly discuss systems $\{\mathcal{L}, \mathcal{C}, K\}$ for which the previous condition $\bar{\mathcal{B}} \supset \mathcal{B}_\infty$ is not satisfied. As we have seen in Chapter 1, famous models with constraints such as the KDP and F model fall into this category. For such systems, a covering of the total set of bonds $\mathcal{B}'$ is needed. As already remarked, Property 1 of Sec. 5.2.1 remains valid to determine coverings of $\mathcal{B}'$. Since now infinite bonds can also undergo contractions, we have to associate a complex variable z_{B_∞} with any infinite bond $B_\infty \in \mathcal{B}' / \mathcal{B}$ which undergoes a contraction. But the polynomial $\mathcal{M}(z_{\mathcal{B}'})$ is only associated with $Z^{red}(\Lambda, \mathcal{C}_\Lambda, K)$ if one can take after performing all contractions $z_{B_\infty} = +1$ for all $B_\infty \in \mathcal{B}'/\mathcal{B}$, i.e. if $z_{B_\infty} = 1$ belongs to the analyticity domain for all $B_\infty \in \mathcal{B}'/\mathcal{B}$. This condition is by no means easy to satisfy. As we shall see on the example of the KDP model, the analyticity domain in the variable z_{B_∞} can be increased at the cost of the analyticity domains of the finite bond variables. However the same idea fails in the case of the F model and it is impossible to have $z_{B_\infty} = 1$ in the domain free of zeroes of the small polynomials. The applicability of the above idea depends strongly on the structure of the small polynomials and a conclusion can only be obtained after a careful study of the small polynomials.

Application to the KDP model :

The free energy density of the KDP model below the critical temperature is constant; hence showing analyticity properties in $z_B = e^{-2K(B)} = e^{-\beta\frac{\varepsilon_1}{2}}$ only implies that the free energy density remains constant for some complex temperatures. Despite of this not very exiting result, we compute a domain free of zeroes of $\mathcal{M}^{KDP}(z_{\mathcal{B}'})$ in order to show how to deal with systems $\mathcal{B} \not\supset \mathcal{B}_\infty$.

Taking for $\{\mathcal{K}_i^1\}$ the subgroups of $\mathcal{K}^a$ defined on the unit squares of the lattice, we find a covering satisfying the conditions (a) and (b) of Property 1, Sec. 5.2.1. The discussion of the small polynomials yields for any a < 1

$$\mathcal{M}^{KDP}(z_{\mathcal{B}'}) \neq 0 \quad \text{if} \quad \begin{cases} |z_B| < \sqrt{a}\,(\sqrt{2}-1) & B \in \mathcal{B} \\ |z_B| < \sqrt{\frac{1-a}{a}}\,(\sqrt{2}-1) & B \in \mathcal{B}'/\mathcal{B}\,,\ B = \text{horiz. bond.} \end{cases}$$

For the horizontal bonds belonging to $\mathcal{B}_\infty$, we have to show that $z_B = 1$ belongs to the analyticity domain. This is the case if $1 < \sqrt{\frac{1-a}{a}}\,(\sqrt{2}-1)$, which implies $a < \frac{\sqrt{2}-1}{2\sqrt{2}}$. Thus the free energy density of the KDP model is analytic in the domain

$$|e^{-\frac{1}{2}\varepsilon_1}| < (\sqrt{2}-1)\sqrt{\frac{\sqrt{2}-1}{2\sqrt{2}}} = 0.15\ldots$$

The F model :

To obtain an analyticity domain for this model at low temperature would be of more interest than for the KDP model. The covering and the small polynomials needed for a discussion of the F model in the variable $z_B = e^{\frac{1}{2}\beta\varepsilon_2}$ are identical with those of the KDP model. For the F model however infinite bond variables undergo contraction. (*) For these bonds, the small polynomial can vanish if $|z_B| = 1$ $B \in \mathcal{B}_\infty$ and no analyticity property can be derived for $\mathcal{M}^F(z_{\mathcal{B}'})$.

(*) all n.n. bonds belong to $\mathcal{B}_\infty$.

APPENDIX

The use of Ruelle's theorem involves the computation of power sets of complex domains :

$$\mathfrak{D}^{*n} = \prod_{j=1}^{n} \mathfrak{D}_j \equiv \{ z ; \; z = \prod_{j=1}^{n} z_j \; , \; z_j \in \mathfrak{D}_j \, , \; j = 1,2,..,n \} \tag{A.1}$$

In most applications, the domains $\mathfrak{D}_j$ belong to one of the two classes of domains defined below.

Definitions :

(A) A closed domain $\mathfrak{D}$ is a star domain with respect to the origin 0 if $z \in \mathfrak{D}$ and $\alpha \geq 1$ implies $\alpha z \in \mathfrak{D}$.

(B) A closed domain $\mathfrak{D}$ is convex with respect to the origin 0 if $z, z' \in \mathfrak{D}$ and $\arg z = \arg z'$ implies $\alpha z \in \mathfrak{D}$ for any $\alpha \in [1, \frac{|z'|}{|z|}]$.

We shall now prove that the boundary $\partial \mathfrak{D}^{*n}$ of products of star domains, respectively convex domains with respect to the origin, is a subset of

$$\{ z ; \; z = \prod_{j=1}^{n} z_j \; , \; z_j \in \partial \mathfrak{D}_j \, , \; j = 1,2,\ldots, n \}$$

moreover, in some cases to be specified below, this result yields an explicit expression for the boundary of power sets.

Theorem A

Let $\mathfrak{D}_j \subset \mathbb{C}, j=1,..,n$ be a family of star domains in the complex plane whose complement $\mathfrak{D}_j^c$ containts a neighbourhood of the origin 0 . Moreover, let $\rho_j(\varphi_j)$ be the boundary function of $\mathfrak{D}_j$ defined by :

$$\rho_j(\varphi_j) = \begin{cases} \min\limits_{\substack{z \in \mathfrak{D}_j \\ \arg z = \varphi_j}} |z| \\ + \infty \quad \text{otherwise} \end{cases} \tag{A.2}$$

then the following properties hold :

i) $\mathfrak{D}^{*n}$ is a star domain with respect to O and $(\mathfrak{D}^{*n})^c$ contains a neighbourhood of O .

ii) If $\rho_j(\varphi_j)$ is of class $\mathscr{C}^1$, $j = 1, \ldots, n$ then the boundary of $\mathfrak{D}^{*n}$ is defined by the function :

$$\rho(\varphi) = \min_{\{\bar{\varphi}_j\}} \prod_{j=1}^{n} \rho_j(\bar{\varphi}_j)$$

where the minimum is taken over all solutions $\{\bar{\varphi}_j\}_{j=1,\ldots,n}$ of the set of equations

$$\frac{\partial}{\partial\varphi_j} \ln \rho_j = \frac{\partial}{\partial\varphi_k} \ln \rho_k \qquad \forall\ j,k \tag{A.3}$$

$$\sum_{j=1}^{n} \bar{\varphi}_j = \varphi + 2m\pi \qquad m = 0, 1, \ldots, n-1$$

iii) If the domains $\mathfrak{D}_j$ are similar and connected, i.e. with boundaries defined by $\rho_j(\varphi) = b_j\, \rho_0(\varphi)$, $b_j > 0$, satisfying :

a) $\rho_0(\varphi)$ is a strictly positive function of period 2π ;

b) $\rho_0(\varphi)$ is of class $\mathscr{C}^2$ in the domain where it is finite;

c) there is an interval $[\varphi_0, \varphi_0 + 2\pi]$ where $\ln \rho_0(\varphi)$ is strictly convex.

Then the boundary of $\mathfrak{D}^{*n}$ is given by the function of period 2π defined by :

$$\rho(\varphi) = \left[\rho_0\left(\frac{\varphi}{n}\right)\right]^n \prod_{j=1}^{n} b_j \ , \quad \varphi \in [n\bar{\varphi}, n\bar{\varphi} + 2\pi] \tag{A.4}$$

where $\bar{\varphi} \in [0, 2\pi]$ is the solution of :

$$\rho_0(\bar{\varphi}) = \rho_0\left(\bar{\varphi} + \frac{2\pi}{n}\right) \tag{A.5}$$

For the proof of this Theorem see [122] .

Applications

a) A typical star domain (cf. Eq. (5.25)) is $\mathfrak{D}_b = \{z ; \operatorname{Re} z \geqslant b\}$ in this case $\rho_0(\varphi) = \frac{b}{\cos\varphi}$ if $\varphi \in [-\frac{\pi}{2}, \frac{\pi}{2}]$ and $\rho_0(\varphi) = \infty$ otherwise. Moreover $\ln \rho_0(\varphi)$ is strictly convex in the interval $[-\pi, +\pi]$, hence the boundary of $\mathfrak{D}_b^{*n}$ is defined by

$$\rho_0(\varphi) = \left(\frac{b}{\cos\frac{\varphi}{n}}\right)^n \qquad \varphi \in [-\pi, \pi] \qquad (A.7)$$

It should be remarked that only those points on the boundaries of $\mathfrak{D}_b$ with $-\frac{\pi}{n} \leqslant \varphi \leqslant \frac{\pi}{n}$ will give points of the boundary of $\mathfrak{D}^{*n}$; moreover, this result confirms a conjecture given in [119].

b) The boundary of the domain $M^{(3)}$ given by Runnels and Hubbard [119] satisfies the conditions of iii), hence Eq. (A.4) immediately gives the shape of all its power sets.

Theorem B

Let $\mathfrak{D}_j \subset \mathbb{C}$ $j=1,..,n$ be a family of connected domains, convex with respect to 0, whose complements contain a neighbourhood of the origin 0, and such that there exists r_j with $r_j e^{i\varphi} \in \mathfrak{D}_j$ for all $\varphi \in [\psi_j, \vartheta_j]$, where ψ_j, ϑ_j are the two angles limiting the domain $\mathfrak{D}_j$. The boundary of $\mathfrak{D}_j$ is then defined by the functions (Fig. 6) :

$$\rho_j^{min}(\varphi) = \min_{\substack{z \in \mathfrak{D}_j \\ \arg z = \varphi}} |z| \quad \text{if } \varphi \in [\psi_j, \vartheta_j] \qquad \rho_j^{max}(\varphi) = \max_{\substack{z \in \mathfrak{D}_j \\ \arg z = \varphi}} |z| \quad \text{if } \varphi \in [\psi_j, \vartheta_j]$$

$$\rho_j^{min}(\varphi) = +\infty \quad \text{otherwise} \qquad \rho_j^{max}(\varphi) = 0 \quad \text{otherwise}$$

then the following properties hold :

i) $\mathfrak{D}^{*n}$ is connected, convex with respect to the origin 0, and $(\mathfrak{D}^{*n})^c$ contains a neighbourhood of the origin; moreover, there exists r such that $r e^{i\varphi} \in \mathfrak{D}^{*n}$ for all $\varphi \in [\sum_{j=1}^{n} \psi_j, \sum_{j=1}^{n} \vartheta_j]$.

ii) If $\rho_j^{min}(\varphi)$ and $\rho_j^{max}(\varphi)$ are of class $\mathscr{C}^1$, $j=1..n$ then the boundary of $\mathfrak{D}^{*n}$ is defined by the two funtions :

$$\rho^{\left[\substack{max\\min}\right]}(\varphi) = \left[\substack{max\\min}\right]_{\{\varphi_j^{\left[\substack{max\\min}\right]}\}} \prod_{j=1}^{n} \rho_j^{\left[\substack{max\\min}\right]}\left(\bar\varphi_j^{\left[\substack{max\\min}\right]}\right)$$

where the extremas are taken over all solutions $\{\bar\varphi_j^{\left[\substack{max\\min}\right]}\}_{j=1,\dots,n}$ of the set of equations :

$$\frac{\partial}{\partial\varphi_j} \ln \rho^{\left[\substack{max\\min}\right]}(\varphi_j) = \frac{\partial}{\partial\varphi_k} \ln \rho^{\left[\substack{max\\min}\right]}(\varphi_k) \qquad \text{(A.9)}$$

$$\sum_{j=1}^{n} \varphi_j^{\left[\substack{max\\min}\right]} = \varphi + 2m\pi \quad , \quad m = 0, 1, \dots, n-1$$

iii) If the domains $\mathfrak{D}_j$ are similar, i.e. with boundaries defined by :

$$\rho_j^{\left[\substack{max\\min}\right]}(\varphi) = b_j \; \rho_0^{\left[\substack{max\\min}\right]}(\varphi) \quad , \quad b_j > 0$$

satisfying the following two conditions :

a) $\rho_0^{\left[\substack{max\\min}\right]}(\varphi)$ are strictly positive functions of class $\mathscr{C}^2$ in the interval $[\Psi_0, \vartheta_0]$ and of period 2π.

b) There is an interval $[\varphi_0, \varphi_0 + 2\pi]$ where $\ln \rho_0^{max}(\varphi)$ (resp. $\ln \rho_0^{min}(\varphi)$) is strictly concave (resp. convex).

Then the boundary of $\mathfrak{D}^{*n}$ is given by the functions of period 2π defined by :

$$\rho^{\left[\substack{max\\min}\right]}(\varphi) = \left(\rho_0^{\left[\substack{max\\min}\right]}\left(\frac{\varphi}{n}\right)\right)^n \prod_{j=1}^{n} b_j \qquad \text{(A.10)}$$

$$\text{for} \quad \varphi \in \left[n\,\bar\varphi^{\left[\substack{max\\min}\right]} \, , \; n\,\bar\varphi^{\left[\substack{max\\min}\right]} + 2\pi\right]$$

where $\bar\varphi^{\left[\substack{max\\min}\right]} \in [0, 2\pi]$ are the solutions of

$$\rho_0^{max}(\bar\varphi^{max}) = \rho_0^{max}\left(\bar\varphi^{max} + \frac{2\pi}{n}\right)$$

$$\rho_0^{min}(\bar\varphi^{min}) = \rho_0^{min}\left(\bar\varphi^{min} + \frac{2\pi}{n}\right) \qquad \text{(A.11)}$$

For the proof we refer to [122] .

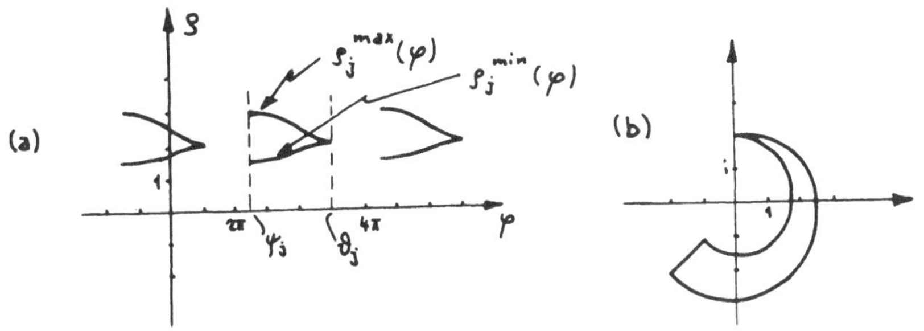

Fig. 6 : (a) Boundary functions $\varrho_j^{min}(\varphi)$, $\varrho_j^{max}(\varphi)$ $\varphi \in [\psi_j, \vartheta_j]$ for a simple connected convex domain with respect to 0 .

(b) Corresponding domain.

Application

In the discussion of lattice gas systems [106a] or anti-ferromagnetic systems [111] , power sets of the domain $\mathfrak{D} = \{ z \; ; \; |z-a| = \sqrt{a^2-1} \}$ where $a = e^{-2K(B)} > 1$ have been considered without giving its precise shape. In this case :

$$\varrho_0^{\left[\substack{max \\ min}\right]}(\varphi) = a \cos\varphi \pm \sqrt{a^2\cos^2\varphi - 1} \quad , \; \varphi \in [-\operatorname{arc\,cos}\sqrt{1-a^{-2}}, \operatorname{arc\,cos}\sqrt{1-a^{-2}}]$$

Since
$$\frac{\partial^2}{\partial\varphi^2} \ln \varrho_0^{\left[\substack{max \\ min}\right]}(\varphi) = \mp \frac{a^2(a^2-1)\cos\varphi}{(a^2\cos^2\varphi - 1)^{3/2}} \left[\varrho_0^{\left[\substack{max \\ min}\right]}(\varphi)\right]^2$$

the conditions of part iii) of theorem B are satisfied, therefore the boundary of $\mathfrak{D}^{*n}$ is limited by :

$$\varrho^{\left[\substack{max \\ min}\right]}(\varphi) = \left(a \cos\frac{\varphi}{n} \pm \sqrt{a^2\cos^2\frac{\varphi}{n} - 1} \right)^n$$

and only those points of $\mathfrak{D}$ with $\varphi \in [-\frac{1}{n}\operatorname{arc\,cos}\sqrt{1-a^{-2}}, \frac{1}{n}\operatorname{arc\,cos}\sqrt{1-a^{-2}}]$ contribute to the boundary of $\mathfrak{D}^{*n}$.

PART III

ARBITRARY SPIN LATTICE SYSTEMS

CHAPTER 1 - GENERAL FRAMEWORK OF HIGHER SPIN SYSTEMS

1.1. Definition of Arbitrary Spin Systems; Physical Picture and Group Picture [125].

In the preceding parts I and II it was shown that a group structure appears naturally in the study of classical spin $\frac{1}{2}$ lattice systems; in this last part we shall extend this formalism to discuss general properties of higher spin system , multicomponent systems, diluted systems, or any other system constituted of subsystems (molecules) which may exist in a finite number q of configurations. We shall call such systems simply "Arbitrary Spin Systems" whatever be the physical system we have under consideration. It should already be remarked that some of the methods and results which we shall describe can be extended to continuous "spin" variables as well as to "quantum" systems; we shall not go in detail into such generalisations but the notation will be such that the extension is straightforward. For general informations concerning arbitrary spin systems the reader is referred to [126] and references cited therein.

Arbitrary spin systems are defined by a lattice $\mathcal{L}$; at each site x of $\mathcal{L}$ is associated a variable s_x which is allowed to take q values $s_x \in \{s_x^1, \dots, s_x^q\}$ and defines the physical configurations at the site x ; it is natural to label these q values by the set of numbers $\{0,1,\dots,q-1\}$ and therefore if the system is homogeneous (i.e. all subsystems are identical) a configuration of the system is defined either by

$$\underline{s} = \{s_x\}_{x \in \mathcal{L}} \quad \in \quad G_{\mathcal{L}} = \prod_{x \in \mathcal{L}} \{s_x^1, \dots, s_x^q\} \tag{1.1}$$

or by

$$\underline{n} = \{n_x\}_{x \in \mathcal{L}} \quad \in \quad \mathcal{G}_{\mathcal{L}} = \prod_{x \in \mathcal{L}} \{0, 1, \dots, q-1\} \tag{1.2}$$

The hamiltonian of the finite system Λ is defined by a function on the configuration space and we have :

$$H_\Lambda^{(n)}(\underline{n}) = H_\Lambda^{(s)}(\underline{s}) \quad \text{where} \quad \underline{n} = \phi(\underline{s}) \tag{1.3}$$

As we have seen in parts I and II it is convenient to introduce a group structure on the configuration space, which makes it possible to apply standard results of harmonic analysis. In the case of spin $\frac{1}{2}$ systems we did not have any choice since each subsystem had only two possible configurations; in the present case however we could choose a priori any group of order q .

In the following we shall identify the set of q numbers $\{0, 1, \dots, q-1\}$ with the additive group $\mathbb{Z}_q$ of integers modulo q ; we thus have a "Group of Configurations $\mathcal{G}_{\mathcal{L}}$ " defined by:

$$\begin{aligned} \mathcal{G}_{\mathcal{L}} &= \prod_{x \in \mathcal{L}} \mathcal{G}_x \qquad\qquad \mathcal{G}_x \cong \mathbb{Z}_q \\ &\ni \\ \underline{n} &= \{n_x\}_{x \in \mathcal{L}} \\ \underline{m} \cdot \underline{n} &= \{m_x + n_x \,,\ \mathrm{mod}\ q\}_{x \in \mathcal{L}} \end{aligned} \tag{1.4}$$

which we call the "Group Picture" for the physical system under consideration. (*)

(*) Note that for $q = 2$ this group picture corresponds to the lattice gas language.

As usual the algebra of local observables is defined by means of the algebra $\mathfrak{A}_\Lambda$ of observables for the finite system $\Lambda \subset \mathcal{L}$, where $\mathfrak{A}_\Lambda$ is the algebra of continuous functions $A : \underline{n} \mapsto A(\underline{n})$ of $\mathcal{G}_\Lambda = \prod_{x \in \Lambda} \mathcal{G}_x$ into the complex numbers; using the group structure we can choose the elements of the dual group $\hat{\mathcal{G}}_\Lambda$ as basis for $\mathfrak{A}_\Lambda$ and we have [4] :

$$\hat{\mathcal{G}}_\Lambda = \prod_{x \in \Lambda} \hat{\mathcal{G}}_x \tag{1.5}$$

$$\cup\!\!\!\!|$$

$$\chi = \prod_{x \in \Lambda} \chi_x$$

i.e. for each $\underline{n} \in \mathcal{G}_\Lambda$, $\chi(\underline{n}) = \langle \chi ; \underline{n} \rangle_\Lambda = \prod_{x \in \Lambda} \langle \chi_x ; n_x \rangle_\Lambda$ (*)

With any $\underline{m}$ in $\mathcal{G}_\Lambda$ we can associate the function $\chi_{\underline{m}}$ on $\mathcal{G}_\Lambda$ defined by :

$$\chi_{\underline{m}}(\underline{n}) = \prod_{x \in \Lambda} \exp\left(\frac{2 i \pi}{q} m_x n_x\right) = \langle \underline{m} ; \underline{n} \rangle_\Lambda \tag{1.6}$$

The mapping $\underline{m} \mapsto \chi_{\underline{m}}$ is then the natural isomorphism between $\mathcal{G}_\Lambda$ and its dual group $\hat{\mathcal{G}}_\Lambda$; moreover for any χ in $\hat{\mathcal{G}}_\Lambda$ we defined the "support of χ " as the subset of Λ such that $\chi_x \neq \mathbb{1}_x$; in the same manner the support of any configuration is the subset of $\mathcal{L}$ where $n_x \neq 0$.

(*) As in parts I and II we shall make use of the bicharacter notation $\langle\, ; \,\rangle$ which we shall index when necessary to recall the group to which it refers; moreover for any $\Lambda' \supset \Lambda$ we shall use the same notation to denote either the element of $\hat{\mathcal{G}}_\Lambda$ or the element of $\hat{\mathcal{G}}_{\Lambda'}$.

Any observable $A \in \mathfrak{A}_\Lambda$ admit then a Fourier decomposition :

$$A(\underline{n}) = \sum_{\chi \in \mathcal{G}_\Lambda^\wedge} \tilde{A}(\chi)\ \chi(\underline{n}) = \sum_{\underline{m} \in \mathcal{G}_\Lambda} \tilde{A}(\underline{m})\ \chi_{\underline{m}}(\underline{n}) \quad (1.7.)$$

where $\tilde{A}(\chi) = q^{-|\Lambda|} \sum_{\underline{n} \in \mathcal{G}_\Lambda} A(\underline{n})\ \chi(\underline{n}^{-1})$; $\tilde{A}(\underline{m}) = \tilde{A}(\chi_{\underline{m}})$

The hamiltonian H_Λ is assumed to be an observable in $\mathfrak{A}_\Lambda$ and therefore can be decomposed as :

$$-\beta H_\Lambda = \sum_{\chi \in \mathcal{G}_\Lambda^\wedge} K(\chi)\cdot\chi = \sum_{\underline{m} \in \mathcal{G}_\Lambda} K(\underline{m})\ \chi_{\underline{m}} \quad ; \quad \beta = \frac{1}{kT} \ ; \ K = \beta J \quad (1.8)$$

where the Fourier coefficient $K(\chi)$ is a complex function on $\mathcal{G}_\Lambda^\wedge$ which describes the "interactions" between the sites x where $\chi_x \neq 1\!\!1_x$; moreover for real hamiltonian

$$\overline{K(\chi)} = K(\chi^{-1})$$

We are thus led to introduce as before the "support $\mathcal{I}$ of the interactions", subset of $\mathcal{G}_\Lambda^\wedge$ defined by:

$$\mathcal{I} = \{ \chi ;\ K(\chi) \neq 0 \}$$

From the above "group picture" we can discuss a large class of physical systems by means of the mapping $\phi^{-1} : \underline{n} \mapsto \underline{s}$ of the group $\mathcal{G}_\mathcal{L}$ onto the configuration space $G_\mathcal{L}$. For example,

- for systems of spin $s = \frac{q-1}{2}$, we have $G_x = \{-\frac{q-1}{2}, -\frac{q-1}{2}+1, \dots, \frac{q-1}{2}\}$ and the mapping ϕ can be taken as $\phi(s_x) = s_x + \frac{q-1}{2}$.
- for systems consisting of a mixture of k different species of atoms $A_1, \dots, A_k$ with at most one atoms per site, we have $G_x = \{0, A_1, A_2, \dots, A_k\}$ and the mapping ϕ can be taken as $\phi(0) = 0$, $\phi(A_i) = i$.

- for systems such that $p_x = \exp\left(\frac{2i\pi}{q} n_x\right)$ the mapping is given by $\phi(p_x) = n_x$.

In any case we would like to stress that the introduction of this group structure on the configuration space is not unique; in particular we have $q!$ distinct mapping ϕ between G_x and $\mathbb{Z}_q$ corresponding to the group of permutation of q objects $\mathcal{S}_q$. This symmetry under permutations will be reflected as new symmetry properties for physical quantities such as the free energy or the correlation functions to be introduced in the next section; however in the following we shall usually choose the mapping ϕ in such a way that $H_\Lambda(\underline{n}) - H_\Lambda(\underline{0}) \geq 0$ for all $\underline{n}$ in $\mathcal{G}_\Lambda$, and any configuration such that $H_\Lambda(\underline{n}) - H_\Lambda(\underline{0}) = 0$ will be called a "Ground State".

1.2. Thermodynamic and Equilibrium States of Finite Systems

The partition function and the correlation functions of the finite system $\{\Lambda, \mathcal{G}_\Lambda, K\}$ are defined by :

$$Z(\Lambda, K) = \sum_{\underline{n} \in \mathcal{G}_\Lambda} e^{-\beta H_\Lambda(\underline{n})} \tag{1.9}$$

$$\omega_{(\Lambda,K)}[\chi] = \langle \chi \rangle_{(\Lambda,K)} = Z^{-1} \sum_{\underline{n} \in \mathcal{G}_\Lambda} \chi(\underline{n})\, e^{-\beta H_\Lambda(\underline{n})} \tag{1.10}$$

i.e. for any observables A in $\mathfrak{A}_\Lambda$,

$$\omega_{(\Lambda,K)}[A] = \langle A \rangle_{(\Lambda,K)} = \sum_{\chi \in \mathcal{G}_\Lambda^\wedge} \tilde{A}(\chi)\, \omega_{(\Lambda,K)}[\chi]$$

and $\omega_{(\Lambda, K)}$ is by definition the "Gibbs State" of the finite system $\{\Lambda, \mathcal{G}_\Lambda, K\}$.

We are going to analyse $Z(\Lambda, K)$ and $\omega_{(\Lambda, K)}$ using the expansion of H_Λ in terms of characters; however it is more convenient to introduce first a family $\{\chi_b\}_{b \in \mathcal{B}}$ of characters in $\mathcal{G}_\Lambda^{\wedge}$ which generates $\mathcal{T}$, i.e. for all $\chi \in \mathcal{T}$ $\exists$ $b \in \mathcal{B}$ and $\ell \in \{1, 2, \dots, \alpha_b - 1\}$ such that $\chi = \chi_b^{\ell}$, with α_b order of χ_b.

The hamiltonian H_Λ of the finite system has thus the general form :

$$-\beta H_\Lambda = \sum_{b \in \mathcal{B}} \sum_{\ell = 0}^{\alpha_b - 1} K(b; \ell)\, \chi_b^{\ell} = -\beta \sum_{b \in \mathcal{B}} H^{(b)} \qquad (1.11)$$

where for all b in $\mathcal{B}$ $K(b; \ell) \neq 0$ for at least one $\ell \in \{1, \dots, \alpha_b - 1\}$. For physical system the hamiltonian is real and this condition will be satisfied in particular if we impose :

$$\overline{K(b; \ell)} = K(b; \alpha_b - \ell) \qquad (1.12)$$

which we shall do in the following.

Let us remark that we could take as family $\{\chi_b\}_{b \in \mathcal{B}}$ the set $\mathcal{T}$ itself as was done by W. Greenberg [127] ; this choice, which is in fact the only possibility for spin ½ systems, is however not very convenient in most applications and in most cases we shall choose the smallest family which generates $\mathcal{T}$. In conclusion the support of the interactions is specified by a "rectangular matrice" $\{M_{b,x}\}_{\substack{b \in \mathcal{B} \\ x \in \Lambda}}$ where :

$$\chi_b(\underline{n}) = \prod_{x \in \Lambda} \exp\left(\frac{2i\pi}{q} M_{b,x}\, n_x\right) \qquad M_{b,x} \in \{0, 1, \dots, q-1\}$$

Remark :

For real hamiltonian, the correlation functions satisfy $\overline{\omega[\chi]} = \omega[\chi^{-1}]$.

Instead of introducing first the general group structure and then show its interest for physical applications, as was done in parts I and II, we shall proceed in the opposite direction to show how this structure appears naturally as soon as one has adopted the expansion of the hamiltonian Eq. (1.11).

1.2.1. Low Temperature Expansion for the Partition Function

From the definition Eq. (1.9) and the decomposition Eq. (1.11) we obtain :

$$Z(\Lambda, K) = \sum_{\underline{n} \in \mathcal{G}_\Lambda} \prod_{b \in \mathcal{B}} exp\left[\sum_{\ell=0}^{\alpha_b - 1} K(b; \ell)\, \chi_b^{\ell}(\underline{n}) \right]$$

It is then clear that we can replace the sum on $\underline{n}$ in $\mathcal{G}_\Lambda$ by the sum on equivalence classes $[\underline{n}] = \{ \underline{m};\ \chi_b(\underline{m}) = \chi_b(\underline{n})\ \forall b \in \mathcal{B} \}$ and therefore with :

$$\mathcal{S} = [\underline{o}] = \{ \underline{s} \in \mathcal{G}_\Lambda\ ;\ \chi_b(\underline{s}) = 1\ \ \forall b \in \mathcal{B} \}$$

we obtain the so-called "Low Temperature Expansion" for the partition function

$$Z(\Lambda, K) = |\mathcal{S}| \sum_{[\underline{n}] \in \mathcal{G}_\Lambda / \mathcal{S}} \prod_{b \in \mathcal{B}} \omega(b; [\underline{n}])$$

$$\omega(b; [\underline{n}]) = exp[-\beta H^{(b)}(\underline{n})] = exp\left[\sum_{\ell=0}^{\alpha_b - 1} K(b;\ell)\, \chi_b^{\ell}(\underline{n}) \right] \tag{1.13}$$

which we shall further discuss in Sec. 2.1.

1.2.2. High Temperature Expansion for the Partition Function

To derive the high temperature expansion, we consider first the expansion :

$$\exp\left[\sum_{\ell=0}^{\alpha_b-1} K(b;\ell)\, \chi_b^{\ell}(\underline{n})\right] = \exp\left[\sum_{\ell=0}^{\alpha_b-1} K(b;\ell)\, e^{\frac{2i\pi}{\alpha_b}\ell r}\right] = \sum_{\ell=0}^{\alpha_b-1} f(b;\ell)\, \chi_b^{\ell}(\underline{n})$$

where $r = \sum_{x\in\Lambda} M_{b,x}\, n_x$ and

$$f(b;\ell) = \frac{1}{\alpha_b}\sum_{k=0}^{\alpha_b-1} e^{-\frac{2i\pi}{\alpha_b}k\ell} \exp\left[\sum_{\ell'=0}^{\alpha_b-1} K(b;\ell')\, e^{\frac{2i\pi}{\alpha_b}\ell' k}\right] \qquad (1.14)$$

is the Fourier transform of the Boltzmann factor $\omega(b) = e^{-\beta H^{(b)}}$.

Using the orthogonality property of the characters, we obtain the "High Temperature Expansion" :

$$Z(\Lambda, K) = |\mathcal{G}_\Lambda| \sum_{\underline{\ell}}{}^{*} \prod_{b\in\mathcal{B}} f(b;\ell_b) \qquad (1.15)$$

where the sum on $\underline{\ell} = \{\ell_b\}_{b\in\mathcal{B}}$ $\ell_b \in \{0,1,\dots,\alpha_b-1\}$ is restricted to those $\underline{\ell}$ satisfying the condition :

$$\sum_{b\in\mathcal{B}} \ell_b\, M_{b,x} = 0 \mod q \quad \text{for all } x \text{ in } \Lambda .$$

We shall further discuss this expansion in Sec. 2.2.

1.3. Group Structure for Finite and Infinite q-Component Systems

The group structure introduced in Sec. 1.3.4 of Part I was sufficiently general to make the extension to arbitrary spin systems straightforward; moreover this extension appears very natural in view of the results described in the preceding section. In this section we shall consider only the case of q component systems and we shall give the extension to arbitrary systems in the next section.

Arbitrary q-component lattice systems are defined abstractly by :

$$\{ \mathcal{L} ,\ \mathcal{G}_{\mathcal{L}} = \prod_{x \in \mathcal{L}} \mathcal{G}_x ,\ \mathcal{B},\ \pi,\ K \}$$

where $\mathcal{L}$ is a countable set of point x , called "lattice sites"

$\mathcal{G}_x \cong \mathbb{Z}_q$

$\mathcal{B}$ is a countable set of indices b , called "bonds"

π is a mapping $b \mapsto \chi_b$ from $\mathcal{B}$ into the dual group $\mathcal{G}_{\mathcal{L}}^{\wedge}$ (*) :

$$< \chi_b ; \underline{n} >_{\mathcal{L}} = \prod_{x \in \mathcal{L}} \exp \left[\frac{2 i \pi}{q} M_{b,x}\, n_x \right]$$

K is a complex function $(b, \ell) \mapsto K(b; \ell)$ on $\mathcal{B} \times \mathbb{Z}_q$.

Together with the "Group of Configurations $\mathcal{G}_{\mathcal{L}} = \prod_{x \in \mathcal{L}} \mathcal{G}_x$ ", we consider also the "Group of Graphs $\mathcal{G}_{\mathcal{B}}$ " defined by :

$$\mathcal{G}_{\mathcal{B}} = \prod_{b \in \mathcal{B}} \mathcal{G}_b \qquad \mathcal{G}_b \cong \mathbb{Z}_{\alpha_b} \qquad \alpha_b = \text{order of } \chi_b$$

The group structure is then defined by means of the groups $\mathcal{G}_{\mathcal{L}}$, $\mathcal{G}_{\mathcal{B}}$, their dual groups $\mathcal{G}_{\mathcal{L}}^{\wedge}$, $\mathcal{G}_{\mathcal{B}}^{\wedge}$, together with the following homomorphisms π and γ :

$$\pi : \begin{array}{ccc} \mathcal{G}_{\mathcal{B}} & \longrightarrow & \prod_{x \in \mathcal{L}} \mathcal{G}_x^{\wedge} \cong \mathcal{G}_{\mathcal{L}} \\ \cup & & \cup \\ \underline{\ell} & \longmapsto & \pi(\underline{\ell}) = \prod_{b \in \mathcal{B}} \chi_b^{\ell_b} \end{array} \qquad \gamma : \begin{array}{ccc} \mathcal{G}_{\mathcal{L}} & \longrightarrow & \prod_{b \in \mathcal{B}} \mathcal{G}_b^{\wedge} \cong \mathcal{G}_{\mathcal{B}} \\ \cup & & \cup \\ \underline{n} & \longmapsto & \gamma(\underline{n}) \end{array}$$

(*) We recall that for infinite $\mathcal{L}$, the dual group $\mathcal{G}_{\mathcal{L}}^{\wedge}$ of $\mathcal{G}_{\mathcal{L}}$ is by definition

$\mathcal{G}_{\mathcal{L}}^{\wedge} = \left[\prod_{x \in \mathcal{L}} \mathcal{G}_x^{\wedge} \right]_f = \{ \chi ;$ support of $\chi \in \mathcal{P}_f(\mathcal{L}) \}$

property which remains true for any compact abelian group [4] ; moreover for any locally compact group $\mathcal{G}_{\mathcal{L}}$, we have $(\mathcal{G}_{\mathcal{L}}^{\wedge})^{\wedge} = \mathcal{G}_{\mathcal{L}}$. Following Rudin's terminology $\mathcal{G}_{\mathcal{L}}$ is a "complete direct sum", while $\mathcal{G}_{\mathcal{L},f} = \{ \underline{n} \in \mathcal{G}_{\mathcal{L}} ;$ support of $\underline{n}$ finite $\}$ is a "direct sum".

where $\gamma(\underline{n})$ is defined by the equation :

$$< \gamma(\underline{n}) ; \underline{\ell} >_{\mathcal{B}} = < \pi(\underline{\ell}) ; \underline{n} >_{\mathcal{L}} = \exp\left[\frac{2i\pi}{q} \sum_{\substack{b \in \mathcal{B} \\ x \in \mathcal{L}}} \ell_b M_{b,x} n_x \right] \qquad (1.16)$$

Using the isomorphism Eq.(1.6) we have :

$$\gamma(\underline{n}) \leftrightarrow \left\{ \sum_x \frac{\alpha_b}{q} M_{b,x} n_x \mod \alpha_b \right\}_{b \in \mathcal{B}} \in \mathcal{G}_{\mathcal{B}}$$

(1.17.)

$$\pi(\underline{\ell}) \leftrightarrow \left\{ \sum_b \ell_b M_{b,x} \mod q \right\}_{x \in \mathcal{L}} \in \mathcal{G}_{\mathcal{L}}$$

relation valid for all $\underline{n} \in \mathcal{G}_{\mathcal{L}}$, $\underline{\ell} \in \mathcal{G}_{\mathcal{B},f}$
and for all $\underline{n} \in \mathcal{G}_{\mathcal{L},f}$, $\underline{\ell} \in \mathcal{G}_{\mathcal{B}}$

The Kernel $\mathcal{S}$ of γ and the image $\bar{\mathcal{T}}$ of π restricted to $\mathcal{G}_{\mathcal{B},f}$ are subgroups of $\mathcal{G}_{\mathcal{L}}$ and $\mathcal{G}_{\mathcal{L}}^{\wedge}$ respectively defined by :

$$\mathcal{S} = \{ \underline{s} \in \mathcal{G}_{\mathcal{L}} ; \chi_b(\underline{s}) = 1 \ \forall b \in \mathcal{B} \} = \{ \underline{s} \in \mathcal{G}_{\mathcal{L}} ; \sum_{x \in \mathcal{L}} M_{b,x} s_x = 0 \mod q \ \forall b \in \mathcal{B} \}$$

(1.18)

$$\bar{\mathcal{T}} = \{ \chi \in \mathcal{G}_{\mathcal{L}}^{\wedge} ; \chi = \pi(\underline{\ell}) , \underline{\ell} \in \mathcal{G}_{\mathcal{B},f} \} \cong \{ \underline{n} = \{ \sum_b \ell_b M_{b,x} \}_{x \in \mathcal{L}} ; \underline{\ell} \in \mathcal{G}_{\mathcal{B},f} \}$$

The Kernel $\mathcal{K}$ of π and the image Γ of γ are subgroups of $\mathcal{G}_{\mathcal{B}}$ and $\prod_{b \in \mathcal{B}} \mathcal{G}_b^{\wedge}$ respectively given by :

$$\mathcal{K} = \{ \underline{\ell} \in \mathcal{G}_{\mathcal{B}} ; \prod_{b \in \mathcal{B}} \chi_b^{\ell_b}(\underline{n}) = 1 \ \forall \underline{n} \in \mathcal{G}_{\mathcal{L},f} \} = \{ \underline{\ell} \in \mathcal{G}_{\mathcal{B}} ; \sum_{b \in \mathcal{B}} \ell_b M_{b,x} = 0 \mod q \ \forall x \in \mathcal{L} \}$$

(1.19)

$$\Gamma = \{ \gamma(\underline{n}) ; \underline{n} \in \mathcal{G}_{\mathcal{L}} \} \cong \{ \underline{\ell} = \{ \sum_x \frac{\alpha_b}{q} M_{b,x} n_x \mod \alpha_b \}_{b \in \mathcal{B}} ; \underline{n} \in \mathcal{G}_{\mathcal{L}} \}$$

Property 1

1) $\mathcal{S} = \bar{\mathcal{T}}^{\perp}$ $\qquad$ $\bar{\mathcal{T}} = \mathcal{S}^{\perp}$

2) $(\mathcal{G}_{\mathcal{L}}/\mathcal{S})^{\wedge} \cong \bar{\mathcal{T}} \cong \Gamma^{\wedge}$

$(\mathcal{G}_{\mathcal{L},f}/\mathcal{S}_f)^{\wedge} \cong (\Gamma^{(f)})^{\wedge} \cong \mathcal{I}m\,\pi$ where : $\Gamma^{(f)} = \{\gamma(\underline{n})\,;\ \underline{n} \in \mathcal{G}_{\mathcal{L},f}\}$

$\mathcal{S}_f = \mathcal{S} \cap \mathcal{G}_{\mathcal{L},f}$

Property 2.

1) $\mathcal{K} = \Gamma^{(f)\perp}$ $\qquad$ $\mathcal{K}^{\perp} = \Gamma^{(f)}$

2) $\mathcal{K}_f = \Gamma^{\perp}$, $\mathcal{K}_f^{\perp} = \Gamma$ where : $\mathcal{K}_f = \mathcal{K} \cap \mathcal{G}_{\mathcal{B},f}$

3) $(\mathcal{G}_{\mathcal{B}}/\mathcal{K})^{\wedge} \cong \Gamma^{(f)} \cong (\mathcal{I}m\,\pi)^{\wedge}$

$(\mathcal{G}_{\mathcal{B},f}/\mathcal{K}_f)^{\wedge} \cong \Gamma \cong (\bar{\mathcal{T}})^{\wedge}$

The proofs of these properties follow directly from Eq. (1.16) and lemma 1 of Part I (Sec. 1.3.2) [5].

Summary of the Group Structure

For arbitrary spin lattice systems $\{\mathcal{L}, \mathcal{G}_{\mathcal{L}}, \mathcal{B}, \pi, K\}$, we have the following group structure :

$$\mathcal{G}_{\mathcal{L}} = \prod_{x \in \mathcal{L}} \mathcal{G}_x \xrightarrow{\ \gamma\ } \prod_{b \in \mathcal{B}} \mathcal{G}_b^{\wedge}$$

$\ker\gamma = \mathcal{S} = \bar{\mathcal{T}}^{\perp}$	$\mathcal{I}m\,\gamma = \Gamma = \mathcal{K}_f^{\perp}$
$\mathcal{S}_f = (\mathcal{I}m\,\pi)^{\perp}$	$\Gamma^{(f)} = \mathcal{K}^{\perp}$

In this example the bonds are thus defined by <u>oriented</u> pairs of nearest neighbour sites $b = (x,y)$ and $\chi_b(\underline{\vartheta}) = e^{-i(\theta_x - \theta_y)}$ yields $M_{b,x} = -1$, $M_{b,y} = +1$, $M_{b,z} = 0$ otherwise.

The group of graphs is then defined by :

$$\mathcal{G}_{\mathcal{B}} = \prod_b \mathcal{G}_b \qquad \mathcal{G}_b \cong \mathbb{Z}$$

and the group $\mathcal{K}$ of closed graph is the subgroup of integers $\underline{k}$ associated with each oriented nearest neighbour satisfying the condition :

$$\sum_{y \in \Lambda} k_{(x,y)} = \sum_{y \in \Lambda} k_{(y,x)} \qquad \forall\, x \in \Lambda$$

It is easily seen that this group is generated by the elements shown on Fig. 1.b).

On the other hand the Low Temperature group Γ is a subgroup of $\hat{\mathcal{G}}_{\mathcal{B}} = \prod_{b \in \mathcal{B}} \hat{\mathcal{G}}_b$, $\hat{\mathcal{G}}_b \cong T^{(1)}$, defined by :

$$\Gamma = \{ \underline{\Psi} = (\Psi_1, \ldots, \Psi_{|\mathcal{B}|}) \,;\ \Psi_i \in [0, 2\pi] \ ,\ \exists\, \underline{\vartheta} \in \mathcal{G}_\Lambda \ \text{s.t.}\ \Psi_b = \theta_y - \theta_x \}$$

Again it is easily verified that this group is generated by the elements shown on Fig. 1.c).

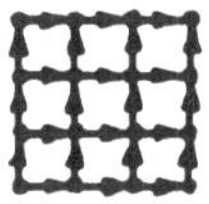

(a)

(b)

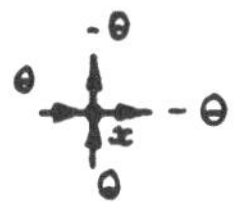

(c)

<u>Fig</u>.1 a) $\mathcal{L} = \{\bullet\}$ $\mathcal{B} = \{\bullet\!\!\rightarrow\!\!\bullet\}$

b) generators for $\mathcal{K}$

c) generators for Γ , $\theta \in [0, 2\pi]$

$$\prod_{x\in\mathcal{L}} \mathcal{G}_x^{\wedge} \xleftarrow{\;\pi\;} \mathcal{G}_{\mathcal{B}} = \prod_{b\in\mathcal{B}} \mathcal{G}_b$$

$\mathrm{Im}\,\pi = \mathcal{S}_f^{\perp}$	$\ker \pi = \mathcal{K} = \Gamma^{(f)\perp}$
$\bar{\mathcal{T}} = \mathcal{S}^{\perp}$	$\mathcal{K}_f = \Gamma^{\perp}$

$$\Gamma \cong \mathcal{G}_{\mathcal{L}}/\mathcal{S} \cong (\mathcal{G}_{\mathcal{B},f}/\mathcal{K}_f)^{\wedge} \cong \bar{\mathcal{T}}^{\wedge}$$

$$\Gamma^{(f)} \cong \mathcal{G}_{\mathcal{L},f}/\mathcal{S}_f \cong (\mathcal{G}_{\mathcal{B}}/\mathcal{K})^{\wedge} \cong (\mathrm{Im}\,\pi)^{\wedge}$$

Remarks

1 - For $q=2$ (spin $\frac{1}{2}$ systems), $\mathcal{G}_{\mathcal{L}} \cong \mathcal{P}(\mathcal{L})$, $\mathcal{G}_{\mathcal{L}}^{\wedge} \cong \mathcal{P}_f(\mathcal{L})$, $\mathcal{G}_{\mathcal{B}} \cong \mathcal{P}(\mathcal{B})$, $\mathcal{G}_{\mathcal{B}}^{\wedge} \cong \mathcal{P}_f(\mathcal{B})$ and we recover the structure introduced in Part I for spin $\frac{1}{2}$; moreover the introduction of the group $\mathcal{G}_{\mathcal{B}} = \prod_{b\in\mathcal{B}} \mathcal{G}_b$ for arbitrary spin is suggested by the decomposition of the hamiltonian and the H.T. expansion Eq. (1.15).

2 - The kernel of the mapping γ coincide with the group $\mathcal{S}$ introduced in Sec. 1.2 which justifies the notation; moreover $\bar{\mathcal{T}}$ coincide the group finitely generated by the support $\mathcal{T}$ of the interactions.

1.4. Group Structure for General Spin Systems

The definitions of Sec. 1.2 and 1.3 can be extended to cover more general cases where at each lattice sites x of $\mathcal{L}$ is associated an arbitrary group $\mathcal{G}_x$ together with a measure μ_x ; the group of configurations is again defined by :

$$\mathcal{G}_{\mathcal{L}} = \prod_{x\in\mathcal{L}} \mathcal{G}_x$$

and for finite systems the partition function and the Gibbs states are defined by :

$$Z(\Lambda) = \int_{\mathcal{G}_\Lambda} \prod_x d\mu_x(n_x) \; e^{-\beta H_\Lambda(\underline{n})} \qquad (1.20)$$

$$\omega_\Lambda[A] = Z(\Lambda)^{-1} \int_{\mathcal{G}_\Lambda} \prod_x d\mu_x(n_x) \; e^{-\beta H_\Lambda(\underline{n})} \; A(\underline{n})$$

To illustrate such a generalisation we consider the case where $\mathcal{G}_x$ is a "Compact Abelian Group",with the Haar measure dn_x ; with the hamiltonian expressed in the form

$$-\beta H_\Lambda = \sum_{b\in\mathcal{B}} \sum_\ell K(b;\ell)\, \chi_b^\ell \qquad \chi_b \in \hat{\mathcal{G}}_\Lambda$$

we introduce again the "group of graphs"

$$\mathcal{G}_\mathcal{B} = \prod_{b\in\mathcal{B}} \mathcal{G}_b \qquad \begin{array}{ll} \mathcal{G}_b = \mathbb{Z}_{\alpha_b} \text{ if } & \alpha_b = \text{order of } \chi_b < \infty \\ \mathcal{G}_b = \mathbb{Z} \text{ if } & \alpha_b = \infty \end{array}$$

and the homomorphisms π and γ as well as the groups $\mathcal{K}$, Γ , $\mathcal{S}$ and $\bar{\mathcal{T}}$, are introduced following the definitions of Sec. 1.3 and have the same properties.

From :
$$e^{-\beta H_\Lambda} = \prod_{b\in\mathcal{B}} \exp\left[\sum_\ell K(b;\ell)\, \chi_b^\ell\right]$$

together with the expansion of the Boltzmann factors

$$\exp\left[\sum_\ell K(b;\ell)\, \chi_b^\ell\right] = \sum_\ell f(b;\ell)\, \chi_b^\ell$$

we obtain the "H.T. - expansion" from the orthogonality of character :

$$Z(\Lambda) = \prod_{b\in\mathcal{B}} f(b;0) \left[\prod_{x\in\Lambda} \int_{\mathcal{G}_x} dn_x\right] \sum_{\underline{\ell}\in\mathcal{K}} \prod_{b\in\mathcal{B}} \frac{f(b;\ell_b)}{f(b;0)}$$

where $\mathcal{K} = \{\underline{\ell} \in \mathcal{G}_\mathcal{B} ;\ \prod_b \chi_b^{\ell_b}(\underline{n}) = 1 \quad \forall\ \underline{n} \in \mathcal{G}_\Lambda\} \subset \mathcal{G}_\mathcal{B}$

Example : "Rotator with N.N. interactions on $\mathbb{Z}^2$"

Let us illustrate this generalisation with the system defined by $\mathcal{G}_x = T^{(1)}$(the 1-torus),

$$\mathcal{G}_\Lambda = \prod_{x\in\Lambda} \mathcal{G}_x = \{\underline{\vartheta} = (\theta_1,\dots,\theta_{|\Lambda|});\ \theta_i \in [0,2\pi]\}$$

and the hamiltonian

$$-\beta H_\Lambda(\underline{\vartheta}) = \sum_{\{x,y\}} \sum_{\ell\geq 1} K(\{x,y\};\ell) \cos \ell(\theta_x - \theta_y)$$

To conclude the discussion of this generalisation we recall the following standard results.

Theorem [4]

1) $\mathcal{G}^{\wedge\wedge} = \mathcal{G}$

2) if $\mathcal{G}$ is discrete abelian, $\mathcal{G}^{\wedge}$ is compact
   if $\mathcal{G}$ is compact abelian, $\mathcal{G}^{\wedge}$ is discrete

3) $\mathcal{G}_x \cong \mathbb{Z} \Rightarrow \mathcal{G}_x^{\wedge} \cong T^{(1)}$; $\chi(\underline{n}) = \prod_{x \in \mathcal{L}} e^{i\varphi_x n_x}$; $\varphi_x \in [0,2\pi]$, $n_x \in \mathbb{Z}$

$\mathcal{G}_x \cong T^{(1)} \Rightarrow \mathcal{G}_x^{\wedge} \cong \mathbb{Z}$

$\mathcal{G}_x \cong \mathbb{R} \Rightarrow \mathcal{G}_x^{\wedge} \cong \mathbb{R}$; $\chi(\underline{n}) = \prod_{x \in \mathcal{L}} e^{ik_x n_x}$; $k_x \in \mathbb{R}$, $n_x \in \mathbb{R}$

CHAPTER 2 - PHYSICAL IMPLICATIONS OF THE GROUP STRUCTURE

The interest of the groups Γ and $\mathcal{K}$ introduced in the preceding chapter lies in the fact that "Low Temperature" (L.T.) and "High Temperature" (H.T.) expansion of the partition function can be written in a natural way in terms of these groups. Moreover these expansions are simply related by means of Poisson formulae for abelian groups [5] .

On the other hand the groups $\mathcal{S}$ and $\overline{\mathcal{G}}$ are related to symmetry properties of the hamiltonian, which yield for finite systems symmetry properties of the free energy and of the state; these properties will then enable us to introduce "order parameters" describing possible breakdown of the symmetry group $\mathcal{S}$ for infinite systems.

As we shall see, for arbitrary spin systems it is not sufficient to consider $\mathcal{S}$ as was done in part I ; there may exist additional symmetry properties which are obtained by means of the permutation group $\mathcal{S}_q$ of q objects and which could also be broken for infinite systems.

Following the philosophy of parts I and II, we shall then derive equation for the correlation functions of finite systems [125] and we shall take the solutions of these equations extended to infinite systems as definition of the equilibrium states; the existence or absence of phase transitions, together with the symmetry properties of the equilibrium states can then be discussed in terms of properties of this equation.

We shall conclude this chapter with a discussion on the relation between the physical picture and the group picture together with the Potts Models as illustration.

As usual we keep the notation Λ instead of $\mathcal{L}$ to denote the lattice sites of a finite system; moreover the hamiltonian is given by :

$$-\beta H_{\Lambda} = \sum_{b\in\mathcal{B}} \sum_{\ell=0}^{\alpha_b-1} K(b;\ell)\,\chi_b^{\ell} = -\beta \sum_{b\in\mathcal{B}} H^{(b)}$$

In particular if Λ is a finite subset of an infinite system and if $\underline{m} \in \mathcal{G}_{\mathcal{L}}$, the hamiltonian $H_{\Lambda,\underline{m}}$ of the finite system Λ with boundary condition $\underline{m}$ is defined by :

$$-\beta H_{\Lambda,\underline{m}} = \sum_{b\in\mathcal{B}_{\Lambda}} \sum_{\ell=0}^{\alpha_b-1} K(b;\ell) < \chi_b^{\ell}; \underline{m} >_{\mathcal{L}/\Lambda} \chi_b^{\ell}$$

where : $\mathcal{B}_{\Lambda} = \{ b \in \mathcal{B}\,;\, < \chi_b ; \underline{n} >_{\Lambda} \neq 1$ for some $\underline{n} \in \mathcal{G}_{\Lambda} \}$.

Finally we shall always keep the notation $\underline{\ell}$ for the elements in $\mathcal{G}_{\mathcal{B}}$ and $\underline{n}$ for the elements in $\mathcal{G}_{\mathcal{L}}$.

2.1. Low Temperature Expansion

Using the subgroup Γ of $\hat{\mathcal{G}_{\mathcal{B}}}$, the "Low Temperature Expansion" Eq. (1.13) becomes :

$$Z(\Lambda, K) = \prod_{b\in\mathcal{B}} e^{\sum_{\ell=0}^{\alpha_b-1} K(b;\ell)} \cdot |\mathcal{S}| \cdot \sum_{\underline{\ell}\in\Gamma} \prod_{b\in\mathcal{B}} z(b;\ell_b) \qquad (2.1)$$

where the generalised activities $z(b;\ell)$ are defined by :

$$z(b;\ell) = exp\left[-\sum_{\ell'=1}^{\alpha_b-1} K(b;\ell')\left(1 - e^{\frac{2i\pi}{\alpha_b}\ell\ell'}\right)\right] ; \; z(b;0) = 1 \qquad (2.2)$$

We remark in particular that for spin ½ systems, $z(b,1) = e^{-2K(b)}$, and we recover the L.T. expansion Eq. (2.1) discussed in part I ; on the other hand for arbitrary spin systems we have the following properties (Eq.(1.12)) :

$$\overline{K(b;\ell)} = K(b;\alpha_b-\ell) \text{ implies } z(b;\ell) > 0$$

As we have already discussed, the group picture can always be chosen such that :

$$\sum_{b\in\mathcal{B}} \sum_{\ell=1}^{\alpha_b-1} K(b;\ell)\left[1 - \chi_b^{\ell}(\underline{n})\right] \geq 0 \qquad \forall\, \underline{n} \in \mathcal{G}_{\Lambda}$$

This condition is not sufficient however to insure that the generalised activies are small at low temperature; we thus introduce the following definition :

Definition

The system is said ferromagnetic of

$$\sum_{\ell'=1}^{\alpha_b-1} K(b;\ell')\left[1-e^{\frac{2i\pi}{\alpha_b}\ell\ell'}\right] > 0 \qquad \forall b \in \mathcal{B}, \quad \ell \in \{1,\dots,\alpha_b-1\}$$

Again we remark that this condition yields $K(b) > 0$ for spin $\frac{1}{2}$ system which is the usual definition; on the other hand for arbitrary spin systems this concept of "ferromagnetic" is related to the decomposition of the hamiltonian Eq. (1.11) and therefore a given system could be ferromagnetic for a given decomposition and not be with respect to another. In particular for the decomposition $\{\chi_b\} = \mathcal{T}$ considered by W. Greenberg [127] the system can never be ferromagnetic. (except if $\chi^2 = 1 \quad \forall \chi \in \mathcal{T}$).

Properties 1

1 - If $\overline{K(b;\ell)} = K(b;\alpha_b-\ell) > 0$ the system is ferromagnetic; this condition is however not necessary.

2 - For ferromagnetic systems, $0 < z(b;\ell) < 1 \quad \forall \ell \in \{1,\dots,\alpha_b-1\}$ and therefore the parameters which appear in the L.T. expansion will be small at low temperature which justifies the expansion "L.T. expansion".

2.2. High Temperature Expansion

Using the subgroup $\mathcal{K}$ of $\mathcal{G}_{\mathcal{B}}$ Eq. (1.19), the "High Temperature Expansion" Eq. (1.15) becomes :

$$Z(\Lambda,K) = |\mathcal{G}_\Lambda| \prod_{b\in\mathcal{B}} f(b;0) \sum_{\underline{\ell}\in\mathcal{K}} \prod_{b\in\mathcal{B}} t(b;\ell_b) \qquad (2.3)$$

where the "generalised tanh functions" $t(b;\ell)$ are defined by :

$$t(b;\ell) = \frac{f(b;\ell)}{f(b;0)}$$

with

$$f(b;\ell) = \frac{1}{\alpha_b} \sum_{k=0}^{\alpha_b-1} e^{-\frac{2i\pi}{\alpha_b}\ell k} \exp\left[\sum_{\ell'=1}^{\alpha_b-1} K(b;\ell')\, e^{\frac{2i\pi}{\alpha_b} k\ell'}\right] \tag{2.4}$$

i.e.

$$t(b;\ell) = \frac{\sum_{k=0}^{\alpha_b-1} e^{-\frac{2i\pi}{\alpha_b}\ell k}\, z(b;k)}{\sum_{k=0}^{\alpha_b-1} z(b;k)} \tag{2.5}$$

For spin $\frac{1}{2}$ systems, $t(b;1) = \tanh K(b)$, and we recover the H.T. expansion Eq. (2.1) discussed in part I ; on the other hand for any lattice system the parameters $|t(b;\ell)|$ will be small at high temperature which justifies the expression "H.T. expansion"; in fact we have

$$t(b;\ell) = K(b;\ell) + O(T^{-2}) \qquad K = \frac{1}{kT} J$$

It should also be remarked that the condition $\overline{K(b;\ell)} = K(b;\alpha_b-\ell)$ is not sufficient to yield real t's and in most cases these parameters will be complex; in fact we have only the following property :

$$\overline{K(b;\ell)} = K(b;\alpha_b-\ell) \geqslant 0 \quad \text{implies} \quad 0 \leq t(b;\ell) < 1$$

while

$$\overline{K(b;\ell)} = K(b;\alpha_b-\ell) \quad \text{implies} \quad \overline{t(b;\ell)} = t(b;\alpha_b-\ell)$$

2.3. Poisson Formulae

The Kramers - Wannier duality relation [7] for the infinite two dimensional Ising model was derived by H.P. McKean [11] by means of Poisson formulae for finite abelian groups. This was then generalised by C.Gruber and A. Hintermann [5] to arbitrary lattice systems where it was shown that the above H.T. and L.T. expansions are simply related by means of the same Poisson formulae. It should be stressed however that Poisson formulae is a transformation related to a given system, which should not be confused with the duality transformation which relates different systems (even though the two concepts will be connected).

Poisson formulae for finite groups was given in Sec. 8.3 of part I ; let then $G = \mathcal{G}_{\mathcal{B}}$, $H = \mathcal{K}$; using property 2 of Sec. 1 we have $(G/H)^{\wedge} = (\mathcal{G}_{\mathcal{B}}/\mathcal{K})^{\wedge} \cong \Gamma$. Moreover with

$$f(\underline{\ell}) = |\mathcal{G}_{\Lambda}| \prod_{b\in\mathcal{B}} f(b;\ell_b)$$

we obtain :

$$\tilde{f}(\chi_{\underline{m}}) = |\mathcal{G}_{\mathcal{B}}|^{-1} |\mathcal{G}_{\Lambda}| \prod_{b\in\mathcal{B}} \left[\sum_{\ell_b=0}^{\alpha_b-1} e^{-\frac{2i\pi}{\alpha_b} m_b \ell_b} f(b;\ell_b) \right]$$

Therefore using the definition of f(b;ℓ) Eq. (2.4) together with the relation $|\mathcal{G}_{\mathcal{B}}|^{-1} |\mathcal{K}| \, |\mathcal{G}_{\Lambda}| = |\mathcal{S}|$ (Property 2) the Poisson formulae applied to the H.T. expansion yields precisely the L.T. expansion.

It should be recalled that Poisson formulae is in fact valid for arbitrary locally compact abelian group and we have :

<u>Theorem</u> [128]

Let G be any locally compact Abelian group and let H be a closed subgroup. Let the Haar measures on G , H and G/H be adjusted so that $\int_G = \int_{G/H} \int_H$ and let f be a function of $[\mathcal{L}^1 \cap \mathcal{P}](G)$ such that $g(y) = \int_H f(xy)\,dx$ is a continuous function (on G/H) of y . Then

$$\int_H f(x)\,dx = \int_{(G/H)^{\wedge}} \tilde{f}(\alpha)\,d\alpha$$

In this theorem $\mathcal{P}$ is the class of positive definite functions and we have

<u>Theorem</u> [128]

If $f \in [\mathcal{L}^1 \cap \mathcal{P}](G)$ then $\tilde{f} \in \mathcal{L}^1(G)$ and

$$f(x) = \int_{G^{\wedge}} \chi(x)\,\tilde{f}(\chi)\,d\chi \quad \text{for almost all } x \in G$$

where $d\chi$ is the Haar measure of $G^{\wedge}$ suitably normalised. Moreover the vector space $[\mathcal{L}^1 \cap \mathcal{P}](G)$ generated by $\mathcal{L}^1 \cap \mathcal{P}$ is dense in $\mathcal{L}^1$.

2.4. Symmetry Properties

2.4.1 Internal Symmetry Group $\mathcal{S}$

Just as in part I, the configuration group $\mathcal{G}_{\mathcal{L}}$ acts as a group of automorphisms of $\mathfrak{A}_{\Lambda}$ by :

$$\forall \; \underline{m} \in \mathcal{G}_{\Lambda} \qquad (\tau_{\underline{m}} A)(\underline{n}) = A(\underline{m}^{-1} \cdot \underline{n}) \qquad (2.6)$$

in particular : $\tau_{\underline{m}} \chi = \chi^{-1}(\underline{m}) \cdot \chi$ for all $\chi \in \hat{\mathcal{G}}_{\mathcal{L}}$

The automorphism $\tau_{\underline{m}}$ induces as usual a transformation on the states defined by :

$$(\tau'_{\underline{m}} \omega)[A] = \omega[\tau^{-1}_{\underline{m}} A]$$

which yields

$$(\tau'_{\underline{m}} \omega)[\chi] = \chi(\underline{m}) \; \omega[\chi]$$

Definition.

A state ω is said invariant under the subgroup $\mathcal{G}_{\omega}$ of $\mathcal{G}_{\mathcal{L}}$ if

$$\tau'_{\underline{m}} \omega = \omega \qquad \text{for all } \underline{m} \in \mathcal{G}_{\omega}$$

Lemma 1

A state ω is invariant under $\mathcal{G}_{\omega} \subset \mathcal{G}_{\mathcal{L}}$ if and only if

$$\omega[\chi] = 0 \qquad \text{for all } \chi \notin \mathcal{G}_{\omega}^{\perp}$$

Indeed $(\tau'_{\underline{m}} \omega)[\chi] = \omega[\chi] \quad \forall \; \underline{m} \in \mathcal{G}_{\omega} \quad \forall \; \chi \in \hat{\mathcal{G}}_{\mathcal{L}}$

implies $\omega[\chi] \, (1 - \chi(\underline{m})) = 0$.

Therefore : if $\chi \notin \mathcal{G}_{\omega}^{\perp} \equiv \{ \chi \in \hat{\mathcal{G}}_{\mathcal{L}} \; ; \; \chi(\underline{m}) = 1 \;\; \forall \, \underline{m} \in \mathcal{G}_{\omega} \}$

then $\omega[\chi] = 0$

The interest of the subgroups $\mathcal{S}$ and $\mathcal{T}$ of $\mathcal{G}_{\Lambda}$ is then reflected by the following properties.

Properties 2

1 - The hamiltonian H_{Λ} is invariant under the internal symmetry group $\mathcal{S}$.

2 - The Gibbs states are, for finite systems, invariant under $\mathcal{S}$ and thus

$$\omega_{(\Lambda,K)}[\chi] = 0 \qquad \text{if} \qquad \chi \notin \bar{\mathcal{T}} = \mathcal{S}^{\perp}$$

i.e. $\omega_{(\Lambda,K)}[\chi] \neq 0$ only if $\chi \in \bar{\mathcal{T}}$

3 - For ferromagnetic systems, the group $\mathcal{S}$ is a group of "Ground States".

These properties follow immediately from Lemma 1 just as were derived the similar properties for spin ½.

2.4.2 Symmetric States and Symmetric Algebra

As in parts I and II, a state ω is said "symmetric" if it is invariant under the internal symmetry group $\mathcal{S}$; it follows from the preceding discussion that a symmetric state can also be defined as a state on the "symmetric algebra $\mathfrak{A}^{Sym}$" defined as

$$\mathfrak{A}^{Sym} = \{ A \in \mathfrak{A} ;\ \tau_{\underline{s}} A = A \quad \forall\ \underline{s} \in \mathcal{S} \}$$

This algebra being the closure of the linear span of the family of observables

$$\pi(\underline{\ell}) = \prod_b \chi_b^{\ell_b} \qquad \underline{\ell} \in \mathcal{G}_{\mathcal{B},f}$$

any symmetric state will be uniquely defined by the function σ on $\mathcal{G}_{\mathcal{B},f}$

$$\sigma(\underline{\ell}) = \omega[\pi(\underline{\ell})]$$

On the other hand $\mathfrak{A}^{Sym.}$ can also be defined by the family of observables

$$\mu_{\underline{\ell}} = \prod_b \exp\Big\{-\sum_{k=0}^{d_b-1} K(b;k)\,[1 - e^{\frac{2i\pi}{d_b}\ell_b k}]\,\chi_b^{k}\Big\} \quad , \quad \underline{\ell} \in \mathcal{G}_{\mathcal{B},f}$$

and any symmetric state will be uniquely defined by the function μ on $\mathcal{G}_{\mathcal{B},f}$

$$\mu(\underline{\ell}) = \omega[\mu_{\underline{\ell}}]$$

Property 3 :

For any finite system the Gibbs state is symmetric and

$$\sigma(\underline{\ell}) = \frac{\sum_{\underline{\ell}' \in \Gamma} \chi_{\underline{\ell}}(\underline{\ell}') \prod_{b \in \mathcal{B}} z(b;\ell'_b)}{\sum_{\underline{\ell}' \in \Gamma} \prod_{b \in \mathcal{B}} z(b;\ell'_b)} \tag{2.7}$$

$$\mu(\underline{\ell}) = \frac{\sum_{\underline{\ell}' \in \mathcal{K}} \chi_{\underline{\ell}}(\underline{\ell}') \prod_{b \in \mathcal{B}} t(b; \ell'_b)}{\sum_{\underline{\ell}' \in \mathcal{K}} \prod_{b \in \mathcal{B}} t(b; \ell'_b)} \tag{2.8}$$

where : $\chi_{\underline{\ell}}(\underline{\ell}') = \prod_{b \in \mathcal{B}} e^{\frac{2i\pi}{\alpha_b} \ell_b \ell'_b}$

This property is the immediate generalisation of proposition 1 Sec. 3.3 of Part I and follows immediately from the definition. Another expansion of the Gibbs state ω is given by :

$$\sigma(\underline{\ell}) = \frac{\sum_{\underline{\ell}' : \pi(\underline{\ell}') = \pi(\underline{\ell}^{-1})} \prod_{b \in \mathcal{B}} t(b; \ell'_b)}{\sum_{\underline{\ell}' \in \mathcal{K}} \prod_{b \in \mathcal{B}} t(b; \ell'_b)}$$

From which we obtain the following result which is a form of "Griffith's first inequality".

Property 4. "Griffith's first inequality"

For ferromagnetic systems such that $\overline{K(b;\ell)} = K(b; \alpha_b - \ell) \geq 0$, the function $\sigma(\underline{\ell}) = \omega_{(\Lambda,K)}[\pi(\underline{\ell})]$ is non negative :

$$\omega_{(\Lambda,K)}[\prod_b \chi_b^{\ell_b}] \geq 0 \qquad \forall\ \underline{\ell} \in \mathcal{G}_{\mathcal{B}}$$

(As we shall see in the next section the condition $K(b;\ell) \in \mathbb{R}$ implies $\sigma(\underline{\ell}) \in \mathbb{R}$)

2.4.3 Permutation Group

For arbitrary spin systems, the group $\mathcal{S}$ does not usually exhaust all possible symmetries of the hamiltonian; moreover physical consequences can be obtained if one considers also the action of the permutation group $\mathcal{S}_q$ of q elements on $\mathfrak{A}_\Lambda$ (for example, to discuss phase transitions Sec. 3.2 or domain of analyticity of the free energy Ch. 5). This action of $\mathcal{S}_q$ can also be viewed as the $q!$ distinct ways of introducing the group structure $\mathbb{Z}_q$ on the physical configuration space G_x .

The permutation group $\mathcal{S}_q$ acts as a group of automorphisms of $\mathfrak{A}_\Lambda$, by the relation :

$$\forall\, p \in \mathcal{S}_q \;,\quad (\tau_p A)(\underline{n}) = A(p^{-1}(\underline{n})) \qquad p^{-1}(\underline{n}) = \{p^{-1}(n_x)\}_{x\in\Lambda}$$

In particular by action of τ_p on the hamiltonian H_Λ, we obtain a new hamiltonian :

$$H_\Lambda^{p} = \tau_p H_\Lambda \tag{2.9}$$

with new interactions K^p given by :

$$-\beta H_\Lambda^{p}(\underline{n}) = \sum_{\chi} K^p(\chi)\,\chi(\underline{n}) = \sum_{\chi} K(\chi)\,\chi(p^{-1}(\underline{n}))$$

i.e. with $\chi(p(\underline{n})) = \sum_{\chi'} a^p(\chi;\chi')\,\chi'(\underline{n})$

we obtain : $a^p(\chi;\chi') = \prod_{x\in\Lambda}\left\{\frac{1}{q}\sum_{n=0}^{q-1}\chi_x(p(n))\,\chi'_x(n^{-1})\right\}$

$$K^p(\chi) = \sum_{\chi'\in\mathcal{G}_\Lambda^*} K(\chi')\,a^{p^{-1}}(\chi';\chi)$$

As before the automorphism τ_p induces a transformation τ'_p on the states defined by :

$$(\tau'_p\,\omega)[A] = \omega[\tau_p^{-1}A] \qquad \tau'_{p_1p_2} = \tau'_{p_1}\cdot\tau'_{p_2}$$

Property 5.

1) The partition function is invariant under the transformation $H_\Lambda \longrightarrow H_\Lambda^{p}$ i.e. $Z(\Lambda,K) = Z(\Lambda,K^p) \quad \forall\, p\in\mathcal{S}_q$

2) If $\omega_{(\Lambda,K)}$ is the Gibbs state of the finite system Λ with hamiltonian H_Λ, then $\tau'_p\,\omega_{(\Lambda,K)}$ is the Gibbs state of the finite system Λ with hamiltonian H_Λ^{p} ; moreover

$$(\tau'_p\,\omega_{(\Lambda,K)})[\chi] = \sum_{\chi'} a^p(\chi;\chi')\,\omega_{(\Lambda,K)}[\chi'] \tag{2.10}$$

We remark that $a^p(\chi;\chi')$ is different from zero iff χ and χ' have same support. From this property 5 we thus obtain symmetry relations for the correlation of finite systems if the hamiltonian is invariant under τ_p

i.e. $\tau_p H_\Lambda = H_\Lambda$ implies $\tau'_p\,\omega = \omega$ for finite Λ and symmetry relations for the free energy if the hamiltonian is not invariant under τ_p.

As example we consider the following subgroup of permutations.

i) $p_I(n) = q - n \qquad p_I^{-1} = p_I$

then: $K^p(\chi) = K(\chi^{-1})$ i.e. $K^p(b;\ell) = K(b;\alpha_b - \ell)$

and $(\tau_I'\omega)[\chi] = \omega[\chi^{-1}]$

moreover the generalised activities for the interaction K^p are given by : $z^p(b;\ell) = z(b;\alpha_b - \ell)$ (2.11)

ii) $p_j(n) = n + j \qquad p_j^{-1}(n) = n - j \qquad j = 1, \dots, q$

then : $K^p(\chi) = \bar{\chi}(\underline{j})\, K(\chi) \qquad \underline{j} = \{ j_x = j \}_{x \in \mathcal{B}}$

i.e. $K^p(b;\ell) = \exp[-\frac{2i\pi}{\alpha_b} \ell r j]\, K(b;\ell) \qquad r = \frac{\alpha_b}{q} \sum_{x \in \Lambda} M_{b,x}$

and : $(\tau_j'\omega)[\chi] = \chi(\underline{j})\ \omega[\chi]$

moreover the generalised activities for the interaction K^p are given by : $z^p(b;\ell) = \dfrac{z(b;\ell - rj)}{z(b;\alpha_b - rj)}$ (2.12)

iii) $p_{(I,j)} = p_I\, p_j \qquad p_{(I,j)}^{-1} = p_{(I,j)}$

then : $K^p(\chi) = \chi(\underline{j})\, K(\chi^{-1})$ i.e. $K^p(b;\ell) = \exp[\frac{2i\pi}{\alpha_b} \ell r j]\, K(b;\alpha_b - \ell)$

and : $(\tau_{(I,j)}'\omega)[\chi] = \bar{\chi}(\underline{j})\ \omega[\chi^{-1}]$

moreover the generalised activities for the interaction K^p are given by : $z^p(b;\ell) = \dfrac{z(b;\alpha_b - \ell - rj)}{z(b;\alpha_b - rj)}$ (2.13)

Conclusion

I) The partition function satisfies the symmetry relation

$$Z_{(\Lambda,K)}(\{z(b;\ell)\}) = Z_{(\Lambda,K)}\left(\left\{\frac{z(b;\ell - rj)}{z(b;\alpha_b - rj)}\right\}\right) = Z_{(\Lambda,K)}\left(\left\{\frac{z(b;\alpha_b - \ell - rj)}{z(b;\alpha_b - rj)}\right\}\right)$$

II) For real hamiltonian, the interactions are

1) Invariant under p_I if $K(\chi) \in \mathbb{R} \Rightarrow \omega_{(\Lambda,K)}[\chi] \in \mathbb{R}$

2) Invariant under p_j if $\left\{ \begin{array}{l} K(\chi) = 0 \\ \text{for } \chi(\underline{j}) \neq 1 \end{array} \right\} \Rightarrow \begin{array}{l} \omega_{(\Lambda,K)}[\chi] = 0 \\ \text{if } \chi(\underline{j}) \neq 1 \end{array}$

3) Invariant under $p_{(I,j)}$ if $\left\{ \begin{array}{l} K(\chi) = g(\chi)\, \chi(\underline{j})^{1/2} \\ g(\chi) = g(\chi^{-1}) \in \mathbb{R} \end{array} \right\} \Rightarrow \begin{array}{l} \omega_{(\Lambda,K)}[\chi] = R(\chi)\, \chi(\underline{j})^{-1/2} \\ R(\chi) = R(\chi^{-1}) \in \mathbb{R} \end{array}$

Remarks

1) The permutation $p_{(I,1)}$ is the standard spin reversal transformation in the spin language.

2) The group $\mathcal{S}$ appears as symmetry group of the hamiltonian, while the permutation group $\mathcal{S}_q$ appears as symmetry group for the measure associate with each lattice site (which is the Haar measure for $\mathcal{G}_x \cong \mathbb{Z}_q$).

3) For general systems such that $\mathcal{G}_x$ is any arbitrary group with a measure μ_x (Sec. 1.4), the kernel $\mathcal{S}$ of Γ is again the "internal symmetry group of the Hamiltonian", i.e. $\mathcal{S} = \{ \underline{s} \in \mathcal{G}_\Lambda \, ; \, H_\Lambda(\underline{n}) = H_\Lambda(\underline{s}^{-1}\cdot\underline{n}) \;\; \forall\, n \in \mathcal{G}_\Lambda \}$

On the other hand we can also define the "symmetry group $\mathcal{S}_\mu$ of the measure μ " as the group of measure preserving bijections $p : n_x \mapsto p(n_x)$ of $\mathcal{G}_x$ onto itself.

Property 5 remains valid for $\mathcal{S}_\mu$, while property 2 will remain valid only for the subgroup $\mathcal{S}' \subset \mathcal{S}$ which leave the measure invariant. In conclusion the symmetry group $\mathcal{S}_\Sigma$ of the system is given by $\mathcal{S}_\Sigma = \mathcal{S}' \times \mathcal{S}_\mu$

2.5. Correlation Functions

In order to derive equation for the correlation functions which remains valid for the more general case where $\mathcal{G}_x$ is any abelian group with a measure μ_x we introduce the notation :

$$q^{-1} \sum_{n_x=0}^{q-1} f(n_x) = \int_{\mathcal{G}_x} d\mu_x(n_x)\, f(n_x)$$

We then have for any χ in $\hat{\mathcal{G}}_\Lambda$ and any $x \in \Lambda$ the following identity :

$$\omega_{(\Lambda,K)}[\chi] = Z^{-1} \int_{\mathcal{G}_\Lambda} \prod_{y \in \Lambda} d\mu_y(n_y)\; e^{-\beta H_\Lambda(\underline{n})}\, \chi(\underline{n}) \;\Rightarrow$$

$$\omega_{(\Lambda,K)}[\chi] = Z^{-1}\int_{\mathcal{G}_{\Lambda/x}} \prod_{y\neq x} d\mu_y(n_y)\ \chi(\underline{n})\,\chi_x^{-1}(n_x)\ \exp\Big[\sum_{\chi':\chi'_x=1_x} K(\chi')\,\chi'(\underline{n})\Big]\ \cdot$$

$$\cdot\int_{\mathcal{G}_x} d\mu_x(n_x)\ \chi_x(n_x)\ \exp\Big[\sum_{\chi':\chi'_x\neq 1_x} K(\chi')\,\chi'(\underline{n})\Big] =$$

$$= Z^{-1}\int_{\mathcal{G}_\Lambda} \prod_y d\mu_y(n_y)\ \chi(\underline{n})\,\chi_x^{-1}(n_x)\, e^{-\beta H_\Lambda(\underline{n})}\cdot \frac{\int_{\mathcal{G}_x} d\mu_x(n_x)\,\chi_x(n_x)\ \exp[\cdots]}{\int_{\mathcal{G}_x} d\mu_x(n_x)\ \exp[\cdots]}$$

Therefore :

$$\omega_{(\Lambda,K)}[\chi] = \omega_{(\Lambda,K)}\Big[\chi\cdot\chi_x^{-1}\ \frac{\int_{\mathcal{G}_x} d\mu_x(n_x)\,\chi_x(n_x)\ \exp\big[\sum_{\chi':\chi'_x\neq 1_x} K(\chi')\chi'\chi_x'^{-1}\,\chi'_x(n_x)\big]}{\int_{\mathcal{G}_x} d\mu_x(n_x)\ \exp[\ ''\ ]}\Big]$$

Equation which is of the form :

$$\omega_{(\Lambda,K)}[1\!\!1] = 1 \qquad\qquad \omega_{(\Lambda,K)}[\chi] = \omega_{(\Lambda,K)}[\mathcal{A}_x(\chi)]$$

Following the general procedure discussed in Parts 1 and 2, we adopt the following

Definition :

Any state solution of the equations

$$\omega[1\!\!1] = 1 \qquad\qquad \omega[\chi] = \omega[\mathcal{A}_x(\chi)] \qquad\qquad \forall\ \chi\in\hat{\mathcal{G}}_{\mathcal{L}}$$

is called an "Equilibrium State with respect to the interaction K ". From this equation, which was derived by several authors [125,129,130] one can then prove the existence of the thermodynamic limit for the correlation functions, the unicity of the equilibrium state, as well as analyticity properts in the High Temperature domain following the lines discussed in Part I, Ch. 6, [78] . We shall not repeat these arguments, but give the following properties which appear as direct consequence of the definition.

Properties 6.

1) If ω is an equilibrium state with respect to K, then $\tau'_{\underline{s}}\,\omega$ is also an equilibrium state with respect to K for all $\underline{s}$ in $\mathcal{S} = \mathcal{T}^{\perp}$ which leaves the measure μ_x invariant. $[d\mu_x(n_x) = d\mu_x(\underline{s}_x^{-1}\cdot n_x)]$

2) Let $\mathcal{G}_x \cong \mathbb{Z}_q$; if ω is an equilibrium state with respect to K, then $\tau'_p\,\omega$ is an equilibrium state with respect to K^p for any permutation p in $\mathcal{S}_q$. (This property can be generalised at once to the case where $\mathcal{G}_x$ is arbitrary, if one defines p to be a bijection of $\mathcal{G}_x$ onto itself).

Proof

The proof of (1) is straightforward and relies only on the following :

$$(\tau'_{\underline{s}}\,\omega)[\chi] = \chi(\underline{s})\;\omega[\chi]$$

$$(\tau'_{\underline{s}}\,\omega)[\mathcal{A}_x(\chi)] = \omega[\tau_{\underline{s}}^{-1}\,\mathcal{A}_x(\chi)] \quad ; \quad [\tau_{\underline{s}}^{-1}\,\mathcal{A}_x(\chi)](\underline{n}) = (\mathcal{A}_x(\chi))(\underline{s}\cdot\underline{n})$$

$$\omega[\tau_{\underline{s}}^{-1}\,\mathcal{A}_x(\chi)] = \chi(\underline{s})\quad\omega[\mathcal{A}_x(\chi)]$$

The proof of (2) uses the same argument together with the remark that $a^p(\chi;\chi')\neq 0$ only if χ and χ' have same support which yields

$$\sum_{\chi':\,\chi'_x \neq 1_x} K(\chi')\;\chi'(p^{-1}(\underline{n})) = \sum_{\chi:\,\chi_x\neq 1_x} K^p(\chi)\;\chi(\underline{n})$$

Remark

This property generalises to infinite systems the properties 2 and 5 given for finite systems. To conclude this discussion let us remark that we could have chosen any set of functions $f_a(n_x) \in \mathcal{L}^1(\mu_x)$ with $f_0(n_x) = 1$ such that

$$-\beta H_\Lambda = \sum_{A\in \prod_{x\in\Lambda}\mathbb{Z}_x^+} \tilde{H}(A)\prod_{x\in\Lambda} f_{a_x}$$

instead of the characters.

We would obtain similarly for any $A \in \prod_{x \in \Lambda} \mathbb{Z}^+_x$ and any $x \in \Lambda$ such that $a_x \neq 0$

$$\omega_{(\Lambda,K)}[\mathbb{1}] = 1$$

$$\omega_{(\Lambda,K)}[\prod_{y \in \Lambda} f_{a_y}] = \omega_{(\Lambda,K)}[\mathcal{A}_x(\prod_{y \in \Lambda} f_{a_y})]$$

where : $$\mathcal{A}_x(\prod_{y\in\Lambda} f_{a_y}) = \prod_{y \neq x} f_{a_y} \frac{\int_{\mathcal{G}_x} d\mu_x(n_x)\, f_{a_x}(n_x) \exp\left[\sum_{B: b_x \neq 0} \tilde{H}(B) \prod_{y \neq x} f_{b_y} f_{b_x}(n_x)\right]}{\int_{\mathcal{G}_x} d\mu_x(n_x) \exp[\ldots]}$$

2.6. Relation between Physical Picture and Group Picture

Usually lattice systems are defined by the configuration space

$$G_{\mathcal{L}} = \prod_{x \in \mathcal{L}} G_x \qquad G_x = \{s_x^1, \ldots, s_x^q\}$$

and the hamiltonian is given as a polynomial in the variable $\underline{s}$,

$$-\beta H_\Lambda^{(s)}(\underline{s}) = \sum_{\underline{m} \in \mathcal{G}_\Lambda} \varepsilon(\underline{m}) \prod_{x \in \Lambda} s_x^{m_x} =$$

$$= \sum_{x \in \Lambda} \sum_{m=1}^{q-1} \varepsilon(x;m) s_x^m + \sum_{(x,y)\subset\Lambda} \sum_{m,m'=1}^{q-1} \varepsilon(xy;m,m') s_x^m s_y^{m'} + \cdots$$

In this section we shall give the explicit transformation from the physical picture into the group picture for the case of spin systems, i.e.

$$G_x = \{-\tfrac{q-1}{2}, -\tfrac{q-1}{2}+1, \ldots, \tfrac{q-1}{2}\}$$

and the mapping of G_x into $\mathbb{Z}_q$, defined by :

$$\phi(s_x) = n_x = s_x + \frac{q-1}{2} \qquad \phi^{-1}(n_x) = n_x - \frac{q-1}{2}$$

From the above expression of the physical hamiltonian, we obtain the expansion in terms of characters

$$-\beta H_\Lambda^{(n)}(\underline{n}) = -\beta H_\Lambda^{(s)}(\phi^{-1}(\underline{n})) = \sum_{\underline{m} \in \mathcal{G}_\Lambda} K(\underline{m}) \chi_{\underline{m}}(\underline{n}) =$$

$$= K(0) + \sum_{x\in\Lambda} \sum_{m=1}^{q-1} K(x;m)\, e^{\frac{2i\pi}{q} m n_x} + \sum_{(x,y)\subset\Lambda} \sum_{m,m'=1}^{q-1} K(xy;m,m') \cdot e^{\frac{2i\pi}{q}(m n_x + m' n_y)} + \cdots$$

by means of a Fourier transformation;

$$K(\underline{m}) = \sum_{x\in\Lambda} \prod_{\substack{\bar{x}\in\Lambda \\ \bar{x}\neq x}} \delta_{m_{\bar{x}},0} \left\{ \sum_{a=1}^{q-1} \varepsilon(x;a) \frac{1}{q} \sum_{n=0}^{q-1} e^{-\frac{2i\pi}{q} m_x n} [\phi^{-1}(n)]^a \right\}$$

$$+ \sum_{(x,y)\subset\Lambda} \prod_{\substack{\bar{x}\in\Lambda \\ \bar{x}\neq x,y}} \delta_{m_{\bar{x}},0} \left\{ \sum_{a,b=1}^{q-1} \varepsilon(xy;a,b) \frac{1}{q^2} \sum_{n,n'=0}^{q-1} e^{-\frac{2i\pi}{q}(m_x n + m_y n')} \cdot [\phi^{-1}(n)]^a [\phi^{-1}(n')]^b \right\} + \cdots$$

which yields for the constant term

$$K(0) = \sum_{x \in \Lambda} \left\{ \sum_{a=1}^{q-1} \varepsilon(x;a) \frac{1}{q} \sum_{n=0}^{q-1} [\phi^{-1}(n)]^a \right\} +$$

$$+ \sum_{(x,y) \subset \Lambda} \left\{ \sum_{a,b=1}^{q-1} \varepsilon(x,y;a,b) \frac{1}{q^2} \sum_{n,n'=0}^{q-1} [\phi^{-1}(n)]^a [\phi^{-1}(n')]^b \right\} + \cdots \quad (2.14)$$

for the one-body interactions at x

$$K(x;m) = \frac{1}{q} \sum_{n=0}^{q-1} e^{-\frac{2i\pi}{q} m n} \sum_{a=1}^{q-1} \varepsilon(x;a) \, [\phi^{-1}(n)]^a +$$

$$+ \frac{1}{q^2} \sum_{n,n'=0}^{q-1} e^{-\frac{2i\pi}{q} m n} \sum_{a,b=1}^{q-1} \left(\sum_{y \neq x} \varepsilon(xy;a,b) \right) [\phi^{-1}(n)]^a [\phi^{-1}(n')]^b$$

$$+ \cdots$$

for the two-body interactions between (x,y),

$$K(x,y;m,m') = \frac{1}{q^2} \sum_{n,n'=0}^{q-1} e^{-\frac{2i\pi}{q}(mn + m'n')} \sum_{a,b=1}^{q-1} \varepsilon(x,y;a,b) \, [\phi^{-1}(n)]^a [\phi^{-1}(n')]^b$$

$$+ \cdots$$

Conversely from the expansion of the hamiltonian in the group picture we obtain at once the physical hamiltonian by means of the expansion :

$$e^{\frac{2i\pi}{q} k n} = e^{\frac{i\pi}{q} k(q-1)} \sum_{m=0}^{q-1} c_m(k) \, s^m$$

$$c_m(0) = \delta_{m,0} \qquad\qquad s = \phi(n)$$

This expression yields also the connection between the correlation functions $\langle \chi \rangle$ and the correlation functions $\langle \prod_{x \in \Lambda} s_x^{m_x} \rangle$.

Remark :

The group of permutation $\mathcal{S}_q$ induces a transformation on the space of coupling constant $\varepsilon(\underline{m})$ defined by :

$$\sum_{\underline{m}} \varepsilon(\underline{m}) \prod_x (p^{-1}(s_x))^{m_x} = \sum_{\underline{m}} \varepsilon^p(\underline{m}) \prod_x s_x^{m_x}$$

however it does not leave invariant the subspace of coupling constant with same support of $\underline{m}$.

2.7. Example : Generalised Potts Model

To illustrate the concept introduced we consider as example those systems on $\mathbb{Z}^\nu$ with one and two body forces $K(x;m)$, $K(x,y;m,m')$, between nearest neighbours, which are such that the hamiltonian is invariant under the subgroup of permutations discussed in Sec. 2.4.

Invariance under $p_I(n) = q-n$ implies : $K(x;m) = K(x;q-m) \in \mathbb{R}$

$K(x,y;m,m') = K(x,y;q-m,q-m') \in \mathbb{R}$

Invariance under $p_j(n) = n+j$ implies : $K(x;m) = 0$

$K(x,y;m,m') = 0 \quad \text{if} \quad m' \neq q-m$

We remark moreover that the symmetry between lattice sites, $K(x,y;m,m') = K(x,y;m',m)$ together with the invariance under p_1 implies the invariance under p_I and p_j. Therefore the <u>hamiltonian is invariant under this subgroup of permutation</u> if and only if the only non zero interactions are between part of sites which are respectively in the configurations $q-m$ and m, $m \in \{1,\dots,q-1\}$, which yields

$$-\beta H_\Lambda(\underline{n}) = \sum_{\langle x,y\rangle \subset \Lambda} \sum_{m=1}^{q-1} K(x,y;m) \cos\left(\frac{2\pi}{q} m (n_x - n_y)\right)$$

where : $K(x,y;m) = K(x,y;m,q-m) = K(x,y;q-m) \in \mathbb{R}$

i) For these systems the <u>bonds</u> $b \in \mathcal{B}$ are defined by ordered pair of sites $\langle x,y \rangle$:

$$b: x \to y \qquad \chi_b(\underline{n}) = e^{\frac{2i\pi}{q}(n_x - n_y)} \; ; \; M(b;z) = \begin{cases} 1 & z = x \\ q-1 & z = y \\ 0 & \text{otherwise} \end{cases}$$

ii) $\partial(\underline{\ell})_x = -\sum_y \ell_{\langle y,x\rangle} + \sum_z \ell_{\langle x,z\rangle}$

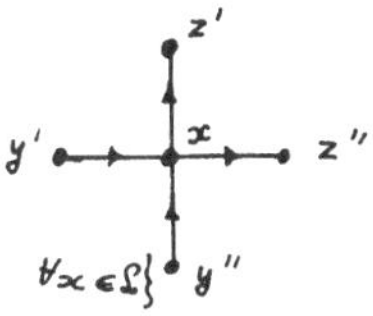

$$\Rightarrow \mathcal{K} = \left\{ \underline{\ell} \in \mathcal{G}_{\mathcal{B}} \; ; \; \sum_y \ell_{\langle y,x\rangle} - \sum_z \ell_{\langle x,z\rangle} = 0 \mod q \quad \forall x \in \mathcal{L} \right\}$$

i.e. $\mathcal{K}$ is the subgroup of $\mathcal{G}_{\mathcal{B}}$ such that there exists "conservation of ℓ at each lattice sites". Moreover $\mathcal{K}$ is generated by the elements shown on Fig. 1 (b).

iii) $\delta(\underline{n})_{\langle x,y\rangle} = n_x - n_y$

$\ell_{(xy)} = n_x - n_y \qquad \underline{\ell} \in \Gamma$

The group Γ is generated by the elements shown on Fig. 1 (a) and the internal symmetry group is given by $\mathcal{S} = \{ \underline{s} \in \mathcal{G}_{\mathcal{L}} \; ; \; s_x = s_y \}$.

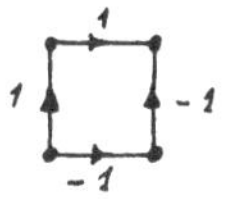

Fig. 1 (a) Generators of Γ (b) Generators of $\mathcal{K}$

iv) Using the results of Sec. 2.4, the correlation functions of the finite system will satisfy the following identities :

i) $\omega_{(\Lambda,K)}[\chi_x^n] = 0 \qquad \forall x \in \Lambda \qquad n \in \{1,\dots,q-1\}$

ii) $\omega_{(\Lambda,K)}[\chi_x^n \chi_y^m] = 0 \qquad \text{if } n+m \neq q$

iii) $\mathrm{Im}\, \omega_{(\Lambda,K)}[\chi_x^n \chi_y^{-n}] = 0$

On the other hand for infinite systems some of the symmetries might be broken :

1) if ω is invariant under τ_j' then : $\omega[\chi_x^n] = 0$ for $jn \neq 0$
$\omega[\chi_x^n \chi_y^m] = 0$ for $j(n+m) \neq 0$

2) if ω is invariant under τ_I' then : $\omega[\chi_x^n] = \omega[\chi_x^{-n}]$

To conclude this section we remark that this model can be considered as a "Generalised Potts Model", model usually defined by the two body hamiltonian :

$$H_{(xy)}(n_x, n_y) = \tilde{e}(x,y; n_x - n_y)$$

which yields :

$$-\beta H_\Lambda(\underline{n}) = \sum_{(x,y)\subset\Lambda} \sum_{m=0}^{q-1} K(xy;m) \cos\left[\frac{2\pi}{q} m (n_x - n_y)\right]$$

$$K(x,y;m) = -\frac{\beta}{q}\sum_{k=0}^{q-1} \tilde{e}(xy;k) \cos\left(\frac{2\pi}{q} m k\right)$$

In particular "Standard Potts Model" [131] are defined by the condition that the only non-zero interactions are between pair of sites which are in the same configuration, i.e.

$$H_{(x,y)}(n_x, n_y) = -J_{xy}\,\delta_{n_x, n_y}$$

i.e. $$-H_\Lambda(\underline{n}) = \sum_{(x,y)\subset\Lambda} J_{xy}\,\delta_{n_x,n_y} = \sum_{(x,y)\subset\Lambda} J_{xy}\,\frac{1}{q}\sum_{m=0}^{q-1}\cos\left[\frac{2\pi}{q} m (n_x - n_y)\right]$$

and the "Vector Potts Model" [131] are defined by the interaction energies :

$$H_{(x,y)}(n_x, n_y) = -J_{(xy)} \cos\frac{2\pi}{q}(n_x - n_y)$$

i.e. $$-H_\Lambda(\underline{n}) = \sum_{(x,y)\subset\Lambda} J_{(xy)} \cos\frac{2\pi}{q}(n_x - n_y)$$

We shall discuss those models in more detail for the particular case $q=3$ in the next chapter; we remark only here that for $q=3$ the only non-zero interaction is $K(xy;12)$ which gives using the technique of Sec. 2.5.

$$\mathcal{E}(x;1) = 0 \qquad \mathcal{E}(x;2) = -2\sum_y \mathcal{E}(xy;11)$$

$$\mathcal{E}(xy;22) = 3\,\mathcal{E}(xy;11)$$

$$\mathcal{E}(xy;11) = \frac{3}{2}K(xy;12) \qquad \mathcal{E}(xy;12) = 0$$

i.e. $$\begin{cases} -\beta H_\Lambda(\underline{\rho}) = \sum_{(x,y)\subset\Lambda} \mathcal{E}(xy)\left[\rho_x\rho_y + 3\rho_x^2\rho_y^2\right] + \sum_{x\in\Lambda}(h_x\rho_x + \mu_x\rho_x^2) \\ \text{with } h_x = 0 \qquad \mu_x = -2\sum_y \mathcal{E}(xy) \end{cases}$$

For $q=4$, the only non-zero interactions are $K(xy;13)$ and $K(xy;22)$ which are real.

Standard Potts Model : $-H = \sum_{(x,y)} J_{xy}\,\frac{1}{4}\left[2\cos\left(\frac{\pi}{2}(n_x - n_y)\right) + \cos\pi(n_x - n_y)\right]$

Vector Potts Model : $-H = \sum J_{xy} \cos\frac{\pi}{2}(n_x - n_y)$

Generalised Potts Model : $-H = \sum J_{xy}\, 2\cos\frac{\pi}{2}(n_x - n_y) + \bar{J}_{xy}\cos\pi(n_x - n_y)$

CHAPTER 3 - SPIN 1 LATTICE SYSTEMS

The first step beyond the discussion of spin $\frac{1}{2}$ given in part I is clearly the investigation of 3 component systems, such as diluted spin $\frac{1}{2}$ systems, ternary alloys, or spin 1 systems. In this chapter we shall consider the case of spin 1 systems with one and two body interactions only; the generalisation to many body interactions is however straightforward.

In the following we shall be more particularly interested in the symmetry properties related to the permutation group $\mathcal{S}_3$; as we shall see, the symmetry relation discussed by D. Kim and R.I. Joseph [132] for a spin 1 model appears as direct consequence of the general discussion of Sec. 2.4. On the other hand, in connection with a result of J. Bernasconi and F. Rys[133] , we use the action of the permutation group $\mathcal{S}_3$ to derive exact phase diagrams for two spin 1 models. Other applications of the permutation group $\mathcal{S}_3$ will be given in chapter 5; in particular it will be used to extend the analyticity domain of the free energy in the variables (e^{-2h} , $e^{-2\mu}$). The set of bonds are thus given by :

$$\mathcal{B} = \{ (x;1) , \langle xy;11\rangle , \langle xy;12\rangle \;;\; x,y \in \Lambda \;\; \text{p.t.} \;\; K \neq 0 \}$$

and the mapping $\pi : b \mapsto \chi_b$ is simply given by χ_x , $\chi_x \chi_y$, $\chi_x \chi_y^2$. and $\alpha_b = 3$. (see sec. 2.1.)

3.1. General Group Structure

We consider a spin 1 system defined as a finite lattice $\Lambda \subset \mathbb{Z}^\nu$ interacting by means of translationally invariant one and two body interactions only; the most general hamiltonian for such system is then given in the "spin picture" and the "group picture" respectively by :

$$
\begin{aligned}
-\beta H_\Lambda &= \sum_{x\in\Lambda} \left(h s_x + \mu s_x^2\right) + \sum_{(xy)\subset\Lambda} \sum_{m,m'=1}^{2} \varepsilon(xy;ab)\, s_x^a s_y^b \\
&= K(0) + \sum_{x\in\Lambda} \sum_{\ell=1}^{2} K(\ell)\chi_x^\ell + \sum_{(xy)\subset\Lambda} \sum_{m,m'=1}^{2} K(xy;m,m')\chi_x^m \chi_y^{m'}
\end{aligned}
\qquad (3.1)
$$

where : $s_x \in \{-1,0,+1\}$, $n_x \in \{0,1,2\}$, $\chi_x(\underline{n}) = \exp(\frac{2i\pi}{3} n_x)$

and $K(2) = \bar{K}(1) \in \mathbb{C}$; $K(xy;11) = \bar{K}(xy;22) \in \mathbb{C}$; $K(xy;12) = K(xy;21) \in \mathbb{R}$

h, μ, ε real, i.e. we have "5" parameters.

The transformation formulae are directly obtained by means of Eq. (2.14) of Sec. 2.5 :

$$
\begin{aligned}
K(1) &= \tfrac{1}{9}\left\{3(\mu-h) + 2z(\varepsilon_{22}-\varepsilon_{12}) + e^{\frac{2i\pi}{3}}\left[3(\mu+h) + 2z(\varepsilon_{22}+\varepsilon_{12})\right]\right\} \\
K(11) &= \tfrac{1}{9}\left\{-4\varepsilon_{12} + e^{\frac{2i\pi}{3}}\left[\varepsilon_{22} - 3\varepsilon_{11} - 2\varepsilon_{12}\right]\right\} \\
K(12) &= \tfrac{1}{9}\left\{3\varepsilon_{11} + \varepsilon_{22}\right\}
\end{aligned}
\qquad (3.2)
$$

where we have dropped the indices x,y to simplify the notation and z is the co-ordination number for nearest neighbour interactions, otherwise $z\,\varepsilon_{i2}$ represents $\sum_y \varepsilon(xy;i2)$.

The connection between the different correlation functions will be moreover given by means of the identity :

$$
e^{\frac{2i\pi}{3}n} = e^{\frac{2i\pi}{3}}\left[1 + i\tfrac{\sqrt{3}}{2}s - \tfrac{3}{2}s^2\right] = 1 + c_1 n + c_2 n^2 \qquad (3.3)
$$

$$
c_1 = \tfrac{1}{4}[-9+5i\sqrt{3}] \qquad c_2 = \tfrac{3}{4}[1-i\sqrt{3}]
$$

In the rest of this section we shall illustrate on this model the concept introduced in chapter 1; for this model the group of configurations is given by

$$
\mathcal{G}_\Lambda = \prod_{x\in\Lambda} \mathcal{G}_x \; ; \quad \mathcal{G}_x \cong \mathbb{Z}_3
$$

the set $\mathcal{B} = \{b\}$ of bonds is a subset of the set

$$
\{(x,1),\ (xy,11),\ (xy,12)\}_{(xy)\subset\Lambda} \qquad (3.4)
$$

the function $K : (b,\ell) \mapsto K(b;\ell)$ is the function defined by :

$$
\begin{aligned}
K((x,1);1) &= K(1) & K((x,1);2) &= K(2) \\
K((xy,11);1) &= K(xy;11) & K((xy,11);2) &= K(xy;22) \\
K((xy,12);1) &= K(xy;12) & K((xy,12);2) &= K(xy;21)
\end{aligned}
$$

while the mapping $\pi : \; b \mapsto \chi_b$ is naturally defined by :

$$(x,1) \mapsto \chi_x$$
$$(xy,11) \mapsto \chi_x \chi_y$$
$$(xy,12) \mapsto \chi_x \chi_y^2$$

All the χ_b are of order 3 and therefore the group of graphs is defined by :

$$\mathcal{G}_{\mathcal{B}} = \prod_b \mathcal{G}_b \qquad \mathcal{G}_b \cong \mathbb{Z}_3$$

The subgroup $\mathcal{S}$ of $\mathcal{G}_\Lambda$ is then given by :(*)

1) if $\mathcal{B} = \mathcal{B}^{(1)} = \{ (\langle xy \rangle, 12) \}$ $\Rightarrow$ $\mathcal{S} = \{ \{\underline{0}\}, \{\underline{1}\}, \{\underline{2}\} \}$

2) if $\mathcal{B} = \mathcal{B}^{(2)} = \{ (\langle xy \rangle, 11) \}$ $\Rightarrow$ $\mathcal{S} = \{ \{\underline{0}\}, \underline{s}_1, \underline{s}_2 \}$

$(\underline{s}_1)_x = 1, \; (\underline{s}_2)_x = 2$ on sublattice A

$(\underline{s}_1)_x = 2, \; (\underline{s}_2)_x = 1$ on sublattice B

3) otherwise $\mathcal{S} = \{\underline{0}\}$

The subgroup $\bar{\mathcal{T}}$ of $\hat{\mathcal{G}}_\Lambda$ for the three cases considered above is given by :

1) $\bar{\mathcal{T}} = \{ \chi = \pi' \chi_x \chi_y^2 \quad \pi'' \chi_x^2 \chi_y \}$

2) $\bar{\mathcal{T}} = \{ \chi = \pi' \chi_x \chi_y \quad \pi'' \chi_x^2 \chi_y^2 \}$

3) $\bar{\mathcal{T}} = \hat{\mathcal{G}}_\Lambda$

The subgroup $\mathcal{K}$ of $\mathcal{G}_{\mathcal{B}}$ for those three cases is generated respectively by the following elements

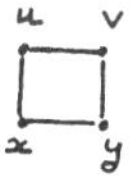

1) $\ell_{(xy,12)} = \ell_{(yv,12)} = 1 \qquad \ell_{(xu,12)} = \ell_{(uv,12)} = 2$

2) $\ell_{(xy,11)} = \ell_{(uv,11)} = 1 \qquad \ell_{(yv,11)} = \ell_{(xu,11)} = 2$

3) i) If $\mathcal{B} = \mathcal{B}^{(1)} \cup \mathcal{B}^{(2)}$

(a) $\ell_{(xy,11)} = \ell_{(xy,12)} = 1 \qquad \ell_{(xu,11)} = \ell_{(xu,12)} = 2 \qquad \ell_b = 0$ otherwise

(b) $\ell'_{(xy,12)} = \ell'_{(yv,11)} = \ell'_{(yv,12)} = 1 \qquad \ell'_{(xy,11)} = 2 \qquad \ell'_b = 0$ otherwise

(c) $\ell''_{(yv,11)} = \ell''_{(uv,12)} = 2 \qquad \ell''_{(yv,12)} = \ell''_{(uv,11)} = 1 \qquad \ell''_b = 0$ otherwise

---

(*) The notation $\{\underline{n}\}$ means $n_x = n \;\; \forall x \in \mathcal{G}_\Lambda$ and $\langle xy \rangle$ that x and y are nearest neighbours.

ii) If $\mathcal{B} = \{x\}_{x\in\Lambda} \cup \mathcal{B}^{(1)}$

$\underline{\ell}_{(xy)}: \ell_{(xy,11)} = 1 \qquad \ell_{(x,1)} = \ell_{(y,1)} = 2 \qquad \ell_b = 0$ otherwise

iii) If $\mathcal{B} = \{x\}_{x\in\Lambda} \cup \mathcal{B}^{(2)}$

$\underline{\ell}'_{(xy)}: \ell'_{(xy,12)} = 1 \qquad \ell'_{(x,1)} = 2 \quad \ell'_{(y,1)} = 1 \qquad \ell'_b = 0$ otherwise

iv) If $\mathcal{B} = \{x\}_{x\in\Lambda} \cup \mathcal{B}^{(1)} \cup \mathcal{B}^{(2)}$

a) $\underline{\ell}_{(xy)}: \ell_{(xy,11)} = 1 \qquad \ell_{(x,1)} = \ell_{(y,1)} = 2 \qquad \ell_b = 0$ otherwise

b) $\underline{\ell}'_{(xy)}: \ell'_{(xy,12)} = 1 \qquad \ell'_{(x,1)} = 2 \quad \ell'_{(y,1)} = 1 \qquad \ell'_b = 0$ otherwise

The generalised activities $z(b;\ell)$ and the generalised tanh functions $t(b;\ell)$ are given by tables 1 and 2.

The system is thus ferromagnetic if :

$1 > z((xy,12);\ell) > 0$ i.e. $\varepsilon_{22} + 3\varepsilon_{11} \geqslant 0$

$1 > z((xy,11);\ell) > 0$ i.e. $\varepsilon_{12} \leqslant 0$

$$3\varepsilon_{11} - \varepsilon_{22} - 2\varepsilon_{12} \geqslant 0$$

$1 > z(x;\ell) > 0$ i.e. $\frac{2}{3} \mathfrak{z}\, \varepsilon_{12} + h \leqslant 0$

$$-\frac{2}{3}\mathfrak{z}(\varepsilon_{22} - \varepsilon_{12}) + h - \mu \leqslant 0$$

3.2 Action of the Permutation Group $\mathcal{S}_3$

$P_I = \begin{pmatrix} 0 & 1 & 2 \\ 0 & 2 & 1 \end{pmatrix} \quad P_{I,1} = \begin{pmatrix} 0 & 1 & 2 \\ 2 & 1 & 0 \end{pmatrix} \quad P_{I,2} = \begin{pmatrix} 0 & 1 & 2 \\ 1 & 0 & 2 \end{pmatrix} \quad P_1 = \begin{pmatrix} 0 & 1 & 2 \\ 1 & 2 & 0 \end{pmatrix} \quad P_2 = \begin{pmatrix} 0 & 1 & 2 \\ 2 & 0 & 1 \end{pmatrix}$

Under the action of $\mathcal{S}_3$, $K \mapsto K^P$ according to formulae of Sec 2.4.3 and the correlation functions for the interactions K^P are related to the correlation functions corresponding to K by :

$$<\chi>^P = (\tau'_P\, \omega)[\chi]$$

In particular, using Eq. (3.3), the one point correlation functions or "order parameters"

$$m_x = <\rho_x> \qquad \xi_x = 3<\rho_x^2> - 2$$

transform according to :

$$\begin{pmatrix} m^P \\ \xi^P \end{pmatrix} = A^P \begin{pmatrix} m \\ \xi \end{pmatrix}$$

where :

$A^I = \frac{1}{2}\begin{pmatrix} 1 & -1 \\ -3 & -1 \end{pmatrix} \qquad A^{I,1} = \begin{pmatrix} -1 & 0 \\ 0 & 1 \end{pmatrix} \qquad A^{I,2} = \frac{1}{2}\begin{pmatrix} 1 & 1 \\ 3 & -1 \end{pmatrix}$

$A^1 = \frac{1}{2}\begin{pmatrix} -1 & -1 \\ 3 & -1 \end{pmatrix} \qquad A^2 = \frac{1}{2}\begin{pmatrix} -1 & 1 \\ -3 & -1 \end{pmatrix}$

In conclusion

1) The "order parameters" transform according to an irreducible representation of $\mathcal{S}_3$.

2) If the hamiltonian is invariant under any of these permutations τ_p , then for finite systems we obtain $m^p = m$, $\xi^p = \xi$ which yields relations between the order parameters.

   In particular : $\tau_p H = H \quad \forall p$ implies $m = 0 \quad \xi = 0$
   in fact : $\tau_{p_1} H = H$ (or $\tau_{p_2} H = H$) implies already $m = \xi = 0$ since A^1 (or A^2) does not have the eigenvalue 1 .
   (as we have already discussed $\tau_{p_1} H = H$ implies $\tau_p H = H$ for all p).

3) If the hamiltonian is invariant under τ_p , then for infinite systems it may happen that the relation $m^p = m$, $\xi^p = \xi$ is not satisfied; this corresponds to a phase transition with spontaneous breakdown of the symmetry under the permutation p , i.e. there exists equilibrium states which are not invariant under p . (Indeed if the hamiltonian is invariant under p , then for each equilibrium state ω , $\tau'_p \omega$ defined by Eq. (2.10) will also be an equilibrium state, see Property 6 Sec. 2.4.3.) Moreover under the action of the permutation group the generalised activities will transform according to Eqs. (2.11-13) (see Table 1).

3.3. Symmetry Properties of the Hamiltonian

As we have seen in the preceding section, if the hamiltonian is invariant under certain permutations, then we may have for infinite system a phase transition associated with a breakdown of this symmetry. In this section we shall discuss the general symmetry properties of the hamiltonian and its consequences. The most general hamiltonian depends upon "5" parameters.

1) The hamiltonian is invariant under P_I if and only if

$K(\chi) = \bar{K}(\chi) \in \mathbb{R}$ i.e. "3" parameters $K(1),\ K(11),\ K(12) \in \mathbb{R}$

which is equivalent to

$\varepsilon_{22} = 3\,\varepsilon_{11} + 2\,\varepsilon_{12}$ $\mu + h = -2z\,(\varepsilon_{11} + \varepsilon_{12})$

and implies for finite system, $\langle \chi \rangle \in \mathbb{R}$, i.e.

$\xi = -m$ i.e. $\langle \rho \rangle = 2 - 3\langle \rho^2 \rangle$

or $\langle n \rangle = \frac{3}{5}\,\langle n^2 \rangle$

2) The hamiltonian is invariant under $P_{I,1}$ (spin reversal transformation) if and only if

$K(\chi) = \chi(1)\,\bar{K}(\chi)$, i.e. "3" parameters $K(1) = s_1 e^{\frac{i\pi}{3}}$ $K(11) = s_{11} e^{-\frac{i\pi}{3}}$

$s_1,\ s_{11},\ K(12) \in \mathbb{R}$

which is equivalent to $\varepsilon_{12} = h = 0$

and implies for finite systems

$m = 0$ i.e. $\langle \rho \rangle = 0$ or $\langle n \rangle = 1$

3) The hamiltonian is invariant under $P_{I,2}$ if and only if

$K(\chi) = \chi(2)\,\bar{K}(\chi)$ i.e. "3" parameters $K(1) = s_1 e^{-\frac{i\pi}{3}}$ $K(11) = s_{11} e^{\frac{i\pi}{3}}$

$s_1,\ s_{11},\ K(12) \in \mathbb{R}$

which is equivalent to $\varepsilon_{22} = 3\,\varepsilon_{11} - 2\,\varepsilon_{12}$ $\mu - h = -2z\,(\varepsilon_{11} - \varepsilon_{12})$

and implies for finite systems

$\xi = m$ i.e. $\langle \rho \rangle = -2 + 3\langle \rho^2 \rangle$

or $\langle n \rangle = \frac{2}{7} + \frac{3}{7}\,\langle n^2 \rangle$

4) The hamiltonian is invariant under P_1 if and only if

$K(\chi) = \bar{\chi}(1)\,K(\chi)$ i.e. one parameter $K(12) \in \mathbb{R}$

which is equivalent to $\varepsilon_{22} = 3\,\varepsilon_{11}$ $\mu = -2z\,\varepsilon_{11}$

$\varepsilon_{12} = 0$ $h = 0$

and implies for finite systems

$m = \xi = 0$ i.e. $\langle \rho \rangle = 0$ $\langle \rho^2 \rangle = \frac{2}{3}$

$\langle n \rangle = 1$ $\langle n^2 \rangle = \frac{5}{3}$

Therefore if the hamiltonian is invariant under ρ_1 it is invariant under $\mathcal{S}_3$, and the model is the 3-component Potts Models (see 2.6.)

In conclusion the possible symmetries for the hamiltonian are the following :

Symmetries	Interactions	"order parameters"
1) $\lvert\mathcal{S}\rvert = 3$ + spin reversal	$\varepsilon_{22} - 3\varepsilon_{11} = 0$ $\quad \varepsilon_{12} = h = 0$ $\quad \mu + 2z\,\varepsilon_{11} = 0$	$m = 0 \quad \xi = 0$
2) $\lvert\mathcal{S}\rvert = 1$ + spin reversal	$\varepsilon_{12} = h = 0$	$m = 0$
3) $\lvert\mathcal{S}\rvert = 1$ + ρ_I	$\varepsilon_{22} - 3\varepsilon_{11} = 2\varepsilon_{12}$ $\quad \mu + 2z\varepsilon_{11} = -h - 2z\varepsilon_{12}$	$\xi + m = 0$
4) $\lvert\mathcal{S}\rvert = 1$ + $\rho_{I,2}$	$\varepsilon_{22} - 3\varepsilon_{11} = -2\varepsilon_{12}$ $\quad \mu + 2z\varepsilon_{11} = h + 2z\varepsilon_{12}$	$\xi - m = 0$

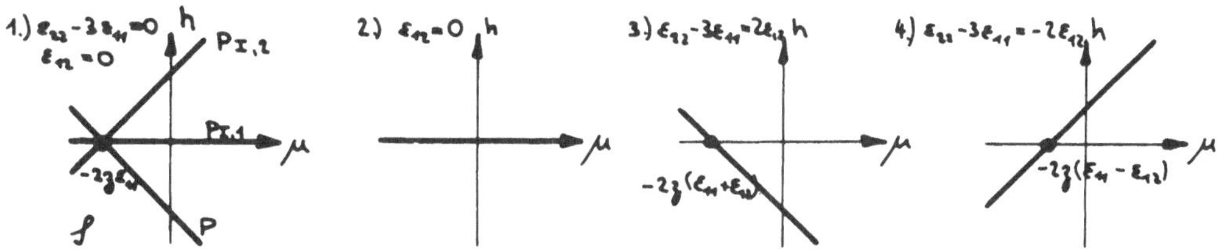

To discuss some consequences of the symmetries, let us first consider the 3-component Potts model (which is invariant under $\mathcal{S}_3$) i.e. $\varepsilon_{22} - 3\,\varepsilon_{11} = 0 \qquad \varepsilon_{12} = 0 \qquad h = 0 \qquad \mu = -2z\,\varepsilon_{11}$

If the equilibrium state of this system is invariant under $\mathcal{S}_3$, then $m = 0, \quad \xi = 0$.

It may happen that for infinite systems and for certain values of temperature the order parameters take non zero values. However since A^1 and A^2 do not have eigenvalue 1 , the only possible situations are the following :
either there exists 3 coexisting phases with order parameters (m, ξ) respectively given by

$$(0, \xi_0) \qquad (-\tfrac{1}{2}\xi_0, -\tfrac{1}{2}\xi_0) \qquad (\tfrac{1}{2}\xi_0, -\tfrac{1}{2}\xi_0)$$

or there exists at least 6 consisting phases with order parameters given by :

$$(m_0, 0) \;,\; (\tfrac{1}{2}m_0, -\tfrac{1}{2}m_0) \;,\; (-m_0, 0) \;,\; (\tfrac{1}{2}m_0, \tfrac{1}{2}m_0) \;,\; (-\tfrac{1}{2}m_0, -\tfrac{1}{2}m_0) \;,\; (-\tfrac{1}{2}m_0, \tfrac{1}{2}m_0)$$

Let us then consider the model defined by :

$$\varepsilon_{22} - 3\,\varepsilon_{11} = 0 \quad , \quad \varepsilon_{12} = 0 \quad , \quad h, \mu \in \mathbb{R}$$

For this model the 2 body interactions are invariant under $\mathcal{S}_3$ while the parameter $(h, \bar{\mu})$ will transform according to

$$\begin{pmatrix} h^P \\ \bar{\mu}^P \end{pmatrix} = A^P \begin{pmatrix} h \\ \bar{\mu} \end{pmatrix}$$

where $\qquad \bar{\mu} = \mu + 2\,z\,\varepsilon_{11}$

and therefore the free energy satisfies the symmetry relation

$$\beta f(h^P, \bar{\mu}^P) = \beta f(h, \mu) + h + \bar{\mu}$$

To conclude this section we remark that this symmetry relation for the model was first obtained by D. Kim and R.I. Joseph using combinatorial arguments and identities valid for spin 1 [132] ; moreover these authors also discussed the implications of the symmetry of the Potts Model on the coexistence surface. On the other hand, the mean field investigation of spin 1 models by D. Mukamel and M. Blume[134] give evidence that even in the mean field approximation, the model $\varepsilon_{22} = 3\,\varepsilon_{11}$ $\quad \varepsilon_{12} = 0$ has very special properties, namely that there exist 3 tricritical points. Evidently the question arises whether this is an exact result or just a consequence of the mean field approximation. However it is clear that the coexistence surface must reflect the symmetry relation which were discussed; in the last chapter we shall use the symmetries to discuss analycity properties of this model and give explicit domain in the (h, μ) plane which must contain the coexistence surface providing there is one.

3.4. Some Exact Phase Diagrams of Spin 1 Models

A simple application of the permutation group $\mathcal{S}_3$ is obtained in connection with a result due to J. Bernasconi and F.Rys[133] . They investigated another physical picture of a 3 component system, namely the one of a magnetic alloy,and were able to find the exact phase diagram for a particular model. Their main idea was to reduce the partition function of the

$q=3$ model by means of local traces to the partition function of a nearest neighbour Ising spin $\frac{1}{2}$ model with ferromagnetic interactions K and a magnetic field h^*. For the particular Hamiltonian

$$-\beta H_\Lambda = \sum_{\langle x,y\rangle} \varepsilon \rho_x^2 \rho_y^2 + \sum_x (h\rho_x + \mu\rho_x^2)$$

these authors obtained

$$Z(\Lambda,\varepsilon,h,\mu) = e^{\frac{|\Lambda|}{2}(3\frac{\varepsilon}{4}+\mu+\ln 2\cosh h)} Z^{Ising}(\Lambda,\ K=\tfrac{\varepsilon}{4},\ h^* = -\tfrac{3\varepsilon}{4} - \tfrac{\mu}{2} - \tfrac{1}{2}\ln 2\cosh h)$$

hence if $\varepsilon > 0$ the phase diagram of this spin 1 model is described by

$$h^* = 0 = \frac{3\varepsilon}{4} + \frac{\mu}{2} + \frac{1}{2}\ln 2\cosh h \tag{3.5}$$

$\mathcal{S}_3$ acts on the hamiltonian as

$$-\beta H^{P_{I,1}} = -\beta H$$

$$-\beta H^{P_I}(\rho) = -\beta H^{P_2}(\rho) = \sum_{\langle xy\rangle} \varepsilon'[\rho_x\rho_y + \rho_x^2\rho_y + \rho_x\rho_y^2 + \rho_x^2\rho_y^2] + \sum_x (h'\rho_x + \mu'\rho_x^2)$$

$$-\beta H^{P_{I,2}}(\rho) = -\beta H^{P_1}(\rho) = \sum_{\langle xy\rangle} \varepsilon'[\rho_x\rho_y - \rho_x^2\rho_y - \rho_x\rho_y^2 + \rho_x^2\rho_y^2] + \sum_x (h'\rho_x + \mu'\rho_x^2)$$

for the last two hamiltonians, the exact phase diagrams are respectively given by (provided that $\varepsilon' > 0$) :

$$2\mathfrak{z}\,\varepsilon' + h' + \mu' - \ln(1+e^{-h'+\mu'}) = 0$$

$$2\mathfrak{z}\,\varepsilon' - h' + \mu' - \ln(1+e^{+h'+\mu'}) = 0$$

Up to the transformation $h \to -h$ the shape of these two phase diagrams is identical. Defining $\varepsilon' = \frac{J}{kT}$, $h' = \frac{\mathcal{H}}{kT}$, $\mu' = -\frac{\Delta}{kT}$ the form of these phase diagrams in the intensive variable space T, $\mathcal{H}$, Δ is shown in Fig. 1.

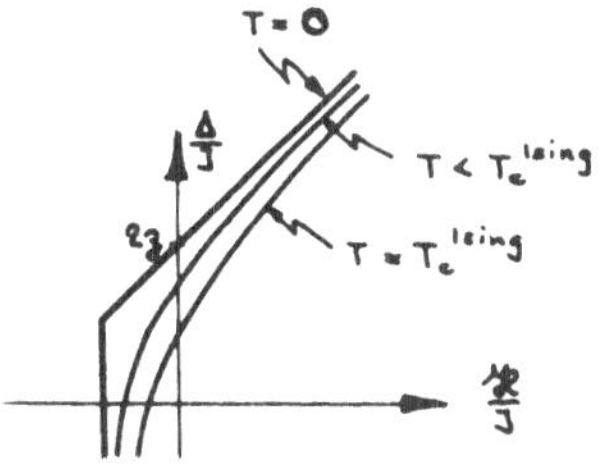

Fig. 1 Exact phase diagram for spin 1 model.

For a discussion of the critical behaviour of the model given by Eq. (3.5) we refer to [133]. Similar discussions can be carried out for the other two reductible models. However one would be more interested in an exact phase diagram exhibiting a tricritical point.

$z(b,\ell)$ \ b	$b = (x;1)$	$b = (xy;11)$	$b = (xy;12)$
$z(b,1)$	$e^{-\frac{2z}{3}(\varepsilon_{22}-\varepsilon_{12})+h-\mu}$	$e^{\frac{4}{3}\varepsilon_{12}}$	$e^{-\varepsilon_{11}-\frac{1}{3}\varepsilon_{22}}$
$z(b,2)$	$e^{\frac{4z}{3}\varepsilon_{12}+2h}$	$e^{\frac{\varepsilon_{22}}{3}-\varepsilon_{11}+\frac{2}{3}\varepsilon_{12}}$	$e^{-\varepsilon_{11}-\frac{1}{3}\varepsilon_{22}}$

Table 1 : The generalized activities $z(b,\ell)$ for $q=3$.

$b=(x;1)$	$t(b,1) = \dfrac{e^{\mu-h+\frac{2}{3}z(\varepsilon_{22}-\varepsilon_{12})}+e^{-\frac{2i\pi}{3}}+e^{\frac{2i\pi}{3}}\cdot e^{\mu+h+\frac{2}{3}z(\varepsilon_{22}+\varepsilon_{12})}}{e^{\mu-h+\frac{2}{3}z(\varepsilon_{22}-\varepsilon_{12})}+1+e^{\mu+h+\frac{2}{3}z(\varepsilon_{22}+\varepsilon_{12})}}$
$b=(xy;11)$	$t(b,1) = \dfrac{e^{-\frac{4}{3}\varepsilon_{12}}+e^{-\frac{2i\pi}{3}}+e^{\frac{2i\pi}{3}}\cdot e^{\frac{1}{3}\varepsilon_{22}-\varepsilon_{11}-\frac{2}{3}\varepsilon_{12}}}{e^{-\frac{4}{3}\varepsilon_{12}}+1+e^{\frac{1}{3}\varepsilon_{22}-\varepsilon_{11}-\frac{2}{3}\varepsilon_{12}}}$
$b=(xy;12)$	$t(b,1) = \dfrac{e^{\varepsilon_{11}+\frac{1}{3}\varepsilon_{22}}-1}{e^{\varepsilon_{11}+\frac{1}{3}\varepsilon_{22}}+2}$

Table 2 : The generalized tanh functions $t(b,\ell)$ for $q=3$
$(\,t(b,2)=\overline{t(b,1)}\,)$.

CHAPTER 4 - THE DUALITY TRANSFORMATION

4.1. Introduction

As we have seen in Part I, the duality transformation plays an important role in the study of spin $\frac{1}{2}$ systems, especially in the discussion of symmetry or "duality" relations for the free energy as well as for the correlation functions, the unicity property of the invariant equilibrium states at low temperature, the location of the critical temperature, the coexistence of phases,...

Soon after the introduction of the duality concept by Kramers and Wannier in 1941 [7], several generalisations to multicomponents were investigated. In 1943 Ashkin and Teller [84] considered a 4-component model on a square lattice; using topological arguments they proved the existence of a duality transformation and located in this way the transition temperature for a special choice of the interaction parameters. Potts [131] proposed in 1952 several generalisation of the spin $\frac{1}{2}$ Ising Model, later called "scalar Potts", resp. "vector Potts", which are special cases of what we have introduced in Sec. 2.7 as "Generalised Potts Models" ; using the Kramers Wannier matrix approach he located the critical temperature of the scalar Potts model for all q and of the vector Potts model $q = 3, 4$. Later Mittag and Stephen [135] investigated the duality transformation for many component models defined on a square lattice; in particular they derived duality transformations using either topological or algebraical arguments. Further progress were then made by Wegner in 1973 [64] who introduced a very general transformation containing the duality transformation for arbitrary components and the weak graph theorem as special cases. Recently an interesting algebraic formulation of the duality transformation for the q component generalised Potts models in $\mathbb{R}^{\nu}$ with non-crossing bonds was given by Wu and Wang[136]; they proved that the Boltzmann factors of the dual model are the eigenvalues of the matrix defined by means of the Boltzmann factors of the original model.

It shall be remarked that the generalisations which we have mentionned, have usually introduced graphical arguments similar to those of the Ising model; it was thus expected that a generalisation of the duality transformation using the group structure should appear naturally. The first attempt to formulate duality in terms of the groups associated with higher spin system was made by Greenberg[127].

In this chapter we shall discuss the general concept of duality for arbitrary lattice systems using the group structure introduced in Sec. 1.3 and following the same ideas as those used in Chap. 2 of Part I. The method will be illustrated by means of the generalised Potts model.

4.2 Duality Transformation and Duality Relation for the Partition Function

From the L.T. expansion Eq. (1.13)

$$Z(\Lambda, K) = |\mathcal{B}| \sum_{\underline{\ell} \in \Gamma} \prod_{b \in \mathcal{B}} \omega(b; \ell_b)$$

and the H.T. expansion Eq. (1.15)

$$Z(\Lambda, K) = |\mathcal{G}_\Lambda| \sum_{\underline{\ell} \in \mathcal{K}} \prod_{b \in \mathcal{B}} f(b; \ell_b)$$

we are led to introduce the following definition.

Definition :

The system $\{\Lambda^*, \mathcal{B}^*, \Pi^*, K^*\}$ is called a HT-LT dual for $\{\Lambda, \mathcal{B}, \Pi, K\}$ if then exists a bijection $\mathcal{D} : \underline{\ell} = \{\ell_b\}_{b \in \mathcal{B}} \mapsto \mathcal{D}\underline{\ell} = \{\ell^*_{b^*}\}_{b^* \in \mathcal{B}^*}$ of $\mathcal{K}$ onto Γ^*, such that for all $\underline{\ell}$ in $\mathcal{K}$

$$\prod_{b \in \mathcal{B}} f(b; \ell_b) = c \prod_{b^* \in \mathcal{B}^*} \omega^*(b^*; \ell^*_{b^*})$$

Similarly LT-HT, HT-HT, LT-LT duals are defined by replacing the pair $(\mathcal{K}, \Gamma^*)$ respectively by $\{\Gamma, \mathcal{K}^*\}$, $\{\mathcal{K}, \mathcal{K}^*\}$, $\{\Gamma, \Gamma^*\}$ with corresponding changes in the f and ω functions.

Proposition :

If $\{\Lambda^*, \mathcal{B}^*, \pi^*, K^*\}$ is a HT-LT, respectively LT-HT, HT-HT, LT-LT dual for $\{\Lambda, \mathcal{B}, \pi, K\}$ then the partition functions satisfy the following Duality Relations :

$$Z(\Lambda, K) = |\mathcal{G}_\Lambda| \, |\mathcal{S}^*|^{-1} \, Z(\Lambda^*, K^*) \qquad \text{HT-LT}$$

$$Z(\Lambda, K) = |\mathcal{S}| \, |\mathcal{G}_{\Lambda^*}|^{-1} \, Z(\Lambda^*, K^*) \qquad \text{LT-HT}$$

$$Z(\Lambda, K) = |\mathcal{G}_\Lambda| \, |\mathcal{G}_{\Lambda^*}|^{-1} \, Z(\Lambda^*, K^*) \qquad \text{HT-HT}$$

$$Z(\Lambda, K) = |\mathcal{S}| \, |\mathcal{S}^*|^{-1} \, Z(\Lambda^*, K^*) \qquad \text{LT-LT}$$

(This proposition is the generalisation to arbitrary spin of proposition 1 of Sec. 2.3 of Part I).

4.3 General Method to Construct a Dual Lattice

Let G be a subgroup of $\mathcal{G}_{\mathcal{B}}$ and $g_{j^*} = \{g_{j^*,b}\}_{b\in\mathcal{B}}, j^*\in\Lambda^*$, be any subset of G which generates G. With each g_{j^*} we associated a lattice site j^* of a dual lattice Λ^* and define a new system

$$\{\Lambda^*, \mathcal{G}_{\Lambda^*} = \prod_{j^*\in\Lambda^*} \mathcal{G}_{j^*}, \mathcal{B}^*, \pi^*\}$$

where : $\mathcal{G}_{j^*} \cong \mathbb{Z}_{q_{j^*}}$ $\qquad q_{j^*}$ = order of g_{j^*}

$\mathcal{B}^* = \mathcal{B}$ (or: $\mathcal{B}\to\mathcal{B}^*$ is then a bijection)

$$\pi^* : b \mapsto \chi_b^*(\underline{n}^*) = \prod_{j^*\in\Lambda^*} e^{\frac{2i\pi}{\alpha_b} g_{j^*,b}\, n^*_{j^*}} = \prod_{j^*\in\Lambda^*} e^{\frac{2i\pi}{q_{j^*}} M^*_{b,j^*} n^*_{j^*}}$$

i.e. the support of the new interactions are defined by

$$M^*_{b,j^*} = \frac{q_{j^*}}{\alpha_b}\, g_{j^*,b} \tag{4.1}$$

By definition $G = \{\underline{g} = \{g_b = \sum_{j^*\in\Lambda^*} g_{j^*,b}\, n^*_{j^*}, \bmod \alpha_b\}_{b\in\mathcal{B}}; \underline{n}^*\in\mathcal{G}_{\Lambda^*}\}$;

on the other hand, with α_b^* the order of χ_b^* we have by construction

$$\Gamma^* \cong \{\underline{\ell}^* = \{\ell_b^* = \frac{\alpha_b^*}{\alpha_b}\sum_{j^*\in\Lambda^*} g_{j^*,b}\, n^*_{j^*}, \bmod \alpha_b^*\}_{b\in\mathcal{B}}; \underline{n}^*\in\mathcal{G}_{\Lambda^*}\}$$

Therefore the mapping of $G\subset\mathcal{G}_{\mathcal{B}}$ into $\mathcal{G}_{\mathcal{B}^*}$ defined by

$$D:\ \underline{g} = \{g_b\}_{b\in\mathcal{B}} \mapsto D\underline{g} = \{\frac{\alpha_b^*}{\alpha_b} g_b = g_b^*\}_{b\in\mathcal{B}}$$

yields an isomorphism between G and Γ^*.

Let $G = \mathcal{K}$; defining $K^*(b;\ell)$ by the relation

$$\omega^*(b;\ell) = \hat{f}(b;\frac{\alpha_b}{\alpha_b^*}\ell) \tag{4.2}$$

i.e. $$\omega^*(b;\ell) = \frac{1}{\alpha_b}\sum_{r=0}^{\alpha_b-1} e^{-\frac{2i\pi}{\alpha_b}\ell r}\,\omega(b;\frac{\alpha_b^*}{\alpha_b}r)$$

we have constructed a HT-LT dual for $\{\Lambda, \mathcal{B}, \pi, K\}$.

Let $G = \Gamma$; defining $K^*(b;\ell)$ by the relation

$$\omega^*(b;\ell) = \omega(b;\frac{\alpha_b}{\alpha_b^*}\ell)$$

we have constructed a LT-LT dual for $\{\Lambda, \mathcal{B}, \pi, K\}$.

4.4 Examples

Example 1 : "Potts Models on $\mathbb{Z}^2$"

To illustrate the Duality Transformation defined in Sec. 4.3 we consider the Generalised Potts Model (Sec. 2.7) on a square lattice. We recall that the set of bonds $\mathcal{B}$ and the mapping $\pi : b \mapsto \chi_b$ can be represented by means of arrows between pair of sites; moreover we can choose as generators of the group $\mathcal{K}$, the elements represented on Fig. 1.

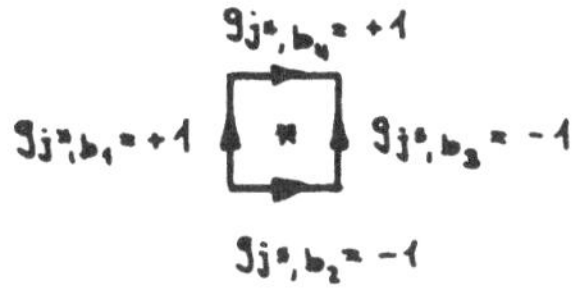

Fig. 1 : Generator $g_{j^*} = \{ g_{j^*,b} \}_{b \in \mathcal{B}}$ of $\mathcal{K}$; $g_{j^*,b_i} = \pm 1$ if $i = 1,2,3,4$, $g_{j^*,b} = 0$ otherwise.

With this choice of generators of $\mathcal{K}$ we have $q_{j^*} = q$ for all $j^* \in \Lambda^*$. Using the general construction we obtain for Λ^* the points situated at the center of those generators; moreover the bonds of the new system as well as $\pi^* : b^* \mapsto \chi^*_{b^*}$ are represented by means of arrows either between pair of sites in Λ^* or containing only one site in Λ^* (boundary sites).

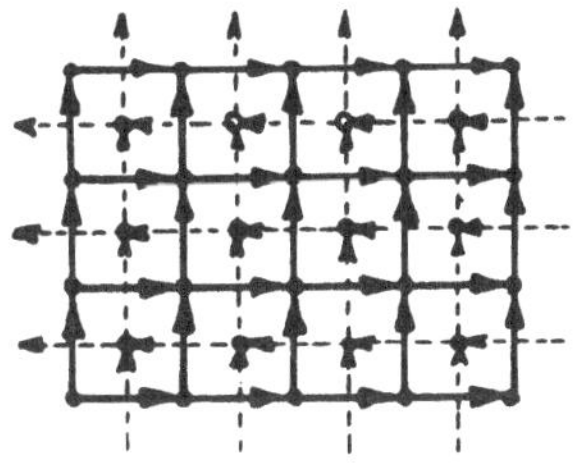

$\{\bullet\}$ lattice site Λ ; $\{ \underset{x}{\bullet} \overset{b}{\longrightarrow} \underset{y}{\bullet} \} = \mathcal{B}$

$\{\circ\}$ lattice site Λ^*; $\{ \underset{y^*}{\circ} \overset{b^*}{\longleftarrow} \underset{x^*}{\circ} \} = \mathcal{B}^*$

$$\chi^*_{b^*}(\underline{n}^*) = e^{\frac{2i\pi}{q}(n^*_{x^*} - n^*_{y^*})}$$

$$n^*_{x^*} = 0 \text{ if } x^* \notin \Lambda^*$$

Fig. 2 : Square Lattice Potts model with free boundary conditions and its dual lattice.

Therefore a dual model for the Potts Model with free boundary condition is a "Potts Model with + boundary condition" (i.e. $n^*_{x^*} = 0 \quad \forall\ x^* \notin \Lambda^*$) and interaction given by

$$\begin{cases} \omega^*(b^*;\ell) = \frac{1}{q}\sum_{r=0}^{q-1} e^{-\frac{2i\pi}{q}\ell r}\ \omega(b;r) \\ \omega(b;r) = \exp\left[\sum_{s=0}^{q-1} K(b;s)\, e^{\frac{2i\pi}{q} r s}\right] \end{cases} \qquad (4.3)$$

In conclusion the Generalised Potts Model on $\mathbb{Z}^2$ is weakly self-dual.

Let us then consider a few special cases as explicit examples.

I) Scalar Potts Model (Potts [131], Mittag-Stephen [135])

$K(b;\ell) = \frac{1}{q} K$

which yields :
$$\begin{cases} \omega(b;0) = e^K \\ \omega(b;\ell) = 1 \qquad \ell \neq 0 \end{cases}$$

$$\begin{cases} \omega^*(b^*;0) = \frac{1}{q}\left(e^K + q - 1\right) = \frac{1}{q}(e^K - 1)\cdot\left(\frac{e^K + q - 1}{e^K - 1}\right) \\ \omega^*(b^*;\ell^*) = \frac{1}{q}(e^K - 1)\cdot 1 \qquad \ell \neq 0 \end{cases}$$

Therefore the scalar Potts Model is weakly self dual with K^* given by :

$$e^{K^*} = \frac{e^K + q - 1}{e^K - 1}$$

and the partition function satisfies the duality relation

$$Z^{sc.\,Potts}_{free}(\Lambda;K) = q^{|\Lambda|} \prod_{b\in B}\left(\frac{e^K - 1}{q}\right) Z^{sc.\,Potts}_{+}(\Lambda^*;K^*)$$

which yields the duality relation for the free energy density of the infinite system

$$F^{sc.\,Potts}(K) = \ln\left[\frac{(e^K - 1)^2}{q}\right] + F^{sc.\,Potts}(K^*)$$

Assuming the existence of a unique phase transition, one can locate the critical point by $K_c^* = K_c$ which gives

$$e^{K_c} = 1 + \sqrt{q}$$

and therefore the critical temperature $T_c \to 0$ as $q \to \infty$, result first given by Potts [131].

II) Generalised 4-components Potts Model (i.e. Special Case of Ashkin-Teller Model)

The most general $q = 4$ model with one and two body forces is characterised by " 9 - parameters". The Ashkin-Teller model [84] (see Sec. 6.4 of Part I) is the most general model whose hamiltonian is invariant under the subgroup of order 4 $\{\phi, P_9, P_{I,1}, P_{I,2}\}$ of $\mathcal{S}_4$ (Sec. 2.4.3); it depends upon 3 parameters and is defined by the following energies $H(n_x, n_y)$ between neighbouring sites (x,y) :

$$H(n,n) = \varepsilon_0 \; ; \quad H(1,2) = H(3,4) = \varepsilon_1 \; ; \quad H(1,3) = H(2,4) = \varepsilon_2 \; ; \quad H(1,4) = H(2,3) = \varepsilon_3 .$$

which yields

$$-\beta H_\Lambda = \sum_{\langle x,y \rangle} \left\{ K_0 + 2 K_1 \cos \frac{\pi}{2}(n_x - n_y) + K_2 \cos \pi (n_x - n_y) - 2 K_3 \sin \frac{\pi}{2}(n_x + n_y) \right\}$$

with

$$K_0 = - \frac{\beta}{4} (\varepsilon_0 + \varepsilon_1 + \varepsilon_2 + \varepsilon_3)$$

$$K_1 = - \frac{\beta}{4} (\varepsilon_0 - \varepsilon_2) = K(xy, 13) = K(b; 1)$$

$$K_2 = - \frac{\beta}{4} (\varepsilon_0 - \varepsilon_1 + \varepsilon_2 - \varepsilon_3) = K(xy, 22) = K(b; 2)$$

$$K_3 = - \frac{\beta}{4} (\varepsilon_1 - \varepsilon_3) = i \; K(xy, 11)$$

Moreover the most general model whose hamiltonian is invariant under the group of order 8 $\{\phi, P_j, P_I, P_{I,j}\}$ is the generalised Potts model discussed in Sec. 2.7; it is a special case of Ashkin -Teller model defined by

$$K_3 = 0 \qquad \text{i.e.} \qquad \varepsilon_1 = \varepsilon_3$$

and depends upon 2 parameters. In fact choosing $\varepsilon_0 = 0$ we have :

$$4 K_1 = \beta \varepsilon_2 \qquad 2 (K_1 + K_2) = \beta \varepsilon_1$$

As special case of Potts models we shall consider the following :

i) Scalar Potts : $\varepsilon_1 = \varepsilon_2 = \varepsilon_3$ i.e. $K_1 = K_2$

ii) Vector Potts : $\varepsilon_0 + \varepsilon_2 = 2 \varepsilon_1$ i.e. $K_2 = K_3 = 0$

iii) Special Potts : $\varepsilon_0 = \varepsilon_2$ i.e. $K_1 = K_3 = 0$

Restricting our attention to the generalised Potts model, we obtain :

$$\omega(b;r) = \exp\left[2K_1 \cos\frac{\pi}{2}r + (-1)^r K_2\right]$$

i.e.

$$\omega(b;0) = e^{2K_1+K_2} \qquad \omega(b;1) = e^{-K_2}$$

$$\omega(b;2) = e^{-2K_1+K_2} \qquad \omega(b;3) = e^{-K_2}$$

which yields using Eq. (4.3)

$$\omega^*(b^*;0) = \frac{1}{4}\left[e^{2K_1+K_2} + 2e^{-K_2} + e^{-2K_1+K_2}\right]$$

$$\omega^*(b^*;1) = \omega^*(b^*;3) = \frac{1}{4}\left[e^{2K_1+K_2} - e^{-2K_1+K_2}\right]$$

$$\omega^*(b^*;2) = \frac{1}{4}\left[e^{2K_1+K_2} - 2e^{-K_2} + e^{-2K_1+K_2}\right]$$

Therefore introducing the dual interactions defined by :

$$e^{2(K_1^*+K_2^*)} = \frac{e^{2K_1} + 2e^{-2K_2} + e^{-2K_1}}{e^{2K_1} - e^{-2K_1}}$$

$$e^{4K_1^*} = \frac{e^{2K_1} + 2e^{-2K_2} + e^{-2K_1}}{e^{2K_1} - 2e^{-2K_2} + e^{-2K_1}} \qquad (4.4)$$

the 4-component generalised Potts model satisfies the duality relation :

$$Z_{free}^{Potts}(q=4;\,K_1,K_2) = q^{|A|}\prod_{b\in B}\left[\frac{1}{4}\left(e^{2K_1}-e^{-2K_1}\right)e^{(K_2+K_2^*)}\right] Z_{+}^{Potts}(q=4;\,K_1^*,K_2^*)$$

and the free energy of the infinite system satisfies :

$$F(K_1,K_2) = -\ln\left[q\left(\frac{1}{4}\left[e^{2K_1}-e^{-2K_1}\right]e^{(K_2+K_2^*)}\right)^2\right] + F(K_1^*,K_2^*)$$

i) For the special case of the <u>Vector Potts Model</u> we have $K(b;2) = 0$

which yields $K_2^* = 0 \qquad e^{-2K_1^*} = \tanh K_1$

therefore the Vector Potts Model $q=4$ is weakly self dual and the interaction transforms as the standard spin ½ Ising Model. It is in fact possible to show that the model factorises into two independent spin ½ Ising Model [137, 138].

ii) For the special case $K(b;1)=0$ $K(b;2)=K$ we obtain

$$K(b;2)^* = \infty \qquad \text{(i.e. model with constraints)}$$

$$e^{-4K(b;1)^*} = \tanh K$$

It should be remarked that in this case one could choose the mapping Π as defined by

$$\Pi : b \mapsto \chi_b(\underline{n}) = e^{i\pi(n_x - n_y)} \quad \text{i.e.} \quad \alpha_b = 2$$

and thus the generators of $\mathcal{K}$ are all of order 2.

Using the general method of Sec. 4.3 the dual model is then a $\frac{1}{2}$ spin system with

$$\chi_b^*(\underline{n}^*) = (-1)^{n^*_{x^*} + n^*_{y^*}}$$

and interaction $K^*(b;\ell)$ defined by

$$\omega^*(b;0) = \cosh K$$

$$\omega^*(b;1) = \sinh K$$

which yields

$$Z^{\text{Potts}}_{\text{free}}(K_1=0, K_2) = 4^{|\Lambda|} \prod_{b\in B} \left(\frac{e^K + e^{-K}}{2}\right) e^{-K^*} Z_+^{\text{Ising}}(K^*) = 2^{|\Lambda|} Z^{\text{Ising}}_{\text{free}}(K)$$

with $e^{-2K^*} = \tanh K.$

To conclude this discussion let us note that the symmetry line, or fixed points of the duality transformation Eq. (4.4) is obtained very simply in terms of the parameters

$$\omega_1 = e^{-\beta\varepsilon_1} = e^{-2(K_1+K_2)}$$

$$\omega_2 = e^{-\beta\varepsilon_2} = e^{-4K_1}$$

The duality transformation yields then

$$\omega_1^* = \frac{1-\omega_2}{1+2\omega_1+\omega_2}$$

$$\omega_2^* = \frac{1-2\omega_1+\omega_2}{1+2\omega_1+\omega_2} \qquad (4.5)$$

and it is immediately seen that the <u>fixed points</u> of Eq. (4.5) are given by :

$$2\,\omega_1 + \omega_2 = 1$$

which is represented on Fig. 3

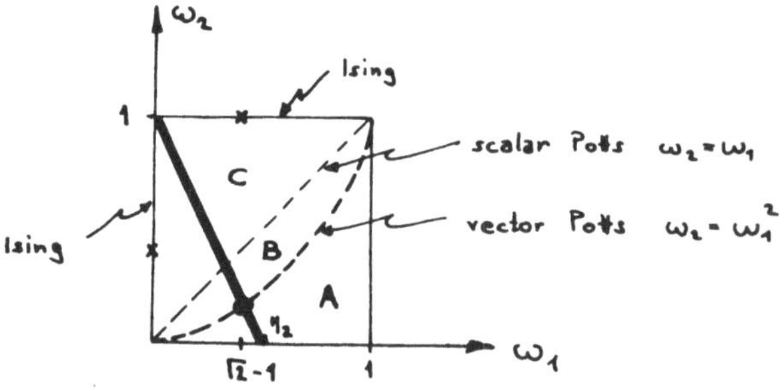

x = critical values for Ising Model

Fig. 3 : Symmetry line for the $q = 4$ generalised Potts models.

Using renormalisation methods Knops [50] has recently shown that there exists no phase transition in A , 1 phase transition in B, and 2 phase transitions in C [49].

Example 2 : Triangular Model with 3 body forces

We consider the model defined by the triangular lattice and 3 body forces between nearest neighbours;

$$-\beta H_\Lambda(\underline{n}) = \sum_{\langle x_1 x_2 x_3 \rangle} \sum_{\ell=0}^{q-1} K(b;\ell) \cos\left[\frac{2\pi}{q}\,\ell\,(n_{x_1} + n_{x_2} + n_{x_3})\right]$$

i.e. the family $\mathcal{B}$ of bonds is defined by the 3 points subsets and the mapping $b \mapsto \chi_b$ is given by

$$b = \langle x_1 x_2 x_3 \rangle \quad \mapsto \quad \chi_b(\underline{n}) = e^{\frac{2i\pi}{q}(n_{x_1} + n_{x_2} + n_{x_3})}$$

As generator of $\mathcal{H}$ we can choose the elements represented by

Using the general construction we obtain for Λ^* the points of Λ and the dual mapping is obtained by

$$b \mapsto b^* = b$$

$$\Pi^*:\ b \mapsto \chi_b^*(\underline{n}) = e^{\frac{2i\pi}{q}(n_1+n_2+n_3)} \quad \text{if} \quad b = \triangledown$$

$$b \mapsto \chi_b^*(\underline{n}) = e^{-\frac{2i\pi}{q}(n_1+n_2+n_3)} \quad \text{if} \quad b = \triangle$$

and therefore the model is <u>weakly self dual</u>.

If the coupling constant are given by (scalar model)

$$K(b;\ell) = \frac{1}{q} K(b) \quad \text{(all energy are zero except } {}_n\overset{n}{\triangle}_n \text{)}$$

we obtain exactly the same result as for the scalar Potts Model; i.e. the scalar Potts Model is weakly self dual with $K^*(b)$ given by

$$e^{K^*(b)} = \frac{e^{K(b)} + q - 1}{e^{K(b)} - 1}$$

Therefore if there exists a unique phase transition the critical temperature is again given by

$$e^{K_c} = 1 + \sqrt{q}$$

CHAPTER 5 - ZEROES OF THE PARTITION FUNCTION

Several discussions of zeroes of the partition function of higher spin systems using Asano contractions combined with Ruelle's theorem [106] have been published . S. Sarbach and F. Rys [139,111] investigated some spin 1 systems and derived bounds on tricritical points. In a first paper, K. Millard and K. Viswanathan [140] derived conditions on some spin 1 systems to have the zeroes on the unit circle. With the help of the Griffith's transformation [141] into equivalent spin $\frac{1}{2}$ system. The general discussion of arbitrary spin systems by means of Griffith's transformation and Ruelle's theorem was made by J. Slawny [98]. In a second paper, K. Millard and K. Viswanathan [142] investigated the zeroes of higher spin systems without using the Griffith's decomposition into an equivalent spin $\frac{1}{2}$ system. However their polynomials were not linear in each variable. Hence these authors had to find a substitute for the theorem of Ruelle. They gave conditions on the interactions to have the zeroes on the unit circle. No result was obtained when these conditions were not satisfied.

Here we present a new method to discuss the zeroes of arbitrary spin systems without using the Griffith's transformation [125, 144]. We shall associate with the partition function a polynomial which is linear in each of its variables and give necessary and sufficient conditions to build up this polynomial by means of Asano contractions. Hence we can apply Ruelle's theorem without modification. These results are valid for any finite q . For $q = 3$, we derive an explicite analyticity domain in the field activities for arbitrary systems with translationally invariant two body interactions and fields in the group picture. For any given model in the spin 1 picture, with the same conditions on the interactions, we find from this domain, using the action of $\mathcal{S}_3$, 6 analyticity domains in the activity variables e^{-2h}, $e^{-2\mu}$. These domains are given explicitely for several well known spin 1 models [141, 143]. The corresponding regions in the (h,μ) plane are shown. In particular we obtain results in regions of the (h,μ) plane where the conditions of K. Millard and K. Viswanathan [142] are not valid. For more detailed informations we refer to [144].

5.1. General Formalism

In the following, we restrict ourselves to a discussion of the polynomial associated with the L.T. expansion of the partition function given in . (2.1). With aid of the generalised activities $z(b;\ell)$ defined in Eq. (2.2) the partition function is a polynomial which is linear in each activity. Since $z(b;0)=1$ we have $\alpha_b - 1$ different activites for each bond $b \in \mathcal{B}$. With Γ subgroup of $\mathcal{G}^{\wedge}_{\mathcal{B}}$ this polynomial becomes

$$M(z_{\mathcal{B}}) = \sum_{\underline{m} \in \Gamma} z^{\underline{m}} \tag{5.1}$$

where $z_{\mathcal{B}} = \{ z(b;\ell) \}$ $\qquad z^{\underline{m}} = \prod_{b \in \mathcal{B}} z(b;m_b)$ (5.2)

The linearity of $M(z_{\mathcal{B}})$ allows us to investigate the zeroes of $M(z_{\mathcal{B}})$ within the same frame as for spin $\frac{1}{2}$ lattice systems discussed in Ch. 8 of Part I, i.e. without any change of the Asano contraction and Ruelle's theorem. Moreover, a straightforward generalisation of theorem 1 of Ch. 8 Part 1 yield coverings satisfying the Asano condition :

Theorem 1

Let $\mathcal{G}^{\wedge} \subset \mathcal{G}^{\wedge}_{\mathcal{B}}$ be any subgroup of $\mathcal{G}^{\wedge}_{\mathcal{B}}$ and $\mathcal{B} = \bigcup_{i=1}^{n} \mathcal{B}_i$ be a finite covering of $\mathcal{B}$, then (*)

$$M(z_{\mathcal{B}}) = \sum_{\underline{m} \in \mathcal{G}^{\wedge}} z^{\underline{m}}$$

(*) In this section we explicitely use the isomorphism $\mathcal{G}^{\wedge}_{\mathcal{B}} \cong \mathcal{G}_{\mathcal{B}}$.

is the Asano contraction of

$$M_i(z_{\mathcal{B}_i}) = \sum_{m_i \in \mathcal{G}_i^\wedge} z^{m_i} \qquad i = 1, 2, \dots, n \tag{5.3}$$

if and only if $\mathcal{G}^{\wedge\perp}$ is equal to the group generated by $\bigcup_{i=1}^{n} \mathcal{G}_i^{\wedge\perp}$ where

$$\mathcal{G}_i^\wedge = \{ \underline{\ell}_i \in \mathcal{G}_{\mathcal{B}_i}^\wedge \; ; \; \ell_{i,b} = \ell_b \quad \text{for some } \underline{\ell} \in \mathcal{G}^\wedge \}$$

Theorem 1 gives the following prescription to find coverings of $\mathcal{B}$ satisfying the Asano condition :

1) given $\mathcal{G}^\wedge$ find $\mathcal{G}^{\wedge\perp}$
2) take $\{\mathcal{G}_i^{\wedge\perp}\}_{i=1,\dots,m}$ a finite family of subgroups of $\mathcal{G}^{\wedge\perp}$ which generates $\mathcal{G}^{\wedge\perp}$
3) then the covering $\mathcal{B} = \bigcup_{i=1}^{n} \mathcal{B}_i$ satisfying the Asano condition is given by $\mathcal{B}_i = \bigcup_{\underline{\ell}_i \in \mathcal{G}_i^{\wedge\perp}} \{ b \; ; \; \ell_{i,b} \neq 0 \} \qquad i = 1, \dots, m$
   and $\bigcup_{i=m+1}^{n} \mathcal{B}_i$ is any covering of $\mathcal{B} / \bigcup_{i=1}^{n} \mathcal{B}_i$.

An application of this general method to find covering sets and small polynomials will be given in the next section where we discuss in detail analyticity domains for the general spin 1 model with two body interactions and fields.

5.2 Applications to Various Spin 1 Models

Following our general strategy, we shall first investigate the L.T. expansion of the group picture for $q = 3$ and derive explicite analyticity domains of $Z(\Lambda, K)$ within the group picture. The change of interaction parameters induced by ϕ^{-1} immediately gives analyticity domains for any physical picture.

The general Hamiltonian Eq. (3.1) has the set of bonds (see eq. (3.4)) $\mathcal{B} = \{(x,1), (xy,11), (xy,12)\}_{(x,y) \subset \Lambda}$. To find a covering of $\mathcal{B}$ satisfying the Asano condition, we follow the prescription of the preceding section :

1) Property 2 of Sec. 1.3 yields $\mathcal{G}^{\perp} = \Gamma^{\perp} = \mathcal{K}$.

2) As for the spin $\frac{1}{2}$ Ising model with 2 body interactions and field [106], we consider the generators of $\mathcal{K}$ (Sec. 3.1). Taking for $\mathcal{G}_i^{\perp}$ the group generated by $\underline{\ell}(xy)$ and $\underline{\ell}'(xy)$, we have

$$\mathcal{B}_i = \{ (xy, 11), (xy, 12), (x,1), (y,2) \} \tag{5.4}$$

and a covering satisfying the Asano condition is

$$\mathcal{B} = \bigcup_{\langle xy \rangle} \{ (xy,11), (xy,12), (x,1), (y,2) \} \tag{5.5}$$

Since we always have $q = \alpha_b = 3$, there are 2 generalised activities associated with any $b \in \mathcal{B}_i$. Hence the small polynomial $M_i(z_{\mathcal{B}_i})$ is a polynomial in 8 variables. To simplify the notation, we define

$$\begin{aligned} &z_1 = z((x,1);1) \quad && z_2 = z((x,1);2) \quad && z_3 = z((y,1);1) \quad && z_4 = z((y,1),2) \\ &z_5 = z((xy,11);1) \quad && z_6 = z((xy,11);2) \quad && z_7 = z((xy,12);1) \quad && z_8 = z((xy,12);2) \end{aligned} \tag{5.6}$$

With this notation, the small polynomials, Eq. (5.3) are all of the form

$$\begin{aligned} M_i(z_1,\dots,z_8) = 1 &+ z_1 z_5 z_7 + z_2 z_6 z_8 + z_3 z_5 z_8 + z_4 z_6 z_7 + \\ &+ z_1 z_3 z_6 + z_1 z_4 z_8 + z_2 z_3 z_7 + z_2 z_4 z_5 \end{aligned} \tag{5.7}$$

To apply Ruelle's theorem, we have to discuss the zeroes of this polynomial. First we remark that with the above choice of the covering of $\mathcal{B}$, only the bonds $\{(x,1)\}_{x \in \Lambda}$ occurs in more than one covering set $\mathcal{B}_i$; therefore only the variables $z_1,\dots,z_4$ can undergo contractions and the variables $z_5,\dots,z_8$ can be kept real. Therefore for fixed values

$$z_5 = a \qquad z_6 = b \qquad z_7 = c \qquad z_8 = d$$

we have to discuss the zeroes of the polynomial M_i' given by

$$M_i'(z_1,z_2,z_3,z_4) = 1 + ac\,z_1 + bd\,z_2 + ad\,z_3 + bc\,z_4 + a\,z_2 z_4 + b\,z_1 z_3 + c\,z_2 z_3 + d\,z_1 z_4 \tag{5.8}$$

This polynomial cannot vanish if all its activities are small enough. In order to use the theorem of Grace [106b] we realise first that $M_i'(z_1,z_2,z_3,z_4)$ is the Asano contraction of the following 6 polynomials :

$$\begin{aligned}
M_{i,1}(z_1,z_2) &= 1 + z_1 + z_2 \\
M_{i,2}(z_1,z_3) &= 1 + a z_1 + a z_3 + \frac{b}{dc} z_1 z_3 \\
M_{i,3}(z_1,z_4) &= 1 + c z_1 + c z_4 + \frac{d}{ab} z_1 z_4 \\
M_{i,4}(z_2,z_3) &= 1 + d z_2 + d z_3 + \frac{c}{ab} z_2 z_3 \\
M_{i,5}(z_2,z_4) &= 1 + b z_2 + b z_4 + \frac{a}{cd} z_2 z_4 \\
M_{i,6}(z_3,z_4) &= 1 + z_3 + z_4
\end{aligned} \tag{5.9}$$

Applying the theorem of Grace, we find

$$\begin{aligned}
M_{i,1} &\neq 0 \quad \text{if} \quad |z_1|,\ |z_2| < \tfrac{1}{2} \\
M_{i,2} &\neq 0 \quad \text{if} \quad |z_1|,\ |z_3| < \frac{cd}{b} \left|\left(a - \sqrt{a^2 - \frac{b}{cd}}\right)\right| \\
M_{i,3} &\neq 0 \quad \text{if} \quad |z_1|,\ |z_4| < \frac{ab}{d} \left|\left(c - \sqrt{c^2 - \frac{d}{ab}}\right)\right| \\
M_{i,4} &\neq 0 \quad \text{if} \quad |z_2|,\ |z_3| < \frac{ab}{c} \left|\left(d - \sqrt{d^2 - \frac{c}{ab}}\right)\right| \\
M_{i,5} &\neq 0 \quad \text{if} \quad |z_2|,\ |z_4| < \frac{cd}{a} \left|\left(b - \sqrt{b^2 - \frac{a}{cd}}\right)\right| \\
M_{i,6} &\neq 0 \quad \text{if} \quad |z_3|,\ |z_4| < \tfrac{1}{2}
\end{aligned} \tag{5.10}$$

To get $\mathcal{M}(z_{\mathcal{B}})$, we have to contract all the polynomials Eq. (5.9) associated with the covering $\mathcal{B} = \bigcup_{i=1}^{n} \mathcal{B}_i$. By definition of the coordination number $\mathfrak{z}$, any bond $(x,1)$ belongs to $\frac{1}{2}\mathfrak{z}$ different covering sets $\mathcal{B}_i$. Applying Ruelle's theorem [106] , we find

$$\mathcal{M}(z_{\mathcal{B}}) \neq 0$$

if for all $x \in \Lambda$ we have :

$$|z((x,1);1)| < \frac{1}{2}\left(a^2cd\right)^3 \left\{ \left| 1 - \sqrt{1-\frac{b}{a^2cd}} \right|^2 \cdot \left| 1-\sqrt{1-\frac{c}{abd^2}} \right| \cdot \left| 1-\sqrt{1-\frac{d}{abc^2}} \right| \right\}^{3/2} \quad (5.11)$$
$$|z((x,1);2)| < \frac{1}{2}\left(b^2cd\right)^3 \left\{ \left| 1 - \sqrt{1-\frac{a}{b^2cd}} \right|^2 \cdot \left| 1-\sqrt{1-\frac{c}{abd^2}} \right| \cdot \left| 1-\sqrt{1-\frac{d}{abc^2}} \right| \right\}^{3/2}$$

The above analyticity domain is independent of Λ hence using standard arguments, it yields an analyticity domain for the free energy in the thermodynamic limit $\Lambda \to \mathcal{L}$.

Now we are interested in analyticity domains in the activity variables e^{-2h} and $e^{-2\mu}$ of the spin 1 picture (Eq. (3.1)). The first step is to map the spin picture by ϕ into the group picture. By Eq. (3.2), the coupling constants $K(b;\ell)$ in the group picture are linear combinations of those of the spin 1 model and therefore the generalised activities are fractional powers of products of the spin 1 activities $e^{-2\varepsilon_{11}}$, $e^{-2\varepsilon_{22}}$, $e^{-2\varepsilon_{12}}$, e^{-2h}, $e^{-2\mu}$. The action of $\mathcal{S}_3$ then gives us 6 different generalised activities for a given spin 1 model, which are explicitely indicated in Table 3. Therefore Eq. (5.11) yields for a given spin 1 model not only 1 but 6 analyticity domains, which are in general different. We notice that $z((xy,1i);\ell)$ do not contain the field variables h , μ ; hence the 2 body coupling constants $\varepsilon_{11}, \varepsilon_{12}, \varepsilon_{22}$ can be kept real and we obtain from Eq. (5.11) analyticity domains in e^{-2h} and $e^{-2\mu}$. Notice that by the structure of the general method introduced in Sec. 5.1., we always conclude simultaneous analyticity in e^{-2h} and $e^{-2\mu}$. In Fig. 1, these analyticity domains are shown in the (h,μ) plane. The domains depend explicitely on the 2 body coupling constants ε_{11} , ε_{12} , ε_{22} . All models discussed below yields similar analyticity domains in the (h,μ) plane. The shaded regions indicate analyticity regions of the free energy. It is interesting to remark the three corridors, which are a direct consequence of the symmetry properties induced by $\mathcal{S}_3$. The only possible localisation of phase transition is within these corridors, hence in particular, we have explicit bounds on tricritical points [125, 144].

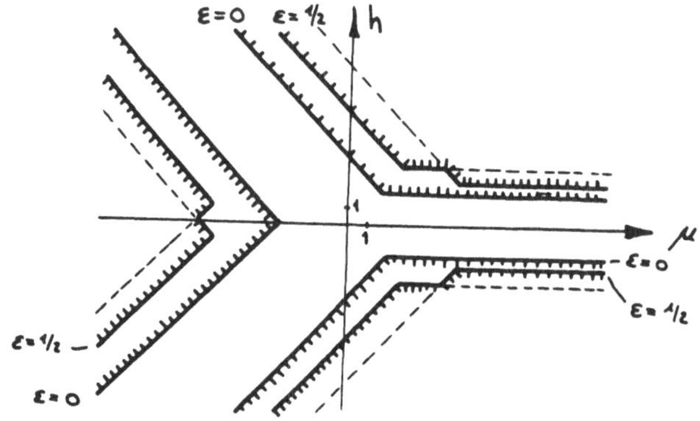

Fig. 1 analyticity domains in the (h,μ) plane

To compute explicite analyticity domains for spin 1 models with $\varepsilon_{12} = 0$, we first observe that it is enough to find domains corresponding to $p_0 = 1$, p_I , p_2 , the other three domains are obtained from the latter by $h \leftrightarrow -h$. In the following two examples, we illustrate the content of Eq. (5.11)

example 1 : (usual spin 1 Ising model) $\varepsilon_{11} = \varepsilon$ $\varepsilon_{12} = \varepsilon_{22} = 0$

$\mathcal{M}^p(z_B) \neq 0$ if

	$\varepsilon > 0$	$\varepsilon < 0$
a) $p = p_0$	$\vert z((x,1);1)\vert < \frac{1}{2} e^{-\frac{\varepsilon}{2}\zeta}$	$\vert z((x,1);1)\vert < \frac{1}{2}\left[e^{-\varepsilon}\left(1-\sqrt{1-e^{\varepsilon}}\right)\left(e^{-\varepsilon}-\sqrt{e^{-2\varepsilon}-1}\right)\right]^3$
	$\vert z((x,1);2)\vert < \frac{1}{2} e^{-\varepsilon\zeta}$	$\vert z((x,1);2)\vert < \frac{1}{2}\left[e^{-3\varepsilon}\left(1-\sqrt{1-e^{4\varepsilon}}\right)\left(e^{-\varepsilon}-\sqrt{e^{-2\varepsilon}-1}\right)\right]^3$
b) $p = p_I$	$\vert z((x,1);1)\vert < \frac{1}{2} e^{-\varepsilon\zeta}$	$\vert z((x,1);1)\vert < \frac{1}{2}\left[e^{-3\varepsilon}\left(1-\sqrt{1-e^{4\varepsilon}}\right)\left(e^{-\varepsilon}-\sqrt{e^{-2\varepsilon}-1}\right)\right]^3$
	$\vert z((x,1);2) < \frac{1}{2} e^{-\frac{\varepsilon}{2}\zeta}$	$\vert z((x,1);2)\vert < \frac{1}{2}\left[e^{-\varepsilon}\left(1-\sqrt{1-e^{\varepsilon}}\right)\left(e^{-\varepsilon}-\sqrt{e^{-2\varepsilon}-1}\right)\right]^3$

c) $p = p_2 \qquad |z((x,1);\ell)| < \frac{1}{2}\left(\frac{1-\sqrt{1-e^{-\varepsilon}}}{e^{-\frac{1}{2}\varepsilon}}\right)^3 \quad \Big| \quad |z((x,1);\ell)| < \frac{1}{2}\left[e^{-\frac{1}{2}\varepsilon}\left(1-\sqrt{1-e^{\varepsilon}}\right)\right]^3$

$\ell = 1, 2$

<u>example 2</u> :(model of Lebowitz Gavallotti [143] $\varepsilon_{11} = -\varepsilon_{22} = \varepsilon \qquad \varepsilon_{12} = 0$

$M(z_{\otimes}) \neq 0$ if $[p = p_0]$

$\varepsilon > 0 \qquad |z((x,1);1)| < 2^{-1} e^{-\frac{3\varepsilon}{3}} \qquad |z((x,1);2)| < 2^{-1} e^{-3\varepsilon}$

$\varepsilon < 0 \qquad |z((x,1);1)| < \frac{1}{2}\left[e^{-\frac{4\varepsilon}{3}}\left(1-\sqrt{1-e^{2\varepsilon}}\right)\right]^3$

$\qquad\qquad |z((x,1);2)| < \frac{1}{2}\left[e^{-4\varepsilon}\left(1-\sqrt{1-e^{2\varepsilon}}\right)\left(1-\sqrt{1-e^{4\varepsilon}}\right)\right]^3$

Analogous domains are obtained for the Potts model with fields.

P \ $Z^P(b;\ell)$	$Z^P((x,1);1)$	$Z^P((x,1);2)$	$Z^P((xy,11);1) = a^P$	$Z^P((xy,11);2) = b^P$	$Z^P((xy,12);\ell) = c^P = d^P$, $\ell = 1,2$
P_0	$e^{-\frac{2}{3}\mathfrak{z}\varepsilon_{22}+\frac{2}{3}\mathfrak{z}\varepsilon_{12}+h-\mu}$	$e^{\frac{4}{3}\mathfrak{z}\varepsilon_{12}+2h}$	$e^{\frac{4}{3}\varepsilon_{12}}$	$e^{-\varepsilon_{11}+\frac{1}{3}\varepsilon_{22}+\frac{2}{3}\varepsilon_{12}}$	$e^{-\varepsilon_{11}-\frac{1}{3}\varepsilon_{22}}$
P_I	$e^{\frac{4}{3}\mathfrak{z}\varepsilon_{12}+2h}$	$e^{-\frac{2}{3}\mathfrak{z}\varepsilon_{22}+\frac{2}{3}\mathfrak{z}\varepsilon_{12}+h-\mu}$	$e^{-\varepsilon_{11}+\frac{1}{3}\varepsilon_{22}+\frac{2}{3}\varepsilon_{12}}$	$e^{\frac{4}{3}\varepsilon_{12}}$	$e^{-\varepsilon_{11}-\frac{1}{3}\varepsilon_{22}}$
P_2	$e^{\frac{2}{3}\mathfrak{z}\varepsilon_{22}+\frac{2}{3}\mathfrak{z}\varepsilon_{12}+h+\mu}$	$e^{\frac{2}{3}\mathfrak{z}\varepsilon_{22}-\frac{2}{3}\mathfrak{z}\varepsilon_{12}-h+\mu}$	$e^{\varepsilon_{11}-\frac{1}{3}\varepsilon_{22}-\frac{2}{3}\varepsilon_{12}}$	$e^{\varepsilon_{11}-\frac{1}{3}\varepsilon_{22}+\frac{2}{3}\varepsilon_{12}}$	$e^{-\varepsilon_{11}-\frac{1}{3}\varepsilon_{22}}$
$P_{(I,2)}$	$e^{\frac{2}{3}\mathfrak{z}\varepsilon_{22}-\frac{2}{3}\mathfrak{z}\varepsilon_{12}-h+\mu}$	$e^{\frac{2}{3}\mathfrak{z}\varepsilon_{22}+\frac{2}{3}\mathfrak{z}\varepsilon_{12}+h+\mu}$	$e^{\varepsilon_{11}-\frac{1}{3}\varepsilon_{22}+\frac{2}{3}\varepsilon_{12}}$	$e^{\varepsilon_{11}-\frac{1}{3}\varepsilon_{22}-\frac{2}{3}\varepsilon_{12}}$	$e^{-\varepsilon_{11}-\frac{1}{3}\varepsilon_{22}}$
P_1	$e^{-\frac{4}{3}\mathfrak{z}\varepsilon_{12}-2h}$	$e^{-\frac{2}{3}\mathfrak{z}\varepsilon_{22}-\frac{2}{3}\mathfrak{z}\varepsilon_{12}-h-\mu}$	$e^{-\varepsilon_{11}+\frac{1}{3}\varepsilon_{22}-\frac{2}{3}\varepsilon_{12}}$	$e^{-\frac{4}{3}\varepsilon_{12}}$	$e^{-\varepsilon_{11}-\frac{1}{3}\varepsilon_{22}}$
$P_{(I,1)}$	$e^{-\frac{2}{3}\mathfrak{z}\varepsilon_{22}-\frac{2}{3}\mathfrak{z}\varepsilon_{12}-h-\mu}$	$e^{-\frac{4}{3}\mathfrak{z}\varepsilon_{12}-2h}$	$e^{-\frac{4}{3}\varepsilon_{12}}$	$e^{-\varepsilon_{11}+\frac{1}{3}\varepsilon_{22}-\frac{2}{3}\varepsilon_{12}}$	$e^{-\varepsilon_{11}-\frac{1}{3}\varepsilon_{22}}$

Table 3 : Generalised activities in terms of the coupling constants $\varepsilon_{11}, \varepsilon_{22}, \varepsilon_{12}, h, \mu$ of the spin 1 picture for all $p \in \mathcal{S}_3$.

REFERENCES

1. Ruelle D. — Statistical Mechanics, W.A. Benjamin N.Y., 1970

2. Gruber C., Merlini D. — J. Math. Phys. 13, 1814 (1972)

3. (a) Kelly D.G., Sherman S. — J. Math. Phys. 9, 466 (1968)

(b) Ginibre J. — Cargèse Lecture in Physics, Gordon and Breach, N.Y. p, 95 (1970)

4. Rudin W. — Fourier Analysis on groups, Interscience Publishers, J.Wiley & Sons, N.Y. 1967

5. Gruber C., Hintermann A. — H.P.A. 47, 67 (1974)

6. Holsztynski W., Slawny J. — Phase transitions in ferromagnetic Systems at low temperatures, Preprint 1976

7. Kramers H.A., Wannier G.K. — Phys. Rev. 60, 252 (1941)

8. Van der Waerden B.L. Z. Phys. 118, 473 (1952)

9. Montroll W.E., Newel G.F. — Rev. Mod. Phys. 25, 2 (1953)

10. Onsager L. — Phys. Rev. 65, 117 (1944)

11. Mc Kean H.P. J. Math. Phys. 5, 775 (1964)

12. Wegner F.J. J. Math. Phys. 12, 2259 (1971)

13. Hintermann A. Phys.Lett. 39 A, 243 (1972)

14. Greenberg W.
Gruber C.
Merlini D. Physica 65, 28 (1973)

15. Slawny J. Comm. Math. Phys. 34, 271 (1973)

16. Slawny J. Private communication

17. Benettin G.
Lasinio G.J.
Stella A. Lettere al Nuovo Cimento 4, 443 (1972)

18. Hintermann A. (1971) unpublished

19. Gallavotti G. La Rivista del Nuovo Cimento 2, 133 (1972)

20. Lee T.D.
Yang C.N. Phys. Rev. 87, 404 (1952)

21. Syozi I in Phase transition and critical phenomena, Ed. C. Domb and M.S. Green, Vol. I Academic Press (1972) pp. 285,288,322

22. Wu F.Y. Phys. Letters 46A, 7 (1973)

23. Baxter R.J. (a) Phys. Rev. Lett. 26, 832 (1971)
(b) Ann.Phys.N.Y. 70, 193 (1972)

24. Hintermann A.
Merlini D. Phys. Lett. 41A, 208 (1972)

25. Gruber C. in Colloquium on group theoretical method in Physics, University of Nijmegen, The Netherland (1973) B. 120

26. Gruber C. Journ. of Stat. Phys. 14, 81 (1976)

27. Gallavotti G. Comm. Math. Phys. 27, 103 (1972)

28. Abraham D.B.
Gallavotti G.
Martin-Löf A. Physica 65, 73 (1973)

29. Gallavotti G.
Martin-Löf A.
Miracle Sole S. in Statistical Mechanics and Mathematical Problems, Battelle Seattle 1971, Rencontres, Springer N.Y. 1973.

30. Ceva H.
Kadanoff L.P. Phys. Rev. B3, 3918 (1971)

31. Slawny J.
Gruber C. (unpublished)

32. Gruber C.
Hintermann A.
Messager A.
Miracle Sole S. "On the uniqueness of the invariant equilibrium state and surface tension"(Preprint 1976).

33. Baxter G. J. Math. Phys. 8, 399 (1967)

34. Mc Coy B.M.
Wu T.T. Phys. Rev. 155, 438, (1967)

35. Merlini D. Lettere al Nuovo Cimento 9, 100 (1974)

36. (a) Fisher M. In Lectures in Theoretical Physics Boulder, Colorado Press, ch. 1 (1964)

(b) Brascamp H.J.
Kunz H. J. Math. Phys. 15, 65 (1974)

37. Griffiths R.B. in Phase Transition and Critical Phenomena Vol. 1, ed. C. Domb and M.S. Green, Academic Press 1972.

38. Martin-Löf A. Comm. Math. Phys. 24, 253 (1972)

39. Merlini D. Lettere al Nuovo Cimento 8, 623 (1973)

40. Gruber C.
Hintermann A.
Merlini D. Lettere al Nuovo Cimento 7, 815 (1973)

41. Peierls R. Proc. Cambridge Phil. Soc. 32 477 (1936)

42. Griffiths R.B. J. Math. Phys. 5, 1215 (1964)

43. Dobrushin R.L. (i) Sov. Phys. Dokl. 10, 111 (1965)
(ii) Theory Probab. and Appl. 10, 193 (1965)

44. (a) Ginibre J.
Grossmann A.
Ruelle D. Comm. Math. Phys. 3, 187 (1966)

(b) Berezin F.A.
Sinaï Ya G. — Proc. Moscow Math. Soc. 17 197 (1967)

(c) Heilmann O.J. — Lett. al Nuovo Cimento 3, 95 (1972)

(d) Heilmann O.J.
Praestgard E. — J. Stat. Phys. 9, 23 (1973)
J. Phys. (A) 7, 1913 (1974)

(e) Heilmann O.J.
Moss R.
Praestgard E. — J. Phys. (C) 6 L 403 (1973)

(f) Abraham D.B.
Heilmann O.J. — J. Stat. Phys. 13, 461 (1973)

(g) Sec. ref. 37 — for a review of Peierls argument

45. (a) Ruelle D. — Phys. Rev. Lett. 27, 1040 (1971)

(b) Lebowitz J.L.
Lieb E.H. — Phys. Letters 39A, 98 (1972)

46. Payandeh B. — Travail de diplôme EPF-L (1972) (unpublished)

47. Slawny J. — Comm. math. Phys. 35, 297 (1974)

48. (a) Ruelle D. — in "Cargèse Lectures in Physics", Gordon and Breach (1970)

(b) Lanford O.E.
Ruelle D. — Comm.Math.Phys. 13, 197 (1968)

49. Wu F.Y.
Lin K.Y. J. Phys. (C) 7, 181 (1974)

50. Knops H.J.F. Branch Point in the critical surface of the Ashkin Teller Model in the renormalization group theory (Preprint 1976)

51. Berge C. Graphes et hypergraphes Dunod, Paris (1970)

52. Minlos R.A.
Sinaĭ Y.G. Trans. Moscow Math. Soc. 17, 237 (1967)

53. Dobrushin R.L. (i) Theor. Prob.and Appl. 13,197 (1968)
(ii) Func.Anal. and Appl. 2,292,302 (1968)

54. Ginibre J. "Systèmes à un nombre infini de degrés de liberté"
Colloque internationaux du CNRS (1970)

55. Heilmann O.J. Comm. Math. Phys. 36, 91 (1974)

56. Pirogov S.
Sinaĭ Y.G. (i) Teor. Mat. Fiz. 25, 358 (1975)
(ii) Seminar presented at IUPAP Conference - Budapest 1975.

57. Runnels L.K. Comm. Math. Phys. 40, 37 (1975)

58. Cassandro M
Da Fano A.
Olivieri E. Comm. Math. Phys. 44, 45 (1975)

59. Ruelle D. Ann. Phys. 69, 364 (1972)

60. Gaunt D.S.
Guttmann A. J. — in "Phase Transition and Critical Phenomena" Vol. 3. Ed. C. Domb and M.S. Green - Academic Press 1974

61. Barber M.N.
Baxter R.J. — J. Phys. Solid State 6, 2913 (1973)

62. Griffiths H.P.
Wood D.W. — J. Phys. (C) 6, 2533 (1973)

63. Baxter R.J.
Wu F.Y. — Phys. Rev. Lett. 31, 1294 (1973)
Australian J. of Physics, 27, 357, (1974) and ibid. 27, 369 (1974)

64. Wegner F.J. — Physica 68, 570 (1973)

65. Kadanoff L.P.
Wegner F.J. — Phys. Rev. B4, 3989 (1971)

66. (a) Kadanoff L.P. — (i) "Proc.of Enrico Fermi Summer School of Physics, Varenna, 1970" Academic Press 1971
(ii) in "Phase Transition and Critical Phenomena, vol 5A, ed. C. Domb and M.S. Green, Academic Press (1970)

(b) Jona Lasinio G. — Nuovo Cimento, 26B 99 (1975)

67. Watts M.G. — J. Phys. (A)7, L85, (1974)

68. Baxter R.J.
Sikes M.F.
Watts M.G. — J. Phys. (A) 8, 1469 (1975)

69. Merlini D. H.P.A. 48, 542 (1975)

70. Baxter R.J.
Enting I.G. J. Phys.(A) 9 L 149 (1976)

71. Shermann S. (i) J. Math. Phys. 8 399 (1960)
(ii) J. Math. Phys. 4 1213 (1963)

72. Calinon R.
Merlini D. To be published

73. Benettin G.
Gallavotti G.
Jona-Lasinio G.
Stella A.L. Comm. Math. Phys. 30 45 (1973)

74. Sykes M.F. and al J. Math. Phys. 14, 1071 (1973) (Eq. 4.5.)

75. Gallavotti G. Il modello a goccia. Scuola di perfezionamento in Fisica dell'Università di Padova (1972)

76. Fisher M.E. Physics Vol.3, 45, 255, (1967)

77. Fisher M.E. Phys. Rev. 113, 969 (1959)

78. Gruber C.
Merlini D. Physica 67, 308 (1973)

79. Pastur L.A. Theor. and Math. Physics 18, 165 (1974)

80. Brascamp H.J. Comm. Math. Phys. 18, 82 (1970)

81. Gruber C.
Lebowitz J.L. — Comm. Math. Phys. 41, 11 (1975)

82. Jelitto R.J. — Phys. Lett. 27A, 267 (1968)

83. Suzuki M. — Phys. Lett. 19, 267 (1965)

84. Ashkin J.
Teller E. — Phys. Rev. 64, 178 (1943)

85. Gruber C.
Merlini D. — Phys. Lett. 41A, 245 (1972)

86. Wegner F.J. — J. Phys. 5 L 131 (1972)

87. Thompson C.J. — Comm. Math. Phys. 24, 61 (1971)

88. Gallavotti G.
Lebowitz J.L. — Physica 70, 219 (1973)

89. Greenberg W. — Comm. Math. Phys. 13 335 (1969)

90. Baxter G. — J. Math. Phys. 6 (1965)

91. Domb C. — J. Phys. (C) 7, 2677 (1974)

92. Stauffer D. — J. Phys. (C) 8, L 172 (1975)

93. Temperley HNV — J. Phys. (A) 9, L 113 (1976)

94. Reatto L.
Rastelli E. — J. Phys. (C) 5, 2785 (1972)

95. (a) Lebowitz J.L.
Martin-Löf A. — Comm. Math. Phys. 25, 276 (1972)

(b) Lebowitz J.L. — Coexistence of Phases in Ising Ferromagnets, Preprint (1976)

96. Gallavotti G.
Miracle-Sole S. Phys. Rev. 5B, 2555 (1972)

97. Messager A.
Miracle-Sole S. Comm. Math. Phys. 40, 187 (1975)

98. Slawny J. Comm. Math. Phys. 46, 75 (1976)

99. Gruber C.
Hintermann A. Physica 83 A, 233 (1975)

100. Heilmann O.J
Lieb E.H. a) Phys. Rev.Letters 86, 1412 (1970)
b) Comm. Math.Phys. 25, 190 (1972)

101. Heilmann O.J Stud. in Appl. Math. 50, 385 (1971)

102. Griffith R.B. J. Math. Phys. 10, 1559 (1969)

103. Asano T. Phys. Rev. 24, 1409 (1970)

104. Asano T. J. Phys. Soc. Jap. 29, 350 (1970)

105. Suzuki M
Fisher M.E. J. Math. Phys. 12, 235 (1971)

106. Ruelle D. a) Phys. Rev. Lett. 26, 303 (1971)
b) Comm. Math. Phys. 31, 265 (1973)

107. Messager A
Trottin J.C. a) Nuovo Cimento 9, 239 (1974)
b) Ann. Inst. Henri Poincaré
sect. A. Vol XXIV, 301 (1976)

108. Slawny J. Comm. Math. Phys. 34, 271 (1973)

109. Gruber C.
Hintermann A.
Merlini D. — Comm. Math. Phys. 40, 83 (1975)

110. Holsztynski W. — Precise Generator of A [G] - Modules (Preprint 1975)

111. Sarbach S.
Rys F. — Phys. Rev. B 7, 3141 (1973)

112. Asano T. — Progr. theor. Phys. 43, 1401 (1970)

113. Gruber C.
Hintermann A. — Physica 83A, 233 (1976)

114. Sewell G.L. — Lett. al Nuovo Cim. 10, 430 (1974)

115. Balian R.
Drouffe J.M.
Itzykson — Phys. Rev. D. 10, 3376 (1974)
11, 2098 (1975)
11, 2104 (1975)

116. Lieb E.
Wu F.Y — In phase transition and critical Phenomena Ed. Domb C.and Green M.S. Vol. 1, Academic Press (1972)

117. Katsura S.
Abe Y.
Ohkohchi K. — J. Phys. Soc. Jap. 29, 845 and 854 (1970)

118. Chang K.S.
Wang S.Y
Wu F.Y. — Phys. Rev. B4, 2324 (1971)

119. Runnels K.K.
Hubbard J.B. — J. Stat. Phys. 6, 1 (1972)

120. Runnels L.K. Quantum Statistical Mechanics in the Natural Sciences; Kursunoglu B, Mintz S.L., Widmayer S.M. (eds), Plenum Publishing Corporation N.Y. (1974)

121. Runnels K.K.
Freasier B.C. Phys. Rev. B8, 2126 (1973)

122. Hintermann A.
Gruber C. Physica 84A, 101 (1976)

123. The essential idea for the proof is that $\mathcal{K}^{(a)}$ can be generated by translates of some $\mathcal{K}_0$, problem which was solved for our class of systems by W. Holsztynski 10 .

124. Merlini D. Lett. al Nuovo Cim. 11, 441 (1974)

125. Hintermann A.
Gruber C. in Lecture Notes in Physics Vol. 50, p. 558, Springer-Verlag (1976)

126. Phase Transitions and Critical Phenomena Vol. 1,2,3,4, Domb C., Green M.S. (Eds) N.Y. Academic Press.

127. Greenberg W. Comm. Math. Phys. 29, 163 (1973)

128. Loomis L.H. An Introduction to Abstract Harmonic Analysis, D. Van Nostrand. Comp. Inc.(1953)

129. Greenberg W. Group Structure of Higher Spin Lattice Models Preprint (1975)

130. Israel R.B. Comm. in Math. Phys. 50, 245 (1976)

131. Potts R.B. Proc. Cambridge Phil. Soc. 48, 106 (1952)

132. Kim D.
Joseph R.I. Phys. Lett. 46A, 359 (1974)

133. Bernasconi J.
Rys F. Phys. Rev. B4 3045 (1971)

134. Mukamel D.
Blume M. Phys. Rev. B7, 3134 (1973)

135. Mittag L.
Stephen M.J. J. Math. Phys. 12, 441 (1971)

136. Wu F.Y.
Wang Y.K. J. Math. Phys. 17 439 (1976)

137. Betts D.D. Comm. J. Phys. 42 1564 (1964)

138. Suzuki M. Prog. Theoret. Phys. (Kyoto) 37, 770 (1967)

139. Sarbach S. Diploma Thesis ETG-Zurich (1972) unpublished

140. Millard K.
Viswanathan K. Phys. Rev. B9, 2030 (1974)

141. Griffiths R.B. J. Math. Phys. 10, 1559 (1969)

142. Millard K.
Viswanathan K. J. Math. Phys. 15, 1821 (1974)

143. Lebowitz J.L.
Gallavotti G. — J. Math. Phys. 12, 1129 (1971)

144. Hintermann A. — Zeroes of the Partition Function of Higher Spin Systems (Preprint 1976)

145. Gruber C.
Hirsbrunner B.
Merlini D. — On the uniqueness of the symmetric equilibrium states in the 2-d Ising ferromagnet.

Vol. 570: Differential Geometrical Methods in Mathematical Physics, Bonn 1975. Proceedings. Edited by K. Bleuler and A. Reetz. VIII, 576 pages. 1977.

Vol. 565: Turbulence and Navier Stokes Equations. Proceedings 1975. Edited by R. Temam. IX, 194 pages. 1976.

Vol. 559: J.-P. Caubet, Le Mouvement Brownien Relativiste. IX. 212 pages. 1976.

Vol. 556: Approximation Theory. Bonn 1976. Proceedings. Edited by R. Schaback and K. Scherer. VII, 466 pages. 1976.

Vol. 543: Nonlinear Operators and the Calculus of Variations, Bruxelles 1975. Edited by J. P. Gossez, E. J. Lami Dozo, J. Mawhin, and L. Waelbroeck, VII, 237 pages. 1976.

Vol. 538: G. Fischer, Complex Analytic Geometry. VII, 201 pages. 1976.

Vol. 532: Théorie Ergodique. Proceedings 1973/1974. Edité par J.-P. Conze and M. S. Keane. VIII, 227 pages. 1976.

Vol. 527: M. Denker, Ch. Grillenberger, and K. Sigmund, Ergodic Theory on Compact Spaces. IV, 360 pages. 1976.

Vol. 526: Probability in Banach Spaces. Proceedings 1975. Edited by A. Beck. VI, 290 pages. 1976.

Vol. 525: Structural Stability, the Theory of Catastrophes, and Applications in the Sciences. Proceedings 1975. Edited by P. Hilton. VI, 408 pages. 1976.

Vol. 524: Séminaire Pierre Lelong (Analyse) Année 1974/75. Edité par P. Lelong. V, 222 pages. 1976.

Vol. 523: S. A. Albeverio and R. J. Høegh-Krohn, Mathematical Theory of Feynman Path Integrals. IV, 139 pages. 1976.

Vol. 522: C. O. Bloom and N. D. Kazarinoff, Short Wave Radiation Problems in Inhomogeneous Media: Asymptotic Solutions. V. 104 pages. 1976.

Vol. 518: Séminaire de Théorie du Potentiel, Proceedings Paris 1972–1974. Edité par F. Hirsch et G. Mokobodzki. VI, 275 pages. 1976.

Vol. 543: Nonlinear Operators and the Calculus of Variations, Bruxelles 1975. Edited by J. P. Gossez, E. J. Lami Dozo, J. Mawhin, and L. Waelbroeck, VII, 237 pages. 1976.

Vol. 538: G. Fischer, Complex Analytic Geometry. VII, 201 pages. 1976.

Vol. 527: M. Denker, Ch. Grillenberger, and K. Sigmund, Ergodic Theory on Compact Spaces. IV, 360 pages. 1976.

Vcl. 525: Structural Stability, the Theory of Catastrophes, and Applications in the Sciences. Proceedings 1975. Edited by P. Hilton. VI, 408 pages. 1976.

Vol. 523: S. A. Albeverio and R. J. Høegh-Krohn, Mathematical Theory of Feynman Path Integrals. IV, 139 pages. 1976.

Vol. 522: C. O. Bloom and N. D. Kazarinoff, Short Wave Radiation Problems in Inhomogeneous Media: Asymptotic Solutions. V. 104 pages. 1976.

Vol. 516: M. L. Silverstein, Boundary Theory for Symmetric Markov Processes. XVI, 314 pages. 1976.

Vol. 515: Bäcklund Transformations. Nashville, Tennessee 1974. Proceedings. Edited by R. M. Miura. VIII, 295 pages. 1976.

Vol. 509: D. E. Blair, Contact Manifolds in Riemannian Geometry. VI, 146 pages. 1976.

Vol. 507: M. C. Reed, Abstract Non-Linear Wave Equations. VI, 128 pages. 1976.

Vol. 503: Applications of Methods of Functional Analysis to Problems in Mechanics. Proceedings 1975. Edited by P. Germain and B. Nayroles. XIX, 531 pages. 1976.

Vol. 490: The Geometry of Metric and Linear Spaces. Proceedings 1974. Edited by L. M. Kelly. X, 244 pages. 1975.

Vol. 489: J. Bair and R. Fourneau, Etude Géométrique des Espaces Vectoriels. Une Introduction. VII, 185 pages. 1975.

Vol. 487: H. M. Reimann und T. Rychener, Funktionen beschränkter mittlerer Oszillation. VI, 141 Seiten. 1975.

Vol. 484: Differential Topology and Geometry. Proceedings 1974. Edited by G. P. Joubert, R. P. Moussu, and R. H. Roussarie. IX, 287 pages. 1975.

Vol. 474: Séminaire Pierre Lelong (Analyse) Année 1973/74. Edité par P. Lelong. VI, 182 pages. 1975.

Vol. 470: R. Bowen, Equilibrium States and the Ergodic Theory of Anosov Diffeomorphisms. III, 108 pages. 1975.

Vol. 468: Dynamical Systems – Warwick 1974. Proceedings 1973/74. Edited by A. Manning. X, 405 pages. 1975.

Vol. 464: C. Rockland, Hypoellipticity and Eigenvalue Asymptotics. III, 171 pages. 1975.

Vol. 463: H.-H. Kuo, Gaussian Measures in Banach Spaces. VI, 224 pages. 1975.

Vol. 461: Computational Mechanics. Proceedings 1974. Edited by J. T. Oden. VII, 328 pages. 1975.

Vol. 459: Fourier Integral Operators and Partial Differential Equations. Proceedings 1974. Edited by J. Chazarain. VI, 372 pages. 1975.

Vol. 458: P. Walters, Ergodic Theory – Introductory Lectures. VI, 198 pages. 1975.

Vol. 449: Hyperfunctions and Theoretical Physics. Proceedings 1973. Edited by F. Pham. IV, 218 pages. 1975.

Vol. 448: Spectral Theory and Differential Equations. Proceedings 1974. Edited by W. N. Everitt. XII, 321 pages. 1975.

Vol. 446: Partial Differential Equations and Related Topics. Proceedings 1974. Edited by J. A. Goldstein. IV, 389 pages. 1975.

Vol. 442: C. H. Wilcox, Scattering Theory for the d'Alembert Equation in Exterior Domains. III, 184 pages. 1975.

Vol. 440: R. K. Getoor, Markov Processes: Ray Processes and Right Processes. V, 118 pages. 1975.

Vol. 434: P. Brenner, V. Thomée, and L. B. Wahlbin, Besov Spaces and Applications to Difference Methods for Initial Value Problems. II, 154 pages. 1975.

Vol. 433: W. G. Faris, Self-Adjoint Operators. VII, 115 pages. 1975.

Vol. 431: Séminaire Bourbaki – vol. 1973/74. Exposés 436–452. IV, 347 pages. 1975.

Vol. 430: Constructive and Computational Methods for Differential and Integral Equations. Proceedings 1974. Edited by D. L. Colton and R. P. Gilbert. VII, 476 pages. 1974.

Vol. 427: H. Omori, Infinite Dimensional Lie Transformation Groups. XII, 149 pages. 1974.

Vol. 426: M. L. Silverstein, Symmetric Markov Processes. X, 287 pages. 1974.

Vol. 425: P. R. Chernoff and J. E. Marsden, Properties of Infinite Dimensional Hamiltonian Systems. IV, 160 pages. 1974.

Vol. 417: H. H. Keller, Differential Calculus in Locally Convex Spaces. XVI, 131 pages. 1974.

Vol. 416: M. E. Taylor, Pseudo Differential Operators. IV, 155 pages. 1974.

Vol. 415: Ordinary and Partial Differential Equations. Proceedings 1974. XVII, 447 pages. 1974.

Vol. 32: Particles, Quantum Fields and Statistical Mechanics. Proceedings 1973. Edited by M. Alexanian and A. Zepeda. V, 132 pages. 1975.

Vol. 33: Classical and Quantum Mechanical Aspects of Heavy Ion Collisions. Proceedings 1974. Edited by H. L. Harney, P. Braun-Munzinger, and C. K. Gelbke. VII, 311 pages. 1975.

Vol. 34: One-Dimensional Conductors GPS Summer School Proceedings, 1974. Edited by H. G. Schuster. VII, 371 pages. 1975.

Vol. 35: Proceedings of the Fourth International Conference on Numerical Methods in Fluid Dynamics, 1974. Edited by R. D. Richtmyer. V, 457 pages. 1975.

Vol. 36: R. Gatignol, Théorie Cinétique des Gaz à Répartition Discrète de Vitesses. II, 219 pages. 1975.

Vol. 37: Trends in Elementary Particle Theory. Proceedings 1974. Edited by H. Rollnik and K. Dietz. V, 472 pages. 1975.

Vol. 38: Dynamical Systems, Theory and Applications. Proceedings 1974. Edited by J. Moser. VI, 624 pages. 1975.

Vol. 39: International Symposium on Mathematical Problems in Theoretical Physics. Proceedings 1975. Edited by H. Araki. XII, 562 pages. 1975.

Vol. 40: Effective Interactions and Operators in Nuclei. Proceedings 1975. Edited by B. R. Barrett. XII, 339 pages. 1975.

Vol. 41: Progress in Numerical Fluid Dynamics. Proceedings 1974. Edited by H. J. Wirz. V, 471 pages. 1975.

Vol. 42: H II Regions and Related Topics. Proceedings 1975. Edited by D. Downes and T. L. Wilson. XII, 488 pages. 1975.

Vol. 43: Laser Spectroscopy. Proceedings 1975. Edited by S. Haroche, J. C. Pebay-Peyroula, T. W. Hänsch, and S. E. Harris. X, 466 pages. 1975.

Vol. 44: R. A. Breuer, Gravitational Perturbation Theory and Synchrotron Radiation. VI, 196 pages. 1975.

Vol. 45: Dynamical Concepts on Scaling Violation and the New Resonances in e^+e^- Annihilation. Edited by B. Humpert. VII, 248 pages. 1976.

Vol. 46: E. J. Flaherty, Hermitian and Kählerian Geometry in Relativity. VIII, 365 pages. 1976.

Vol. 47: Padé Approximants Method and Its Applications to Mechanics. Edited by H. Cabannes. XV, 267 pages. 1976.

Vol. 48: Interplanetary Dust and Zodiacal Light. Proceedings 1975. Edited by H. Elsässer and H. Fechtig. XII, 496 pages. 1976.

Vol. 49: W. G. Harter and C. W. Patterson, A Unitary Calculus for Electronic Orbitals. XII, 144 pages. 1976.

Vol. 50: Group Theoretical Methods in Physics. 4th International Colloquium. Nijmegen 1975. Edited by A. Janner, T. Janssen, and M. Boon. XIII. 629 pages. 1976.

Vol. 51: W. Nörenberg und H. A. Weidenmüller. Introduction to the Theory of Heavy-Ion Collisions. IX, 273 pages. 1976.

Vol. 52: M. Mladjenović, Development of Magnetic β-Ray Spectroscopy. X, 282 pages. 1976.

Vol. 53: D. J. Simms and N. M. J. Woodhouse, Lectures on Geometric Quantization. V, 166 pages. 1976.

Vol. 54: Critical Phenomena. Sitges International School on Statistical Mechanics, June 1976. Edited by J. Brey and R. B. Jones. XI, 383 pages. 1976.

Vol. 55: Nuclear Optical Model Potential. Proceedings 1976. Edited by S. Boffi and G. Passatore. VI, 221 pages. 1976.

Vol. 56: Current Induced Reactions. International Summer Institute, Hamburg 1975. Edited by J. G. Körner, G. Kramer, and D. Schildknecht. V, 553 pages. 1976.

Vol. 57: Physics of Highly Excited States in Solids. Proceedings 1975. Edited by M. Ueta and Y. Nishina. IX, 391 pages. 1976.

Vol. 58: Computing Methods in Applied Sciences. Proceedings 1975. Edited by R. Glowinski and J. L. Lions. VIII, 593 pages. 1976.

Vol. 59: Proceedings of the Fifth International Conference on Numerical Methods in Fluid Dynamics. 1976. Edited by A. I. van de Vooren and P. J. Zandbergen. VII, 459 pages. 1976.

Vol. 60: C. Gruber, A. Hintermann and D. Merlini, Group Analysis of Classical Lattice Systems. XIV, 326 pages. 1977.